VISUAL STATISTICS

VISUAL STATISTICS

SEEING DATA WITH
DYNAMIC INTERACTIVE GRAPHICS

FORREST W. YOUNG

University of North Carolina at Chapel Hill, NC, USA

PEDRO M. VALERO-MORA

Universitat de Valencia, Valencia, Spain

MICHAEL FRIENDLY

York University, Toronto, Canada

 WILEY-INTERSCIENCE

A JOHN WILEY & SONS, INC., PUBLICATION

Published by John Wiley & Sons, Inc., Hoboken, New Jersey.
Published simultaneously in Canada.

For general information on our other products and services or for technical support, please contact our Customer Care Department within the United States at (800) 762-2974, outside the United States at (317) 572-3993 or fax (317) 572-4002.

Wiley also publishes its books in a variety of electronic formats. Some content that appears in print may not be available in electronic format. For information about Wiley products, visit our web site at www.wiley.com.

Library of Congress Cataloging-in-Publication Data:

Young, Forrest W.
    Visual statistics : seeing data with dynamic interactive graphics / Forrest W. Young, Pedro M. Valero-Mora, Michael Friendly.
        p. cm.
    Includes bibliographical references and index.
    ISBN-10 0-471-68160-1
    ISBN-13 978-0-471-68160-1
    1. Statistics—Graphic methods.   2. Statistics—Charts, diagrams, etc.   3. Computer graphics.   4. Information visualization.   I. Valero-Mora, Pedro M., 1967–   II. Friendly, Michael.   III. Title.

QA276.3.Y68 2006
001.4'226—dc22   2006044265

Printed in the United States of America.

10 9 8 7 6 5 4 3 2 1

We dedicate this book to our wives,
Patty, Mar, and Martha

St. Valentine's Day
14 February 2006

seeing data

first you see your data for what they seem to be
    then you ask them if it's so
        are you what you seem to be?

you see with broad expanse, you ask with narrow power
    you see and ask and see
        and ask and see ... and ask ...

with brush you paint the possibilities
            with pen you scribe the probabilities
        for in pictures we find insight
            while in numbers we find strength

forrest young

Forrest W. Young died a few months before this book was published. He was 65. An untiring worker, his passion for visual statistics filled him with enthusiasm and joy in the face of chronic illness. Admired, respected, and loved by his family, friends, and coworkers, with him expired a restless mind, a notable thinker, and, above all, a good man.

Contents

Part I
Introduction

2 Examples   45

Part II
See Data—The Process

3 Interfaces and Environments   73

Part III
Seeing Data—Objects

6 Seeing Univariate Data   181

7 Seeing Bivariate Data   215

8   Seeing Multivariate Data   **263**

Preface

This book is based on a 15-year research project focused on improving our ability to understand data by using dynamic interactive graphics. The result of our research is a statistical analysis system that is radically different from the usual system, a system that presents the user with a visual environment for statistical analysis that supports instantaneous interaction and uses dynamic graphics to communicate information about the data: in short, a system which makes statistical analysis fun rather than drudgery.

Our presentation emphasizes a paradigm for understanding data that is visual, intuitive, geometric, and active, rather than one that relies on convoluted logic, heavy mathematics, systems of algebraic equations, and passive acceptance of results. Not that we reject the underlying logic and mathematics. Rather, we build our visualizations on a firm foundation of mathematical statistics, designing the visuals so that they lead the user to "do the right thing" on the basis of visual intuition.

The usual approach to statistical data analysis can provide deep insight and understanding of our data, but is difficult to learn and complicated to practice. In contrast, visual statistics can simplify, ease and improve the data analysis process We are keenly aware of the benefits and difficulties involved in getting Humans, Statistics and Computers to work smoothly together, that being the goal of our work

While we aim to teach the reader the art and practice of visual statistics, this is not a user's guide to any particular software. We do, however, urge the user to use ViSta, the Visual Statistics system, developed by the first two authors, to obtain a better understanding of the issues discussed in the text. ViSta is available free by downloading it from www.visualstats.org.

Rather than striving to create a commercially viable statistical analysis system, our project has emphasized the ongoing development of a testbed for our ideas about the use of dynamic, highly interactive graphics to improve data analysis. Nevertheless, our testbed has become a viable analysis system, although we continue to adhere to the free (semiopen) software model rather than dealing with all the additional issues involved in turning it into a commercially viable system.

About the Authors

Each author has focused his academic career on topics involving people using computers to do statistics. Each author received his degrees in psychometrics, a field of endeavor that emphasizes people, statistics and computers. Psychometrics is focused on the development and application of statistical methods within the field of psychology in order to improve our ability to measure aspects of people, the aspects including knowledge, ability, attitude and personality.

The authors have extensive experience with the development of statistical software, there being numerous examples of software methods in the major statistical systems developed by, or based on work by one or another of the authors. In addition, one of the authors (Young) has considerable post-graduate training in computer graphics, and another (Valero-Mora) has considerable professional experience in the field of human factors. Thus, our graduate and post-graduate training experiences, and our professional careers have emphasized the interaction between humans, statistics and computers, an emphasis that uniquely qualifies us to be working in an area at the intersection of knowledge in these three seemingly disparate areas.

Forrest W. Young. Forrest W. Young is Professor Emeritus of Quantitative Psychology at the University of North Carolina at Chapel Hill, located in Chapel Hill, North Carolina. Forrest is the creator and designer of ViSta, both in terms of its look and feel, and in terms of its internal software architecture. He has also implemented the design and has written much of the documentation. He has worked on the ViSta project since 1990, receiving considerable help from his students and colleagues.

Prof. Young received his Ph.D. in psychometrics from the University of Southern California in 1967. He has been on the faculty of UNC-CH ever since. His teaching interests focus on "seeing what your data seem to say." This visually intuitive approach to statistics helps to clarify the meaning of data. His courses, ranging from introductory undergraduate course on psychological statistics, to his advanced graduate courses on data analysis, visualization and exploration, reflect this focus.

Prof. Young's early research interests focused on multidimensional scaling and nonlinear multivariate data analysis (for which he was elected president of the Psychometric Society and received the American Market Research Association's O'Dell award, both in 1981). Through these research interests, Prof. Young became involved in software development early in his career. Prof. Young has served as a professional consultant on statistical system interface design with SAS Institute, Statistical Sciences (the S-Plus system), and BMDP Inc. He has written or designed data analysis modules for the SAS, SPSS, and IMSL systems. He is a member of the American Statistical Association's sections on computational and graphical statistics.

Pedro M. Valero-Mora. Pedro M. Valero-Mora is Professor Titular at the University de Valencia (Spain). He received his Ph.D. in psychology from this same university in 1996. He has worked in the Department of Methodology of the Behavioural Sciences of the University of València, Spain, since 1990, teaching introductory and multivariate Statistics, data processing and computational statistics.

Pedro M. Valero-Mora's research interests have combined graphics, statistical visualization, and human computer interaction, being concerned with the design of computer interfaces for statistical graphics systems to make them more useful, friendly and, rewarding to use. This research has lead to his work on the development of innovative computer interfaces for statistical analysis and visualization. This work has been guided by the principles of direct manipulation, instant output, and graphic visualization, making the assumption that statistical systems will be easier to use if the user can directly manipulate visual representations of statistical techniques that respond immediately to the user's actions. This should help the user more easily understand the consequences of the choices at hand, with the user feeling more confident to experiment with the system, thereby obtaining a deeper understanding of the statistical concepts underlying the software.

The development of this type of software has been made possible by the Lisp-Stat language and the ViSta system. Using these tools, Prof. Valero has developed methods for missing data imputation, visually controlled transformations, and visually oriented log-linear analysis, as well as, with Prof. Rubén Ledesma from the University of Mar del Plata, in Argentina, an adaptation of a homogeneity analysis module by Prof. Jan de Leeuw of UCLA.

Michael Friendly. Michael Friendly received his Ph.D. in psychometrics and cognitive psychology from Princeton University, where he held a Psychometric Fellowship awarded by the Educational Testing Service. He is Professor of Psychology at York University, Canada, and has been associate coordinator and director of the Statistical Consulting Service since 1985. He is the author of *The SAS System for Statistical Graphics* and *Visualizing Categorical Data* published by SAS Institute, an Associate Editor of the *Journal of Computational and Graphical Statistics*, and author of numerous research papers. Current research includes graphical methods for data analysis, analysis of categorical data, and the history of data visualization. He teaches graduate and undergraduates in multivariate data analysis, and intermediate and advanced statistics, and has taught many short courses on a wide variety of statistical topics.

Acknowledgments

Forrest W Young. This book represents the major publication of my entire research career, research that has been the focus of my professional life for the last 40 years. As such, hundreds of people have contributed to my work in one way or another. But looking back over that period of time, certain people clearly stand out. I thank them in the order in which their presence made itself known in my life.

First and foremost, in the literal sense of being the person who shaped my professional career before anyone else did, is my mentor, Norman Cliff. I deeply appreciate

Norm's gentle influences. As my mentor throughout graduate school, and as the chairman of my dissertation committee, Norm had a clear and major influence on my career's beginnings. But of even more importance, Norm's appreciation of the simple approach, his gentle strength, quiet certitude, constant civility, and strong sense of ethics, have had a continuing impact throughout my entire career. Thanks Norm. You have been a great role model.

To Lyle Jones, and to his continued support and belief in me, despite my best efforts to show myself to be unworthy, I owe my 35 years of employment at UNC. Without Lyle this crazy 60's hippy might well have ended up down and out. To you, Lyle, I owe my eternal gratitude: For offering me the best job in the world; For making the offer in the airport waiting lounge; for providing a work environment that nurtured and excited my brain; for overlooking my sudden and unexplained "unpaid leave" from which I resurfaced months later half way round the world in Teheran, Iran, narrowly escaping the bowels of the beast; and especially for overlooking all that and supporting me anyway. Your levelness was the perfect antidote to my turbulence.

I will never forget when Jan de Leeuw and I together came up with the concept at the heart of the work we did together, driving way too fast on the back roads of North Carolina, living life way too hard, way too close to the edge. We had an exciting, fun, and productive decade, and I'm sure he enjoyed as much as I did our roles as the Wild Men of Psychometrics, the Hippy Professors. The showers of sparks that exploded from his brain, and the ease with which he let them fall where they may, and yes, even the carelessness with which he let these sparks darken and grow cold, all this set my imagination on fire, igniting and continuing to energize my research long after we had gone our separate ways. Thanks, Jan, it's been fun.

And to John Sall, whose silent determination served him well during the initial ride on the bucking bronco and then over the years of the subsequent internal wars; to John, who I admire more than anyone for his ability to focus; to John, of whom I stand in awe at the influence he has had on the work that all statisticians do; to John, who clearly knew that I would continue to grow if I stayed where I was but would become lost in the labyrinth if he helped me go where I wanted: to John I express my appreciation for standing in my way.

Pedro M. Valero-Mora, of course, has been of central importance in my life for the last 15 years. Without him I would not have written this book. It is he who came up with the basic structure of the book, and it is he who had faith in me, believing, despite all evidence to the contrary, that I could stop fiddling with code and actually come out of my cave and write about what I had been doing. "Mucho gracias," Pedro, for your time, energy, enthusiasm, and last but not least, your writing, both in English and in Lisp. Pedro conceived, created, and implemented all of the missing data features, all of the log-linear modelling features, and the visually based Box–Cox and folded–power transformations, including the visual transformation and imputation methods. His creativity, as well as his sweat, are greatly appreciated. And, finally, to Pedro for his care and concern as my health has declined, from living a life not unlike that of the candle that one burns at both ends.

This book also would not have happened without Michael Friendly who knew everything that was in my head and much much more, being able to fill in my intuition with the necessary details, having a rare ability to work cooperatively in parallel, and on whom I knew I could rely for emergency backup, or total restart when I crashed, who could boot me up no matter how lost I was in the programmer's cave.

For keeping me motivated I must thank the undergrads who unknowingly uttered those magic words "Hey, this is Fun!" And for the same reasons I thank my granddaughter who, on seeing on the computer for the first time what it is that I do said "Cool!" And again for the same reasons I thank Pedro for saying this stuff is his favorite video game -- Hearing such remarks I know I'm really on the right track.

Thanks to Luke Tierney for creating and providing the XLisp-Stat statistical development system, which is the foundation on which ViSta is built. No other system supports the development of dynamic interactive statistical graphics, which is in our opinion a major weakness of all other statistical analysis system.

David Lubinsky, John B. Smith, Richard A. Faldowski and Carla M. Bann all made very fundamental conceptual and architectural contributions early in the project. The concept and realization of the workmap, guidemap and spreadplot was a joint effort on all of our parts.

"Merci Beaucop" to Louis Le Guelte for the French translation. and "mucho gracias" to Maria F. Rodrigo, J. Gabriel Molina, and Pedro M. Valero-Mora for the original Spanish translation of my software, and to Ruben Ledesma for his recent translation efforts and for his new analysis plug-ins.

Many people at the University of North Carolina contributed directly or indirectly to the project. Bibb Latane, of Social Science Conferences, Inc., funded development of the Excel-ViSta connection, multipane windows, and the developer's conference. Angel Beza, Associate Director of UNC's Odum Institute for Research in the Social Sciences, was instrumental in obtaining the SSCI funds. Jose Sandoval, also at UNC's Odum Institute, developed the first public courses for ViSta. Doug Kent and Fabian Camacho, both funded by UNC's Office of Information Technology, implemented critical MS-Windows features. Jenny Williams of UNC's Office of Information Technology, was my Windows systems guru. Mosaic plots and bar charts are based on an algorithm by Ernest Kwan. Frequency polygons and histograms are based on algorithms by Jan de Leeuw and Jason Bond. Many of the model objects were written by Carla Bann, Rich Faldowski, David Flora, Ernest Kwan, Lee Bee Leng, Mary McFarland, and Chris Weisen. The C-to-Lisp parser used in ViVa was written and copyrighted by Erann Gat, who reserves all rights, and which is used under the terms of the GNU General Public License. I also enjoyed support from the Universitat de Valencia, in Spain, the National University of Singapore, and from SAS Institute, Cary, NC, USA.

Finally, I express my deepest thanks and appreciation to my wife, Patricia Young. During much of the last two decades she has stood watch while I wandered lost in the developer's cave, until others could see on the screen what I saw in my mind's eye. Patty's patience, understanding, forbearance, support, and steady love have been crucial to the realization of my professional dreams. Her artistic talent, abilities, and spirit have shaped those dreams.

Pedro M. Valero-Mora. I trace my interest in dynamic-interactive graphics back to 1995, when I experienced, for the first time, dynamic graphics techniques such as linking, interactive selecting, and focusing while running the DataDesk software on a Macintosh SE. I was so fascinated by this program that I carried out all of the statistical analyses for my doctoral thesis with it alone, in overt challenge to the common view that statistics without typing was necessarily inconsequential. Actually, I became so engaged with dynamic-interactive graphics that doing analysis and reanalysis of my data became my favorite video game!

However, once I got my Ph.D., being just a user of the software become progressively less satisfactory. Fortunately, a few years later I discovered the tools that would allow me to go beyond simply playing with data. Those tools were ViSta and Lisp-Stat. As it happened, their discovery was followed by a time of enormous excitement, enjoyment and countless hours in front of the computer. Although many people find Lisp an alien language, I found it so natural from the beginning that I learned many of its features before acquiring a reference book, simply by looking at code programmed by others. Suddenly, I was on the other side of the fence, designing and writing about dynamic-interactive graphics instead of merely using them.

Therefore, my thanks go first to Forrest Young and Luke Tierney, Forrest for starting the ViSta project and for allowing me to get on board, and Luke for developing Lisp-Stat. Without their talent as developers of such wonderful pieces of software, and their generosity in putting them in the public domain, I would probably still be playing the video game and not helping to make it.

My deepest appreciation goes to Jaime Sanmartín. As my Phd supervisor and long-time colleague, Jaime has been not only a major influence on my academic career, but also an example of tolerance, good manners and civility that has influenced me more than all the Psychology that I could have read.

My thanks also go to my colleagues and friends in the department of Methodology of the Behavioural Sciences of the Universitat de València, who have provided a supportive environment for carrying out my work. In particular, Gabriel Molina, and María Rodrigo helped me very significantly with comments, expertise, and friendly support on innumerable occasions.

Many thanks also go to my students Rubén Ledesma and Noelia Llorens. Rubén. Ruben, now at the Universidad de Mar del Plata in Argentina, himself has become involved in ViSta programming. He has been a steady source of code, insight, and ideas. Noelia, contributed documentation and testing during the time she spent in our department.

I also wish to express my gratitude to Maria Teresa Anguera for kindly hosting me during a research visit to the Universitat de Barcelona, and to Juan Luis Chorro of the Universitat de València and Manuel Ato of the University of Murcia, who helped me on several occasions without ever asking anything in return. My appreciation also goes to the Ministerio de Ciencia y Tecnología for providing support in order to work on this book (grant number BSO2002-02220)

I have retained a continued research interest in human computer interaction (HCI), an interest that is easy to link to my attraction for the interactive and visual side of statistical software. As a consequence of this interest, I visited the HCI group of the University of York several times, having the opportunity of enjoying the friendly work atmosphere with Andrew Monk, Leon Watts (now at the University of Bath), and Owen Daily-Jones. I also spent a very useful time with Penelope Sanderson (then at the Swinburne University of Technology in Melbourne and now at the University of Queensland) learning about applications of exploratory data analysis to sequential data.

Nowadays, due to the popularity of the Internet and of free software projects, a programmer has easy access to the code of other programmers. Of the many people that I have contacted in search of advice and code over the years of using the Lisp-Stat language, I especially wish to thank the following people for their generosity and help: Frederic Udina of the Universitat Pompeu Fabra, Didier Josselin of the Université d'Avignon, Jan de Leeuw of the University of California–Los Angeles, Chris Brunsdom of the University of Newcastle Upon Tyne, Sanford Weisberg of the University of Minnesota, and Luke Tierney of the University of Iowa. Thanks to all of you for contributing, often unknowingly, to my programming efforts.

Finally, my deepest thanks to my wife, Mar, and my daughter, Lucía, for their love and support.

Michael Friendly. It has been a great pleasure to collaborate with Forrest and Pedro on this book. I have been working on developing graphical methods for data analysis for the past 15 years, more recently concentrating on methods for the analysis of categorical data, but most of the latter were designed and implemented as static displays. Better, I thought initially, to create informative visual tools to help see relations and patterns in data first, then think about how best to integrate them into the real-time process of data analysis. The opportunity to do just that arose in 1998–1999, when my student, Ernest Kwan, went to UNC to study with Forrest, and Forrest invited me to give some lectures in the winter on visualizing categorical data. The discussions we had over several days led to several joint projects with Forrest and Pedro and ultimately to my participation in this book.

Those who work with graphics know well that doing graphics well is hard work: The devil is in the details. With most modern statistical software, it is relatively easy to get a quick, standard graph, but to produce one that speaks to the eyes with clarity, precision, and elegance is far more difficult. This is certainly so for static graphics; it is much more so when it comes to designing dynamic, interactive graphics because the basic "palettes" from which dynamic graphs are composed are not at all obvious and often have to be redesigned many times. It has been highly rewarding to work with two friends and colleagues who care deeply about the details of graphics and were not reluctant to consign initial or even current designs to the bit bucket for not-quite-good-enough ideas.

I am grateful to the National Science and Engineering Research Council of Canada for continuing research support (Discovery Grant 138748 RGPIN) and to York University for a sabbatical leave fellowship grant on dynamic graphics for categorical data.

Finally, my deepest appreciation to my wife, Martha, for continued love and support.

PART I

INTRODUCTION

1 Introduction

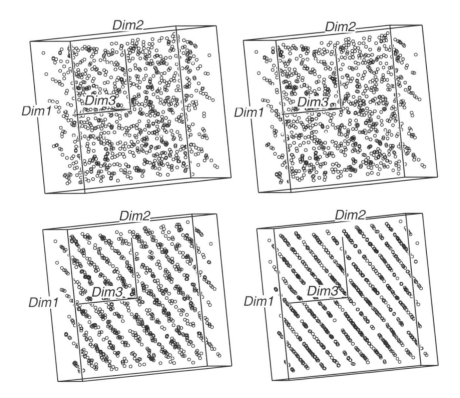

1 Introduction

D*ata* are the lifeblood of science, for it is through data that we obtain scientific understanding of the world around us. Data are also the lifeblood of many other aspects of the modern world. To the financial system, money is data. To the political system, votes are data. To teachers, test scores are data. To marketers, purchases are data. To jet passenger planes, the entire airplane is data. And we could go on and on.

Data are a set of facts suited to the rules of arithmetic that are organized for processing and analysis. Data are often, though not always, numeric.

Methods for processing and analyzing data have evolved, over the past 200 years, to become the bedrock of the mathematical discipline called *statistics*. These methods, collectively called *statistical data analysis*, can provide deep insight and understanding about the processes and phenomena underlying data, can increase the value and usefulness of data, and can suggest how to proceed in the future to obtain additional data that could be even more useful.

However, statistical data analysis is an immensely complicated enterprise that most of us find difficult to learn and to practice. Therefore, recent years have seen a concerted effort by many statisticians to develop new techniques that simplify, ease, and improve the analysis process. Visual statistics is one such effort.

Visual statistics transforms algebraic obscurity into geometric clarity by translating mathematical statistics into dynamic interactive graphics. This can simplify, ease and improve data analysis. When the graphics are mathematically, computationally, perceptually, and cognitively appropriate, they can induce intuitive visual understanding that is simple, instantaneous and accurate.

Dynamic interactive graphics are graphics that can move smoothly and change in response to the data analysts's actions, the changes being computed and presented in real time. The same technology creates the video game's highly dynamic, instantly interactive visual experience, an experience that is fun and exciting.

Seeing data refers to the process—and to the result—of our visual search for meaning in our data. When we interact with a statistical graphic to better understand our data, we are *seeing data*. And when we suddenly see structure that just a moment ago lay hidden within—when we are "hit between the eyes"—we are seeing data.

1.1 Visual Statistics

Visual statistics has evolved from the seminal work by John Tukey on exploratory data analysis (Tukey, 1977). Showing that graphics could be used to generate hypotheses about data, he stated, on page v of the preface:

> Exploratory data analysis is about looking at data to see what it seems to say—It regards whatever appearances we have recognized as partial descriptions, and tries to look beneath them for new insights.

In this quote Tukey reminds us that the data analyst's job is to understand data but that we must always remember that we are only seeing "what the data seems to say," not "what the data say." We should also pay close attention to the wisdom in the words of Mosteller et al., (1983) who observed that .

> Although we often hear that data speak for themselves, their voices can be soft and sly.

The data have a story to tell, but it may not be easy to hear it. We must attend to nuances. There is much art and craft in seeing what the data seem to say. Altho   we have long understood that graphics can strengthen our ability to generate hypotheses, only recently have we begun to understand that graphics can also strengthen our ability to evaluate hypotheses. We are coming to learn that graphics whose geometry faithfully renders the hypothesis-testing algebra of *mathematical statistics* can augment our hypothesis-testing abilities, leading us toward a methodology for visually intuitive significance testing that is mathematically appropriate.

Thus, increasingly, visual statistics is becoming influenced by mathematical statistics, a branch of statistics that uses the formal machinery of theorems and proofs to provide mathematically based answers to scientifically posed questions. In fact, a requirement of the dynamic interactive graphics of visual statistics is that their geometry be a translation of the relevant algebraic results of mathematical statistics: The "subjective" visual insights of the data analyst who is "seeing data" are firmly, though unknowingly, based on the "objective" rigors of mathematical statistics.

Visual statistics also blends influences from computer science, cognitive science, and human–computer interaction studies. These influences inform the way that visual statistics proposes to ease and improve data analysis, leading to an emphasis on the effect of the data analysis environment on the analyst's search for understanding.

See what your data seem to say—that's the bottom line: You learn about your data by interacting with their images. As you interact with images of your data, and as they respond in kind, the interaction helps you get to know your data better. That's what this book's about: how to learn about your data by playing with their pictures, how to use your hands to move the pictures, how to use your eyes to watch them move, and how to use your brain to learn from what your hands are doing and what your eyes are seeing. That's all there is, there's nothing more: no formulas to solve, no convoluted logic. It's just your hands and eyes and brain.

The graphics at the heart of visual statistics are like those used in video games: They move smoothly and rapidly, responding to your actions instantaneously. Such graphics, when well designed, can increase your visual insight and augment your abilities to understand what you see.

Statistical data analysis provides the most powerful tools for understanding data, but the systems currently available for statistical analysis are based on a 40-year-old computing model, and have become much too complex. What we need is a simpler way of using these powerful analysis tools.

Visual statistics is a simpler way. Its dynamic interactive graphics are in fact an interface to these time-proven statistical analysis tools, an interface that presents the results of the hidden tools in a way that helps ensure that our intuitive visual understanding is commensurate with the mathematical statistics under the surface. Thus, visual statistics eases and strengthens the way we understand data and, therefore, eases and strengthens our scientific understanding of the world around us.

1.2 Dynamic Interactive Graphics

We argue that we learn more from properly constructed graphics than from equally well constructed tables of numbers, and that we learn most when the graphics and the user can interact instantaneously, smoothly and continuously. More precisely, a statistical graphic that is either dynamic—capable of smooth motion—or interactive—capable of instantaneous reaction to the user's action—provides an easier and more effective way to understand data than is provided by an ordinary static, noninteractive statistical graphic. Better yet is a statistical graphic that is both dynamic and interactive.

1.2.1 An Analogy

Dynamic interactive graphics are a bit like movies: Both use a series of static pictures to construct and display, in real time, a smoothly moving picture, and sometimes there are a few frames worth the price of admission because of their visual, emotional or cognitive impact. But there is a crucial difference: Visualizations are interactive, waiting for the analyst to act, reacting instantly when the analyst acts. Movies, on the other hand, move whether or not you want them to (indeed, even if you fall asleep or walk out). Movies are watched by a passive viewer, whereas visualizations need an active viewer to interact with. Actually, dynamic interactive graphics are more like video games. In fact, the technology underlying the interactivity and dynamism of statistical visualization is precisely the same as that used by video games. For both it is this technology that makes the graphics change smoothly, continuously, and rapidly in real time, responding instantly to the user's actions.

Note that statistical visualizations are like video games in another way: Statistical visualizations often display multiple views of the data simultaneously, thereby providing even greater information about your data. It may be that multiple views of the data which are highly interactive, extensively interlinked, and immediately responsive—and which are intimately linked with the actions of the user—provide the best chance for accurate insight. This highly dynamic interactivity can change the data analyst's view of the data by providing new information effortlessly.

But there's a crucial difference between video games and dynamic graphics: In video games, the gamer can never stop and reflect—the computer is in control and the user has to keep up—whereas in statistical visualization the data analyst is in full control. When the analyst stops to reflect, the computer stops, too, then waits patiently for the analyst's next action. When the analyst does take action, the computer has to keep up. With statistical visualization you direct your search for meaning, and the computer shows you what there is to see as you pursue that meaning.

1.2.2 Why Use Dynamic Graphics?

Although we know of no firm empirical basis for our beliefs that dynamic interactive statistical graphics are superior to other kinds of statistical graphics, there are several reasons why we believe that to be so. One reason is that we learn by doing. We believe that a data analyst who is actively engaged in the search for understanding is more likely to gain that understanding than the data analyst who is a passive viewer of a dynamic graphic, let alone of a static graphic. If you manipulate your data to get a good view of it, you will be more aware of details than if you watch a movie of someone else doing the same thing. We also believe that dynamic interactive graphics are more fun to use than other kinds of graphics—when one is using dynamic graphics it feels as though one is "playing" with the data, whereas a static graphic lacks that playfulness. And we assume that when you are having fun playing with your data, you are likely to gain greater insight into the data.

Note that most of the plots that we discuss in this book were proposed as static plots long before the advent of computers made it possible to introduce interactivity. Thus, we cover many plots whose static features are very well known and which have been reviewed in numerous other places, but what we do is try to show these plots at their interactive best. We focus chiefly on the interactive aspects of such plots, showing how these aspects strengthen the abilities of the original static plots to reveal data structure.

Of course, there are also plots that are fundamentally dynamic, although not many. The best known example is probably the spinning three-dimensional scatterplot. It may be surprising to learn that these plots can be noninteractive, just spinning on their own, accepting no user interaction. However, we focus on their interactive as well as their dynamic features.

1.2.3 The Four Respects

For a dynamic interactive graphic to be a truly useful discovery tool, it needs to do more than simply provide a "fun" way of doing statistics. It must do this in a way that helps the user be receptive to visual insight, and it must do this in a way that ensures that the insights, when they occur, are valid. For the visual insights to be valid, a dynamic interactive statistical graphic must always be aware of the "Four Respects":

Respect people. A dynamic interactive statistical graphic must respect the user's perceptual and cognitive capabilities, ensuring that the view of the data shown to the analyst will be understood accurately in a visually intuitive manner. Such a system must also ensure that the dynamic and interactive aspects can be dealt with by the user's physiology, and that they are within the perceptual and cognitive capabilities of the ordinary user.

Respect data. A dynamic interactive statistical graphic must respect the nature of the data. As we will soon see, there are different types of data. For example, a major distinction is made between categorical data (example: the gender of a person) and numerical data (example: the height of a person). Respecting data means that each datatype must have an appropriate visualization.

Respect mathematics. A dynamic interactive statistical graphic must respect the mathematics of the statistical situation. The pictures must be faithful to the basic nature of the data (or to the analysis of the data) and to the algebraic framework of the data (or analysis) built by statisticians. Generally, this means translating the algebra of mathematical statistics into the geometry of statistical visualization.

Respect Computers. A dynamic interactive statistical graphics system must also respect the computer's capabilities, so that the dynamic and interactive aspects remain immediate and smooth for the largest possible datasets. We are transferring the tedious work to the computer, leaving the creative work to the analyst. We can augment the analyst's ability to understand the data, but only if we don't overtax the computer.

1.3 Three Examples

We begin with three very brief examples of dynamic interactive graphics, the first emphasizing dynamic aspects of such graphics, the other two focusing on interactive aspects. These examples are presented in more detail in Chapters 2 and 7.

1.3.1 Nonrandom Numbers

We begin by showing how a dynamic statistical graphic can reveal structure that you'd be hard-pressed to find with a static plot. The dynamic graphic is a plot of spinning points. The example uses numbers generated by Randu, a random number generator widely used during the 1960s and 1970s. Marsaglia (1968) showed that Randu, and others in its family of random number generators, yield numbers that are nonrandom. It turns out that dynamic graphics is particularly suited to reveal the nonrandom structure lurking within Randu's supposedly random numbers. We present this example in more detail in Chapter 2. Randu is supposed to generate uniformly distributed random numbers between 0 and 1. These numbers should be generated so that all equal-width subintervals within the [0,1] interval are equally likely (uniform), so that the next number to be generated is unpredictable from the numbers already generated (random).

Briefly, we used Randu to generate 3000 random numbers that should be distributed uniformly. We use these 3000 supposedly uniform random numbers as coordinates of 1000 points in three dimensions, forming a cube of points in three-dimensional space. We then spin the cube around its center and watch the spinning cube for "structure."

In Figure 1.1 we show four views of the spinning cube of points. These are four views that show the rotation of the space leading up to the moment at which structure is revealed (the order of the views is left to right and top to bottom—the first view is upper left, the last is lower right). Each view differs from the preceding view by a rotation of only 3°. You can get an idea of how small a rotation this is by looking at the orientation of the cube. In the upper-left image in Figure 1.1 we see that the space looks like the featureless space it should be if the data are truly random. Then, as we

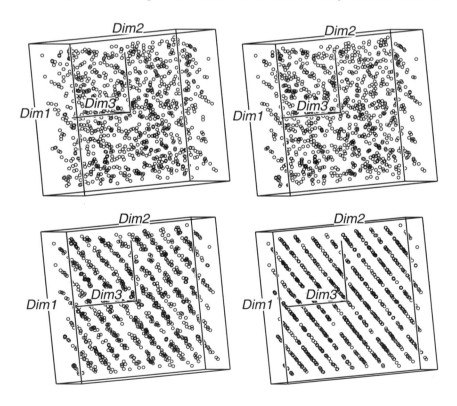

Figure 1.1 Four rotations of the same 3000 uniform random numbers. The 3000 numbers are now displayed as 1000 points in three dimensions.

move to the upper-right image, we begin to see an emerging structure, even though there has been only a $3°$ rotation. The structure, which shows that the points are lined up in parallel planes, emerges completely after two more $3°$ rotations.

Note well: If Randu worked right, we wouldn't see any structure, since there should be no structure in random numbers. However, we do see structure, so something is wrong with the numbers generated by Randu. Also understand that we cannot see this in a one- or two-dimensional plot, since the planes are not lined up with the dimensions of the space. So the question is: How can we find this particular view? The answer is: by using a dynamic graphic which spins the space before our eyes.

Thumb power. To see what this means, take a look at the images shown in the upper-right corner of the right-hand pages of this chapter. These images can be thought of as being frames of a movie that show a cloud of points spinning right in front of you, "before your very eyes," as it were (this point cloud was generated in the same way as the one shown in Figure 1.1). Each figure shows the cloud of points rotated $2°$ from the position of the point cloud in the preceding figure. You "run" the movie by thumb power—simply thumb through these pages and watch for the structure to appear. Go ahead, thumb through these figures and watch for the structure. What you see when you do this is a very rough approximation to what you would see if you were sitting at a computer watching the cloud of points rotate slowly before your eyes. Thus, if you watch the cloud of points slowly rotate or (better, yet) if you use a rotation tool to actively rotate the cloud of points, you would see the nonrandom structure appear for a brief moment and then disappear.

A dynamic graphic, in this case a three-dimensional spinning space, reveals the structure. Although a spinning space was not involved in discovering this structure (it was years before such software was available), having a dynamic graphic that spins the space reveals the structure, and having a dynamic interactive graphic, one that lets you rotate the image by rubbing your cursor on the screen, makes it even easier to find the structure.

1.3.2 Automobile Efficiency

It is very common in statistical data analysis to take actions that involve assuming that the data are *multivariate normal*: that each variable in the data is normally distributed. Because this is such a common situation, we have developed a visualization that uses dynamic interactive graphics to help the data analyst transform data so that they are more nearly multivariate normal. Our approach is based on the fact that when data are multivariate normal, every variable is linearly related to every other variable, a property called *bivariate linearity*. So data that are multivariate normal are also bivariate linear. Thus, if we could transform our variables to be bivariate linear, we could feel comfortable about the question of multivariate normality.

In Chapter 7 we develop a dynamic interactive visualization for transforming variables to improve their bivariate linearity. This visualization is shown in Figure 1.2 as it appears for automobile fuel efficiency data. The data are the values of five variables for 38 automobiles, the variables being the automobile's *Weight*, *MPG* (miles per gallon), *Horsepower, Displacement* (size of engine), and *DriveRatio*. The figure shows the visualization as it appears before any of the variables have been transformed (don't worry about the "fine print" in the figure—it is not important— what is important is the shapes of the distributions, which we wish to linearize).

The left portion of the visualization contains a matrixlike arrangement of plots. This *plotmatrix*, as it is called, has a row and a column of plots for each variable in the data. The right-hand portion of the visualization contains two large scatterplots (there is also a window containing a list of all of the automobiles, but it is not shown).

The diagonal of the plotmatrix, which goes from lowerleft to upperright, shows the transformation for each variable. The off-diagonal plotcells show the bivariate relationships, each with lines (defined in Chapter 7) emphasizing the linearity, or lack thereof, of the relationship. Since the figure shows the initial view, the transformations shown in the diagonal plots are linear.

Imagine that you are the data analyst and that you want to linearize these variables with respect to each other. You interact with the plot in two ways:

- **Click on a plotcell**. This defines the *focal plot* (the one you clicked on) and the *focal variable* (the Y-axis variable of the clicked-on plot).
- **Move the slider**. This transforms the focal variable using the Box–Cox transformation (described in Chapter 7).

When you click on a plot of the plot-matrix the two large plots change. The top one becomes a large version of the plot you clicked on, and the bottom one becomes a

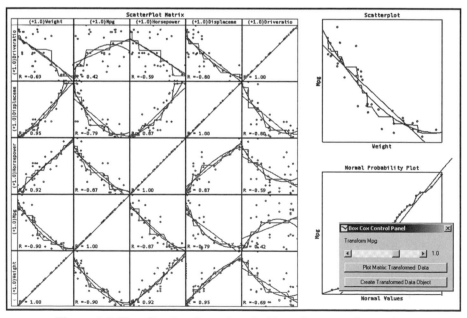

**Figure 1.2 Visualization for using the Box–Cox transformation
to improve bivariate linearity.**

12

**Table 1.1 Transformations That Linearize the Bivariate
Relationships Among Four of the Five Automobile Variables**

Variable	Slider Value	Transformation
Weight	0.5	Square root
MPG	-1.0	Reciprocal
Horsepower	1.0	Linear
Displacement	0.0	Log
Driveratio	—	(not linearizable)

normal-probability plot of the focal variable (which is linear when the focal variable is normal).

When you move the slider you see that the plot you clicked on changes, as do the two large plots. You also see that all of the plots in the focal-plot's row and column also change. They change because they show the bivariate relationships between the focal variable and each of the nonfocal variables. Since several plots change simultaneously, some may get worse. So you should select a position for the slider that shows the best overall change in the plot-matrix, not just the best linearization in the focal plot. When you have done this, you will probably want to focus on another variable to improve the linearity of its bivariate relationships. Gradually, as you focus first on one variable, then on another, the relationships between all pairs of variables may become more linear. At the conclusion of the process you can judge whether the variables are sufficiently bivariate linear to justify the assumption that they are multivariate normal.

In Figure 1.3 we show the plotmatrix we settled on as having the best overall set of linear bivariate relationships. We have "exploded" a 4×4 submatrix from the *Drive-Ratio* row and column to emphasize the fact that we have been, in our judgment, only partially successful in our attempt to linearize all of the bivariate relationships. It is clear that the exploded 4×4 section is mutually bivariate linear. Thus, we judge that these four variables are sufficiently bivariate linear to justify the assumption that they are multivariate normal. We cannot, however, include *DriveRatio* in this decision. This should warn us that, when we continue our investigations of these data, *DriveRatio* needs extra care and attention as the analyses proceed.

As we discuss in Chapter 7, the slider controls an argument of the Box–Cox transformation family, with different transformations resulting from different slider values. We moved the slider to find the value of the argument that produced the most linear transformation. Five of these values are special values that correspond to well known transformations. If the best value was close to a special value, we substituted the special value. The slider values that we choose are shown in Table 1.1. All except the value for *DriveRatio* correspond to special values. *DriveRatio* was not linearizable. It is interesting to note that the transformation of *MPG* not only linearizes with respect to three of the other variables, but also linearizes with respect to the measure used in most of the world, which is L/100 km.

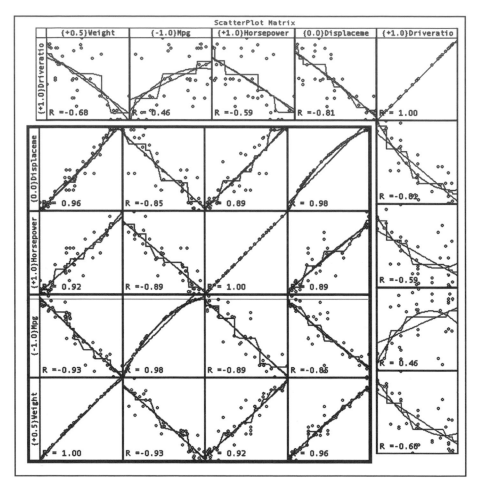

Figure 1.3 Off-diagonal plots show the bivariate relationships after transformation. Diagonal plots show the transformations.

1.3.3 Fidelity and Marriage

This third example shows how being able to interact with a statistical graphic can ease and simplify the interpretation of data. The example is based on a cross-classified table of frequencies calculated from four variables concerning the relationship of divorce to the occurrence of premarital and extramarital sex in men and women. We examine the link between the *Marital Status (M)* and *Gender (G)* of the respondents, and whether they engaged in *Premarital Sex (P)* and/or *Extramarital Sex (E)* in the past (this example is presented in greater detail in Chapter 2). A visualization method for frequency data that has recently gained popularity is the *mosaic display*. This display represents the observed frequencies as well as the residuals from fit of a model of

14

the frequencies (the residuals are the differences between the frequencies and estimates of the frequencies made by the model).

A mosaic plot uses area to represent frequency. Each mosaic plot uses a square to represent the total frequency. The square is subdivided into interlocking rectangles so that a rectangular area corresponds to the cell frequencies in the data table.

Consider the four mosaic plots shown in Figure 1.4. In each of these, the square that forms the perimeter of the plot represents the total frequency. In the top plot the total frequency has been broken down according to the frequencies of the premarital variable. The square is divided vertically into two rectangles called *tiles*. The ratio of the areas of the tiles correspond to the ratio of the frequencies of the table cells. This one-way mosaic tells us that about three times as many people did not have premarital sex as did. We refer to this as the *P mosaic*, since only the premarital (*P*) variable is used.

The second display in Figure 1.4 is a two-way mosaic plot of the frequency cross-tabulation of two of the variables in the study, the *PreMarital* and *ExtraMarital* variables. We refer to this as the *PE mosaic*. Notice that each tile in the upper display has been divided horizontally into two tiles, the division being in proportion to the conditional probability of the second variable given the first. Hence, the area of each tile is proportional to the observed cell frequency: We can tell at a glance that many more people abstained from both premarital and extramarital sex than engaged in either one or both of these behaviors. We can also see, with a second glance, that a larger proportion of those who had premarital sex also had extramarital sex than those who did not have premarital sex.

The third display is a three-way mosaic, where a third variable, in this case *Marital Status*, is used to subdivide each of four tiles of the two-way mosaic. This is the *PEM mosaic*.

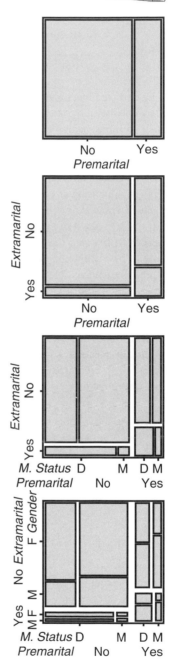

Figure 1.4 Examples of one-way (top) through four-way (bottom) mosaic plots.

Finally, the fourth mosaic is a four-way mosaic, where the fourth variable, namely *Gender*, is used to subdivide the tiles of the three-way mosaic, giving the *PEMG mosaic*.

The mosaic plot always represents the total frequency by a square. For an *n*-way table, the square is first divided into two vertical tiles, their areas being proportional to the cell frequencies of the first of the *n* ways. Then each tile is subdivided horizontally in proportion to the conditional frequencies of the second variable given the first. This continues, each tile being subdivided at right angles to the previous direction, the widths of the divisions being proportional to the frequencies conditioned on the variables used to form the tile. At each level of construction each tile's area is proportional to its observed cell frequency and probability.

By introducing just one more concept, that of shaded tiles, we get the full-fledged mosaic display shown in Figure 1.5. Except for shading, the display is identical to the bottom plot shown in Figure 1.4. Usually, the tiles in the display are shaded in gray or in color to indicate the value and direction of the residuals. Here we use black and white to indicate whether the actual cell frequencies are above or below, respectively, the cell frequencies expected under a given model. Although various models can be used to determine expected cell frequencies, here we use the loglinear equivalent of a logit model. This is done by setting *Marital Status* as the dependent variable.

The mosaic display can help us see patterns in the residuals that help us in our search for good models. Mosaic displays are appealing because they extend naturally to multidimensional tables, a feature uncommon in methods for frequency data. This allows us to visualize several variables using only one figure that summarizes the most important aspects of the data.

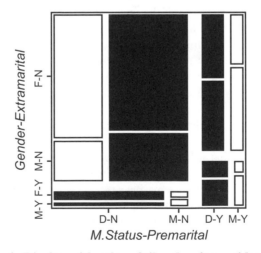

Figure 1.5 Mosaic Display with color of tiles showing residuals from a null logit model (setting *Marital Status* as dependent variable).

Notice that the order in which variables are entered into the mosaic display affects our view and understanding of the data, since the display shows conditional, not unconditional frequency and probability. That is, a four-way mosaic based on entering our four variables in the order *MGPE* gives us a different view than when the variables are entered in the order used in Figure 1.4 and Figure 1.5, which is *PEMG*. Since the view, and therefore our understanding, of the data are different, it is important to explore different variable orders. Furthermore, the different orders should be easily accessible by direct interactive manipulation of the display.

The modelling session, which involves using interactive graphics tools that allow us easily to try various models, begins by default with a model that includes no associations of the dependent variable (marital status) with the independent variables. This model, which is called the *base* or *null logit model*, is surely inadequate, but it is the logical starting point for an interactive modelling session.

The mosaic display in Figure 1.5 shows the variables in the order *PEMG*. Manipulating the order of the variables (as described in Section 2.3), we found that the order *MEPG* put together cells that have the same residual signs, making a more coherent display. The first split, by marital status, separates those still married from those who are divorced. Whereas the residuals were positive (higher observed than expected frequencies) for married people, those for divorced people were negative, and vice versa. This means that we could simplify the interpretation by focusing on either the divorced people or those who are married, seeing the same patterns although with opposite signs. Arbitrarily, we focused on those who were divorced, showing the simplified mosaic in Figure 1.6. This mosaic can be interpreted in a single, simple sentence: People with sexual encounters outside marriage, whether premarital or extramarital, are more likely to divorce than are those without such encounters.

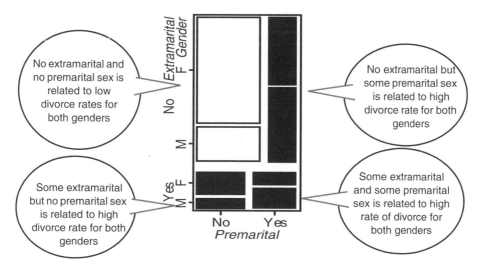

Figure 1.6 Mosaic Display after manipulating the order of the variables and simplification by focusing on *Marital Status*=Divorced.

17

1.4 History of Statistical Graphics

*M*ost modern readers may think of statistical graphics as the new kid on the block, whose immediate family is described in this book. In fact, visual statistics has an extended pedigree, with ancestors reaching back to the earliest visual portraits of information—early maps, navigation charts and tables, geometric and mechanical drawing, the rise of visual thinking, and so on. Much of this history is collected by Friendly and Denis (2004) and described by Friendly (2004). What we see from this history is that data graphics were primarily invented to provide visual solutions to problems (where to build a new railroad, what is the source of a cholera outbreak) by making data "speak to the eyes." As well, important historical graphs were almost invariably created to help tell a story about some significant substance, beyond what mere words and numbers could convey. We illustrate this history with a few short stories from selected periods.

1.4.1 1600–1699: Measurement and Theory

Among the most important problems of the seventh century were those concerned with physical measurement—of time, distance, and space—for astronomy, surveying, map making, navigation, and territorial expansion. This century also saw great new growth in theory and the dawn of practice—the rise of analytic geometry, theories of errors of measurement and estimation, the birth of probability theory, and the beginnings of demographic statistics and "political arithmetic.".

Figure 1.7 (source: Tufte, 1997, p. 15) shows what is believed to be the first visual representation of statistical information, a 1644 graphic by Michael Florent van Langren, a Flemish astronomer to the court of Spain, prepared as an executive summary of a small number of determinations of the distance, in longitude, from Toledo to Rome. At that time, lack of a reliable means to determine longitude at sea hindered navigation and exploration. For navigation, latitudes could be fixed from star inclinations, but longitudes required accurate measurement of time at sea, a problem unsolved until 1765.

The one-dimensional line graph reproduced in Figure 1.7 shows all 12 known estimates of the difference in longitude between Toledo and Rome, and the name of the astronomer (Mercator, Tycho Brahe, Ptolemy, etc.) who provided each observation. What is notable is that van Langren could have presented this information in various tables—ordered by author to show provenance, by date to show priority, or by dis-

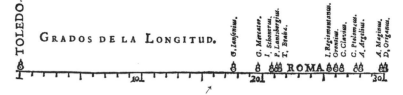

Figure 1.7 Langren's 1644 graph of distance from Toledo to Rome
(From Tufte, 1997, p. 15)

tance. However, only a graph shows the wide variation in the estimates; note that the range of values covers nearly half the length of the scale. Van Langren took as his overall summary the center of the range, where there happened to be a large enough gap for him to inscribe "ROMA." *Unfortunately, all of the estimates were biased upward*; the true distance (16°30') is shown by the arrow. Van Langren's graph is also the earliest-known exemplar of the principle of *effect ordering for data display* (Friendly and Kwan, 2003).

1.4.2   1700–1799: New Graphic Forms and Data

Early in the eighteenth century, map-makers began to try to show more than just geographical position on a map. As a result, new graphic forms (isolines and contours) were invented, and thematic mapping of physical quantities took root, and would later lead to three-dimensional visualization. Toward the end of this century, we see the first attempts at the thematic mapping of geologic, economic, and medical data.

Abstract graphs and graphs of mathematical functions were introduced, along with the beginnings of statistical theory (measurement error) and systematic collection of empirical data. As other (economic and political) data began to be collected, novel visual forms were invented so that the data could "speak to the eyes."

William Playfair (1759–1823) is widely considered the inventor of most of the graphical forms used widely today—first the line graph and bar chart (Playfair, 1786), later the pie chart and circle graph (Playfair, 1801). Figure 1.8 shows a creative combination of different visual forms: circles, pies, and lines (Playfair, 1801).

The use of two separate vertical scales for different quantities (population and taxes) is today considered a sin in statistical graphics (you can easily jiggle either scale to show different things). But Playfair used this device to good effect here to try to show taxes per capita in various nations and too argue that the British were over-

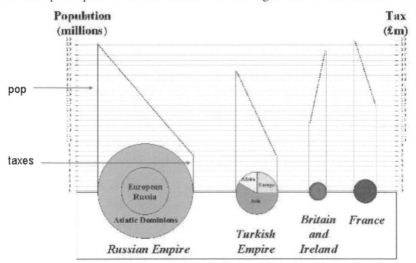

Figure 1.8 Playfair's 1801 pie–circle–line chart.

19

taxed, compared with others. But alas, showing simple numbers by a graph was difficult enough for Playfair (1786), who devoted 21 pages of text describing how to read and understand a line graph. The idea of calculating and graphing rates and other indirect measurements was still to come.

In Figure 1.8 the left axis and line on each circle–pie graph show population, and the right axis and line show taxes. Playfair intended that the *slope* of the line connecting the two would depict the rate of taxation directly to the eye. Playfair's graphic sin can perhaps be forgiven here, because the graph clearly shows the slope of the line for Britain to be in the opposite direction of those for the other nations.

1.4.3 1800–1899: Modern Graphics and the Golden Age

By the early part of the nineteenth century all of the modern forms of data display were invented: bar and pie charts, histograms, line graphs and time-series plots, contour plots, scatterplots, and so on. At least as important, the systematic collection of national data on demographic and economic topics (population distributions, imports/exports, transportation over land and water) became sufficiently widespread in Europe to need a general word to describe it —*statistik*, meaning "numbers of the state.". But there were social issues as well—crime, literacy, poverty (called "pauperism," as if to connote a disease, or conscious individual choice), suicide, and so on, and how these could be understood to affect state policy. Can we better reduce crime by increasing education or by making incarceration more severe? Can we better reduce pauperism by providing some form of social assistance or by building more debtors' prisons? These questions led to the rise of *moral statistics* (Guerry, 1833), now seen as the foundation of modern social science.

By midcentury, official state statistical offices were established throughout Europe, in recognition of the growing importance of numerical information for social planning, industrialization, commerce, and transportation. Together, the successes of a wide variety of graphical methods to convey quantitative information directly to the eyes, along with substantial bodies of data and state interest in outcomes led to an explosive growth in the use of graphical methods, many of unparalleled beauty, and many innovations in graphics and thematic cartography.

To illustrate this period, we choose an 1844 *tableau-figuratif* (Figure 1.9) by Charles Joseph Minard, an early progenitor of the modern mosaic plot (Friendly, 1994). The graphic shows the transportation of commercial goods along the Canal du Centre (Chalon–Dijon). Intermediate stops are spaced by distance, and each bar is divided by type of goods, so the area of each tile represents the cost of transport. Arrows show the direction of transport.

On the surface, mosaic plots descend from bar charts, but Minard introduced two simultaneous innovations: the use of divided and proportional-width bars, so that area had a concrete visual interpretation. The graph shows the transportation of commercial goods along one canal route in France by variable-width, divided bars (Minard, 1844). In this display the width of each vertical bar shows distance along this route; the divided bar segments have height—amount of goods of various types (shown by

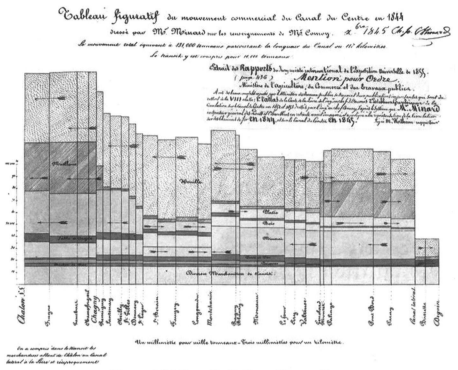

Figure 1.9 Minard's Tableau Figuratif.

shading), so the area of each rectangular segment is proportional to the cost of transport. Minard, a true visual engineer (Friendly, 2000a), developed such diagrams to argue visually for setting differential price rates for partial vs. complete runs. Playfair had tried to make data "speak to the eyes," but Minard wished to make them *calculer par l'oeil* as well.

1.4.4 1900–1950: The Dark Ages of Statistical Graphics— The Golden Age of Mathematical Statistics

If the late nineteenth century was the "golden age" of statistical graphics and thematic cartography, the early twentieth century could be called the "modern dark ages" of visualization (Friendly and Denis, 2001). There were few graphical innovations, and by the mid-1930s, the enthusiasm for visualization that characterized the late nineteenth century had been supplanted by the rise of quantification and formal, often statistical models in the social sciences. Numbers, parameter estimates, and especially, standard errors were precise. Pictures were—well, just pictures: pretty or evocative, perhaps, but incapable of stating a "fact" to three or more decimals. Or so it seemed to statisticians of the time. But it is equally fair to view this as a time of necessary dormancy, application, and popularization rather than one of innovation. In this period, statistical graphics became mainstream. It entered textbooks, the curriculum,

21

and standard use in government, commerce, and science. In particular, perhaps for the first time, graphical methods proved crucial in a number of scientific discoveries (e.g., the discovery of atomic number by Henry Mosely; lawful clusterings of stars based on brightness and color in Hertzprung–Russell diagrams; the Phillips curve in macroeconomics, relating wage inflation and unemployment) (see Friendly and Denis, 2004, for more details).

But the work that became the hallmark of the first half of the twentieth century has come to be known as *confirmatory statistics*, with the main activity being that of hypothesis testing. A statistical *hypothesis* is a statement that specifies a set of possible distributions of the data variable x. In hypothesis testing, the goal is to see if there is sufficient statistical evidence to reject a presumed null hypothesis in favor of a conjectured alternative hypothesis. Confirmatory statistics used the formalisms of mathematical proofs, theorems, derivations, and so on, to provide a firm mathematical foundation for hypothesis testing.

These new mathematically rigorous methods of confirmatory statistics proved to be very useful for research topics that were well enough understood to permit the researcher to construct hypotheses about phenomena of interest. Such a researcher now had a clear-cut way to choose between competing hypotheses: State hypotheses, gather data, and apply hypothesis tests. Thus, confirmatory statistics proved to be very useful and has remained a core topic and fundamental activity within statistics.

But what about those who were working in research areas that had not yet progressed to the point of being able to generate clearly stated hypotheses? For these disciplines, confirmatory statistics was not relevant, and until the groundbreaking work of Tukey (1977), there were no feasible alternatives. Thus, Tukey's work on exploratory data analysis, which provided researchers with a framework for generating hypotheses, was of immense importance.

1.4.5 1950–1975: Rebirth of Statistical Graphics

Still under the influence of the formal and numerical zeitgeist from the mid-1930s on, statistical graphics began to rise from dormancy in the mid-1960s, spurred largely by three significant developments:

1. In the United States, Tukey began the invention of a wide variety of new, simple, and effective graphic displays, under the rubric of *exploratory data analysis*. Tukey's stature as a mathematical statistician, together with his novel graphic methods and pithy philosophy of data analysis (''the greatest value of a picture is when it forces us to notice what we were not prepared to see'') made visual data analysis respectable again.

2. In France, Jacques Bertin published the monumental *Semiologie Graphique* (Bertin, 1967, Bertin, 1983). To some, this appeared to do for graphics what Mendeleev had done for the organization of the chemical elements, that is, to organize the visual and perceptual elements of graphics according to features and relations in data.

3. Finally, widespread computer processing of data had begun, and offered the possibility to construct old and new graphic forms using computer programs. True high-resolution graphics were developed but would take a while to enter common use.

But it was Tukey's work (1962) that had the biggest impact on statistical graphics, influencing how, and why, graphics would be used in statistics. Tukey argued that we should examine the data not only for what they say about a pre-defined hypothesis, but also for what they say about other possibly expected (or completely unexpected) results. One of Tukey's main contributions is the view that carrying out data analysis cannot be reduced to a single set of isolated calculations. In Tukey's view, data analysis is a process where steps are taken that suggest new steps that are based on the knowledge gained from the previous steps. Software that would support this type of freedom in data analysis did not exist at the time of Tukey's book. But today's statistical visualization systems are designed with this philosophy in mind.

By the end of this period, significant intersections and collaborations would begin: computer science research (software tools, C language, UNIX, etc.) at Bell Laboratories (Becker, 1994) and elsewhere would combine forces with developments in data analysis (EDA, psychometrics, etc.) and display and input technology (pen plotters, graphic terminals, digitizer tablets, the mouse, etc.). These developments would provide new paradigms, languages, and software packages for expressing statistical ideas and implementing data graphics. In turn, they would lead to an explosive growth in new visualization methods and techniques, and to statistical visualization, the area of research and development that is the topic of this book.

1.4.6 1975–2000: Statistical Graphics Comes of Age

During the last quarter of the twentieth century, statistical graphics blossomed into a vibrant research discipline. Perhaps the clearest evidence of the maturation of statistical graphics is the development of theoretical frameworks for statistical graphics. Two very important examples are the *viewing pipeline for data analysis* developed by Buja et al., (1988) and the *grammar of graphics* developed by Wilkinson (1999). The viewing pipeline shows how to construct the highly interactive, very dynamic graphics which are the subject of this book. The grammar of graphics presents a theoretical framework for statistical graphics, showing how data flow from their original state to their graphical representation. Surprisingly, Wilkinson's grammar does not include dynamic graphics, although it must have been clear how that could have been done.

The theoretical framework presented in this book, which also shows how data flow from their original state to their graphical representation, is, in fact, very similar to Wilkinson's grammar. However, our framework, being in part based on Buja's viewing pipeline, allows for dynamic graphics. Interestingly, our framework was developed without awareness of Wilkinson's. Such convergence of independent lines of research and development can be taken as further evidence of the maturation of statistical graphics.

The maturation of statistical visualization has also been influenced by work in allied fields outside statistics such as cognitive science, computer science, and human-computer interaction studies. These influences inform the way in which statistical visualization proposes to ease and improve data analysis, leading to an emphasis on the effect of the data analysis environment on an analyst's search for understanding.

Of central importance in this maturation was the concomitant maturation of computational capabilities and the increasing reliance of these capabilities on the use of graphical hardware and software, as emphasized by the National Research Council in their publication on the future of statistical software (Eddy et al., 1991).

It is interesting to note that Tukey, nearly 30 years earlier, was asked upon the occasion of the 125th anniversary of the American Statistical Association to deliver an address on the past, present, and future of technical tools of statistics (Tukey, 1965). His presentation was amazingly prescient: At a time when computers as large as buildings contained less computing power than is available in today's digital wristwatch, Tukey predicted "an increasing swing toward a greater emphasis on graphicality and informality of inference" and "a greater and greater role for graphical techniques as aids to exploration and incisivenes." Indeed, one of the most important challenges for implementing Tukey's ideas came from graphics.

Statistics has long included graphics as an important tool for revealing, seeing, and showing information in data. Statistical graphics have helped scientists explore their data for structure, to confirm the presence of such structure, and to communicate the results of their studies. Tukey himself is famous for introducing graphics that have become very popular, graphics such as the box and whiskers plot and the stem and leaf plot. However, to take full advantage of plots in exploratory data analysis, it was necessary to incorporate ways of interacting with statistical software, ways that were not available until the advent of the microcomputer.

1.5   About Software

At the time this is being written, there are six statistical visualization systems that seem to us to be the most important: JMP, DataDesk (Velleman and Velleman, 1985), Arc (Cook and Weisberg, 1994), ViSta (Young, 1994; Young and Smith, 1991), GGobi (Swayne et al., 1998, 2003) and Manet (Hofman, 2003; Unwin et al., 1996). Each of these systems uses dynamic interactive graphics to simplify data analysis and to display data structure. Of these, two are commercial systems and four are noncommercial. The six statistical visualization systems have evolved from research and development in dynamic interactive statistical graphics that began around 1960, with the first commercial systems becoming available in the late 1980s. Some of the major milestones of progress in visual statistics are:

- A direct manipulation system for controlling a power transformation in real time (Fowlkes, 1971), where we first come across the idea that one could connect a parameter and a statistical method manually with an on-screen controller.

- Prim-9, the first system with 3D data rotations (Fisherkeller et al., 1975), demonstrating dynamic interactive graphics for the first time. Other early systems with these capabilities are ORION (McDonald, 1988) and MacSpin (Donoho et al., 1988).

- The dynamic interactive technique known as *brushing* was developed around 1980. This technique allowed the analyst to select regions or points and to see them linked simultaneously in other plots (McDonald, 1982; Newton, 1978).

- The Grand Tour for visualizing multivariate data sets by creating an animation of data projections by moving a two-dimensional projection plane through n-space (Asimov, 1985). The novel idea here was to help the analyst seek informative views of high-dimensional data by defining various criteria of "interestingness."

- A systematic implementation of interactive statistical graphics, including brushing, linking, and other forms of interaction (Becker and Cleveland, 1988).

1.5.1 XLisp-Stat

XLisp-Stat (Tierney, 1988, 1990), has had considerable impact on the development of statistical visualization systems. XLisp-Stat is not, per se, a statistical system, but is a statistically and graphically oriented programming environment designed to facilitate the creation of statistical systems. It is based on a freely available implementation of the Lisp language called XLisp (Betz, 1985). Tierney (1988, p.4) gave three primary motivations for developing this language: (1) provide a vehicle for experimenting with dynamic graphics and for using dynamic graphics in instruction; (2) explore the use of object-oriented programming ideas for building and analyzing statistical models; and 3) experiment with an environment supporting functional data.

XLisp-Stat provided statisticians with the opportunity to implement ideas related to dynamic graphics in much the way that statistical language *S* had already provided a general statistical programming language (Becker, 1994; Becker and Chambers, 1981; Becker et al., 1988a; Ihaka and Gentleman, 1996). In fact, some of the features of XLisp-Stat were inspired by similar features or functions in *S*. One of the strong points of XLisp-Stat, Tierney and others argued, was the fact that it was based on Lisp, a general-purpose high (and low)-level programming language that was (and is) well known and mature. This guaranteed a solid foundation of technical aspects and a strong set of basic programming tools.

When XLisp-Stat is started by the user, all that is shown to the user is an empty screen with a prompt for typing commands. Using lisp syntax, commands for opening data files and obtaining statistical summaries and plots can be obtained. Also, new functions can easily be written and saved in files, so repeating sessions of analysis or expanding the system could be accomplished.

XLisp-Stat also provides tools for incorporating user interface elements in programs developed by users. In this way, Lisp programs could be turned into direct manipulation systems (Hutchins et al., 1986), the kind of user interaction system popularized by the Macintosh and by Microsoft Windows. These capabilities paved the way to some projects written entirely in XLisp-Stat that provided a user interface for manipulating data, computing results and showing visualizations. These projects are described by their authors in Stine and Fox (1996).

Note that there is some concern that XLisp-Stat is not a healthy, growing statistical system. Indeed, there is a fairly widespread opinion held by many computation statisticians that it has died, an opinion not shared by the current authors. We have recently published articles about the problems that currently exist (Valero-Mora and Udina, 2005; Molina et al., 2005), as have others intimately involved in the development, popularization, and then demise of the system (de Leeuw, 2005; Tierney, 2005). Perhaps the most important point is made by Weisberg (2005) who states that "solutions to applied statistical problems are framed by the limitations imposed by statistical computing packages and languages." Weisberg goes on to point out that the kinds of solutions supported by XLisp-Stat are very different from those supported by other languages, and that a variety of statistical languages can only benefit, not hurt, the field of statistics.

1.5.2 Commercial Systems

The two commercial systems are DataDesk and JMP, which now have a long history that has made them very solid and complete products that can be used for many everyday computing needs of data analysts as well as for visual statistics.

DataDesk. The DataDesk system (Velleman and Velleman, 1985) included from the very beginning, rotating plots, linking and brushing, and dynamic transformations, among other features. It also provided a visual metaphor for managing datasets, variables, analyses, and so on, and for representing the analysis process.

JMP. JMP (JMP, 2002), which is pronounced "jump," first became available in 1989, providing many of the dynamic interactive features introduced previously in the literature. It did not provide as extensive an environment for data analysis as that provided by DataDesk, and there was no vivid metaphor for the data analysis process itself. Yet JMP provided a system in which every analysis was accompanied automatically by graphs, without having to beg them to show themselves.

1.5.3 Noncommercial Systems

Of the four noncommercial projects, ViSta and Arc were developed using XLisp-Stat. They share the potential of being extended to incorporate new methods or techniques. Also, since XLisp-Stat is available behind the scenes, even though each system is based on a direct manipulation interface, command typing is also available.

Arc. Arc (originally named R-code, but was renamed to avoid confusions with the R software system) (Cook and Weisberg, 1994, 1999) is a regression analysis system that emphasizes regression graphics, many of which are dynamic. *Arc* is one of the reference programs for regression analysis.

ViSta. Vista, the Visual Statistics System (Young and Smith, 1991; Young, 1994), was also developed using the tools provided by XLisp-Stat. Since we use ViSta throughout this book, we describe it more extensively in the next section.

Manet. Manet is one of several statistical visualization programs developed by the computer-oriented statistics and data analysis group of the University of Augsburg (Unwin et al., 1996). Manet originally focused on visual estimation of missing values. It now incorporates many other visualization techniques and is particularly outstanding for visualizing categorical data (Hofman, 2003).

XGobi and GGobi. XGobi, and its recent sibling GGobi, are data visualization systems for exploring high-dimensional data that have graphical views that can be brushed and linked (Swayne et al., 1998). These views include dynamic interactive scatterplots, parallel coordinate plots, and scatterplot matrices.

Finally, as the basic ideas of dynamic statistical visualization have become more mainstream, many commercial statistical systems have tried to incorporate them, with, in our opinion, limited success. Programs such as SAS, SPSS, S-Plus, Systat, Statistica, and Minitab have incorporated dynamic interactive graphics techniques, although the complexities involved in integrating these techniques into their parent systems are often problematic and limiting. Also, there have been some attempts to include dynamic plots in R, the free implementation of S.

1.5.4 ViSta

This book uses ViSta (Young, 1994) as the primary system to demonstrate the concepts being described. Originally a research project concerning the use of dynamic interactive graphics, ViSta has become one of the three most widely used statistical visualization systems, along with JMP and DataDesk, with a new version recently released. Whereas JMP and DataDesk are both commercial systems with a closed software distribution model, ViSta is a noncommercial, freely available system using a moderated, partially open software distribution model. ViSta contains the full range of dynamic interactive graphics tools available in other statistical visualization systems. It also includes a visual metaphor for structuring data analysis sessions, a graphical tool for providing the data analyst with expert guidance, and an approach to organizing and coordinating multiple dynamic graphics.

The reason that we have chosen to focus on ViSta is very simple: The first author of this book is the designer and original developer of ViSta, and the second author has made major contributions to the system. As such, we are very familiar with ViSta. We have been using it for visual statistical analysis, for the development of new visualization methods, and for the development of the software architectures underlying such methods. Together, the authors of this book have about 25 years of first hand knowledge in the theoretical basis of highly interactive, very dynamic statistical graphics, as well as literally decades of 20-hour days spent in the trenches doing the immensely difficult and painstakingly demanding software development underlying ViSta's very dynamic, highly interactive graphics.

Although many of the techniques described in this book are well-known methods for data visualization, we have provided new solutions to some of the problems of data visualization that we believe are of fundamental importance. Our implementations of many of the dynamic statistical graphics tools (described later in the book) would have been difficult at best, and more often impossible, had we used only software provided by others.

ViSta focuses on a single thing—visual statistics—and attempts to do that thing as well as possible. As such, we know that ViSta does not include all the statistical visualization techniques, and it is not difficult to find implementations of certain methods that are better than ours. However, we believe that ViSta provides the user with an excellent structured environment for statistical visualization, which simplifies and clarifies statistical visualization. In our view, the most important aspects of ViSta are its graphical interface, its dynamic interactive graphics, and its multiview graphics.

ViSta is operated by a direct manipulation interface that includes menus, dialog boxes, and icons. This type of interface is appropriate for first-time and occasional users because they can use point and click strategies to control the program. A drawback of direct manipulation is that very often there is no method to record the history of actions of the users. ViSta offers the workmap as a way to solve this problem.

ViSta supports a wide range of dynamic and interactive graphics, including nearly all those mentioned in this book. Essentially all plots support brushing, labeling, and linking. The other basic techniques mentioned in Chapter 4 can be added to many plots. Also, many plots are customized for specific analytical situations.

One of the difficulties of having a large number of plots is organizing the windows on the screen so that they can be contemplated without impediments. ViSta provides spreadplots, discussed above. There is a reasonably simple interface that helps the moderately sophisticated developer write new spreadplots.

ViSta includes a basic set of tools for data management and processing. These allow importing and exporting data to text files, merging files, transforming variables, and more. All together, this set of functions guarantee that many data analysis sessions can proceed in ViSta completely without needing help from other programs.

ViSta supports the advanced user who wishes to enhance or extend its capabilities. This can be done by writing plug-ins, applets, or special-purpose programs. Plug-ins are programs that implement major data analysis and visualization techniques and which are interfaced with ViSta so that they become part of its standard capabilities. *Applets* are small bits of code that are served over the Internet to ViSta, which acts as the client in the standard client–server software distribution model. Special-purpose programs may be written on the fly to be used immediately or saved for later use. No restrictions areimposed.

ViSta can be downloaded from www.visualstats.org/. The developer who wishes to extend ViSta's data analysis and visualization capabilities can do so as described above. Those who wish to modify the system have access to portions of the code, with restrictions imposed on critical sections. Even these restrictions may be loosened following procedures defined on the web site. This provides the flexibility of an open system with the reliability of a closed system.

As of the date that this is being written, ViSta 6.4 is the standard version of the software, and the next release, 7.N, is feature-complete. The material in this book is based on version 7.6 which has many capabilities not in 6.4. While essentially all of the figures were prepared using ViSta, some represent specialized programming by the authors and cannot be produced with the point-and-click interface.

But this is not a book about ViSta; it is a book about visual statistics. We focus on the ideas and concepts of visual statistics in and of themselves, not specifically as implemented in ViSta.

1.6 About Data

As used by the ancients, the word *datum* referred to "something given," and its plural, *data*, referred to "a set of given things." Our verbal definition is as follows:

Data are a set of facts suited to the rules of arithmetic that are organized for processing and analysis. Data are often, although not necessarily, numeric.

Commonly, data do not make a clear and unambiguous statement about our world, often requiring tools and methods to provide such clarity. These methods, called *statistical data analysis*, involve collecting, manipulating, analyzing, interpreting, and presenting data in a form that can be used, understood, and communicated to others.

In an unstructured statistical analysis system, the tree of decisions that needs to be considered before selecting an appropriate visualization or statistical model can be large and unwieldy. Negotiating the maze of choices can be confusing and wearing. Many software systems present long lists of options spread across many different dialog boxes or menus, even though many of the choices may be irrelevant for the specific dataset being analyzed. As a consequence, it is not surprising that many data analysts claim that typed commands are the best way to interact with a statistical analysis system. After all, they would argue, when you type commands, you just write what you want to do and you do not need to consider all of the irrelevant choices.

Although we agree that typing commands can be a more efficient technique on many occasions, especially for expert users, we would also argue that interacting with dynamic graphics is usually best accomplished via a well-designed graphical user interface, one that creates a structured data analysis environment.

We suggest that for data analysis, knowing something about the basic nature of the data being analyzed provides an unobtrusive means for structuring the data analysis environment, which simplifies and clarifies the user's data analysis tasks. We call the "basic nature of the data" the *datatype* of the data. By using the data's datatype to structure the data analysis environment, the statistical system and the user, working together, can make educated guesses about what should be done with the data, providing an unobtrusive way of guiding the analyst by limiting the analyst's choices. Thus, using datatype can improve the overall analysis and visualization experience since it limits the number of choices that the user must confront.

As an example, consider the data in Table 1.2, shows all the data for the 104 students who were in the first author's introductory psychological statistics course over a four-year period. The variables are *GPA*, the student's grade-point average (on a 4-point scale, 4 being straight A's); the student's mathematical and verbal *SAT* scores (a standardized university entrance examination, 800 maximum); whether the student was taking the course as part of a major in science (BS) or Arts (BA) (which differ in that the "science" major requires calculus, whereas the "arts" major does not), and the *Gender* of the student (the names are fictitious).

Consider representing graphically a variable such as *"Degree"*, a variable that indicates whether the students were taking the course within the "School of Arts" for a B.A. degree or within the "School of Sciences" for a B.S. degree. Knowing something about the data, such as the fact that we have categories of observation that are BA and BS rather than numbers such as 0 and 1 means that we can simplify the choices provided to the data analyst. For example, categorical variables are usually represented by graphics of the frequencies of the categories in the variable. Bar charts and mosaic plots are both suitable choices for such data, but scatterplots and histograms are not. Therefore, the menu listing the available options can be greatly simplified.

The example just presented uses only one of the five variables in the student dataset. A more realistic example would include all of the variables, especially since some are categorical and some are declared to be numerical. Knowing the dataset's datatype, which is determined by the mix of variable types, helps to determine the most appropriate techniques for the data, thereby reducing unnecessary cognitive effort (why not "let the machine do it" if it can). For these data the datatype is "grouped," as shown in the upper left of the datasheet under the data's name. But that's getting ahead of ourselves. First, we need to discuss the essential aspects of data.

1.6.1 Essential Characteristics

It can be said that data have no innate characteristics that exist independent of human observation, and that data are data only because we organize them in a particular way (de Leeuw, 2004). However, data have essential characteristics that are important to us and that are presented in Table 1.3.

There do not seem to be clearly definable rules that lead to determining which type of organization or which metric is appropriate for a particular set of data. Indeed, we know (Gifi, 1990) that the various ways of organizing data are interchangeable, and that convenience, convention, and common practice are the main reasons for choosing one over another. Nor are there clearly defined empirical aspects that unambiguously determine the data's metric. It is also an annoying fact of a data analyst's life

**Table 1.2 The Students data: *GPA*, *SAT* Scores, *Degree* program, and *Gender* of
102 Students Taking Introductory Psychological Statistics (False Names)**

Student Label	GPA Num	SATM Num	SATV Num	Degr Cat	Gend Cat	Student Label	GPA Num	SATM Num	SATV Num	Degr Cat	Gend Cat
Bennett	3.3	800	770	BS	M	Peg	2.7	610	590	BS	F
Laura	3.5	780	590	BA	F	Laura	3.0	610	550	BS	F
Chris	2.3	770	690	BA	M	Trudy	3.6	610	590	BA	F
Joyce	4.0	760	670	BA	F	Judith	3.0	610	500	BA	F
Nathan	3.7	750	610	BA	M	Abby	2.8	610	610	BA	F
Jean	3.5	730	570	BA	F	Olga	3.3	600	540	BS	F
Toni	3.5	720	590	BA	F	Carol	2.6	600	650	BA	F
Lyle	3.0	720	650	BA	M	Lyris	3.7	600	600	BA	F
Jake	3.7	720	640	BA	M	Jean	2.8	600	600	BA	F
Cindy	3.4	720	650	BA	F	Maria	2.8	600	450	BA	F
Laurie	3.2	710	600	BS	F	Anne	3.5	600	660	BA	F
Mary	2.3	710	560	BA	F	Susanne	3.8	600	630	BA	F
Maurie	2.9	710	700	BA	F	Mary	3.6	590	570	BA	F
Jane	2.9	710	600	BA	F	Chris	3.9	590	690	BA	F
Gary	3.0	700	610	BA	M	Suma	2.9	590	540	BA	F
Edward	3.0	700	660	BA	M	May	3.5	590	610	BA	F
Judy	2.5	700	500	BA	F	Jonathan	3.1	575	575	BA	M
Jeff	3.5	700	540	BA	M	Lena	3.5	560	600	BS	F
Gina	3.8	690	710	BS	F	Maggie	3.0	560	690	BS	F
Debra	3.8	690	570	BS	F	Judith	2.5	560	560	BA	F
Jake	3.3	690	770	BS	M	Lyris	2.9	560	440	BA	F
Bepi	2.8	690	570	BA	F	Alice	3.4	560	600	BA	F
Rick	3.7	690	700	BA	M	Florence	3.1	560	540	BA	F
ALice	3.2	690	710	BA	F	Judy	2.7	560	550	BA	F
David	3.4	680	610	BS	M	Jane	3.2	550	610	BA	F
Sheila	3.0	680	620	BA	F	Rosalie	3.5	550	570	BA	F
Maggie	3.8	680	680	BA	F	Amy	3.0	550	650	BA	F
Dave	3.9	680	660	BA	M	Frederic	3.6	550	560	BA	M
Martina	3.7	680	520	BA	F	Cindy	2.8	540	480	BA	F
Pedro	3.3	680	580	BA	M	Arlette	3.0	540	700	BA	F
Sandy	2.5	670	610	BA	F	Amy	2.5	530	580	BS	F
Thomas	3.8	670	500	BA	M	Becky	2.8	530	540	BS	F
Sue	2.8	670	420	BA	F	Arlene	2.5	530	480	BA	F
Sharon	2.6	660	540	BA	F	John	2.5	530	530	BA	M
Lynette	3.0	650	690	BS	F	Marilyn	3.0	520	700	BA	F
Sharon	2.9	650	550	BA	F	Martha	3.0	520	690	BA	F
Melissa	3.2	650	450	BA	F	Maria	2.8	520	500	BA	F
ALbert	3.1	650	600	BA	M	Ann	3.0	510	510	BA	F
Ellen	3.9	650	630	BA	F	Ian	3.2	500	700	BA	M
Sandy	3.1	650	600	BA	F	Janet	2.5	500	500	BA	F
Patty	3.6	650	640	BA	F	Tammy	2.6	500	500	BA	F
Steve	3.3	650	570	BA	M	Marilyn	2.6	500	560	BA	F
Tony	2.9	640	620	BA	M	Dorothy	2.6	490	550	BA	F
Harriet	3.4	640	580	BA	F	Donald	2.9	490	590	BA	M
Cindy	3.8	640	600	BA	F	Valerie	3.2	480	530	BA	F
Mary	3.6	640	590	BA	F	Valerie	2.3	450	480	BA	F
Sue	3.3	640	620	BA	F	Ian	2.9	440	550	BA	M
Lee	3.1	630	620	BS	M	Ralph	3.1	440	550	BA	M
Charles	3.0	630	670	BA	M	Marie	2.3	430	560	BA	F
Susan	3.2	620	590	BS	F	Kim	2.2	400	580	BA	F
Margaret	2.5	620	480	BA	F	Rose	3.1	390	610	BA	F
Mary	2.7	610	610	BA	F	Lisa	2.8	340	440	BA	F

31

Table 1.3 Essential Characteristics of Data

Dataset Characteristics	Data Element Characteristics
Organization:	**Missingness:**
We organize data into subsets that correspond to one of the following:	Each dataelement has a binary indicator variable that indicates the datum's missingness state:
• Variables	
• Tables	• Observed
• Matrices	• Missing
Metric:	**Activation:**
We define the rules of arithmetic so that the data are one of the following:	Each dataelement has a binary indicator variable that indicates the datum's activation state:
• Categories	
• Ordered categories	• Activated
• Magnitudes	• Nonactivated
• Frequencies	—

that any given set of data may have *missing* elements—elements that should have been observed but for whatever reason, were not.

Finally, the activation state of a datum indicates whether the datum is included or excluded from the ongoing analysis. You may wonder why we need a mechanism to exclude portions of the data from analysis, as it seems that all of the data should always be used. In fact, the flexibility provided by being able to exclude portions of the data temporarily is very important: It lets us obtain a more detailed understanding by focusing on specific portions of the data. With this capability we can shift our attention between various parts of the data as we wish.

1.6.2 Datatypes

As we mentioned above, and as summarized in Table 1.4, the datatype of a dataset is determined by the way it is organized and by the metric of each of its partitions. Although we identified three organizational principles (variable-based, matrix-based, and table-based) and three metrics (category, numerical, and frequency), we cannot simply take all combinations of organization and metric to obtain nine datatypes. There are other considerations. One of these other considerations simplifies the problem of determining datatype. This occurs when we realize that the datatype of (1) variable-based data with one or more frequency variables are always of *frequency* datatype; (2) table-based data are always of *frequency* datatype; and (3) matrix-based data are always of *association* datatype.

Unfortunately, there are also considerations that complicate the problem of determining datatype from the dataset's organization and metric. In particular, variable-based data nearly always have several variables, and the variables can have different metrics. Furthermore, the activation state of the variables can change, meaning that

Table 1.4 The Organization and Metric of Various Datatypes

	Organization		
Datatype	**Type**	**Number**	**Metric**
Frequency			
Table	*n*-way Table	$n=1$	Frequency & Category
Category	Variable	n	Category
Magnitude			
Univariate	Variable	$n=1$	Magnitude
Bivariate	Variable	$n=2$	Magnitude
Multivariate	Variable	$n>2$	Magnitude
Grouped	Variable	$n+c$	Grouped and Magnitude
Missing	Variable	n	Magnitude
Association			
Association	Matrix	n	Associations

the mix of metrics can change. And, since the mix of metrics determines the datatype, the datatype can change when the activation states of the variables changes. Thus, we cannot really view datatype as a property of a dataset. Rather, datatype is a property of the active portion of a dataset.

Regardless, the concept of datatype usefully organizes and distinguishes the various kinds of analyses and visualizations that are possible for a given set of data, and the fact that the datatype can vary as a function of the nature of the active variables in a dataset only serves to increase the importance of the concept.

These considerations lead us, then, to emphasize datatype as a centrally important concept, having great relevance to the task of seeing data. We use the concept to organize the material in this book, just as we used it to organize the functioning of our software. The specific datatypes that we use, and their specific organization and metric characteristics, are given in Table 1.4. The organization number column specifies the number of variables, tables or matrices.

These datatypes classify all datasets that we will encounter. Note, however, that variable-based datasets with a mix of both numerical and frequency variables have a general datatype, and are not analyzable until an active subset of variables is defined which does not have both numerical and frequency variables.

As noted above, the plotting methods discussed in this book reflect the distinctions between the kinds of data just defined. Furthermore, the concept of datatype is one of the concepts used to organize this book, as we discuss in Section 1.7.2.

1.6.3 Datatype Examples

We illustrate datatypes in Figure 1.10 using the students dataset shown in Table 1.2. There are six different panels in the figure. Each panel shows a datasheet. Five of these panels correspond with one of the datatypes in Table 1.4, including the two varieties of frequency data. The remaining panel (c) is a special reformatting of panel e that can only be used for one or two-way tabulations.

Note that three of the panels (a, b, and d) show only an incomplete portion of the entire datasheet, which for all three panels is 104 rows (observations) tall. In these panels the lower edge of the datasheet is missing. The other three panels (c, e, and f) show the entire datasheet, as is evidenced by the fact that you can see the lower edge of the datasheet. Note also that there are three varieties of frequency data. The varieties can be converted into each other and are equivalent except for the fact that students are identified in category variable-based raw frequency data (panel e), but the identity is lost in the other two aggregated codings.

Finally, note that the association data (panel g) comprises correlations computed among the three numerical variables. These could also be other types of derived association measures, such as covariances or distances. Association data can also be measures of distance or association observed directly.

1.7 About This Book

A few words about what this book is (and isn't), how it is organized, who our audience is, and two novel graphic presentation methods (comics and thumb-power).

1.7.1 What This Book Is—and Isn't

This book is about statistical visualization, the goal being to present the concepts of statistical visualization techniques and methods in a way that is independent of specific implementations of such techniques and methods. This book is not a how-to guide. It is not a user's guide for how to do statistical visualization with a specific software system. Admittedly, we walk a fine line, as the book is illustrated by just one software system—ViSta —throughout. But our aim is to use ViSta (Young and Rheingans, 1991) to illustrate points being made about statistical visualization rather than to teach you how to do statistical visualization with the ViSta software system. We do, however, recommend that you access the ViSta website at www.visualstats.org to obtain the system and follow the guidelines presented to generate the figures shown in the book.

1.7.2 Organization

Visual statistics include two related but distinct applications of dynamic interactive graphics. We call them *statistical environment visualization* and *statistical analysis visualization*. These two visualization situations are related to each other in that they

Students Grouped	GPA Num	SATM Num	SATV Num	Degr Cat	Gend Cat
Olga	3.2	600	540	BS	F
Maggie	3.0	560	690	BS	F
Laura	3.0	610	550	BS	F
Benett	3.3	800	770	BS	M
Jake	3.3	690	770	BS	M
...	...	...	...	...	...

(a) Grouped Magnitude Data (variable-based)

Students Magnitude	GPA Num	SATM Num	SATV Num
Olga	3.2	600	540
Maggie	3.0	560	690
Laura	3.0	610	550
Benett	3.3	800	770
Jake	3.3	690	770
...	...	...	...

(b) Magnitude Data (variable-based)

Students Frequency	MALE Frequency	FEMALE Frequency
SCIENCE	4	12
ARTS	22	66

(c) Aggregated Frequency Table Data

This type is not used elsewhere in the book and can only be used for one- or two-way tabulations.

Students Grouped	Degr Cat	Gend Cat
Olga	BS	F
Maggie	BS	F
Laura	BS	F
Benett	BS	M
Jake	BS	M
...	...	...

(d) Raw Frequency Data (category variables)

Students Frequency	Frequency Frequency	Degr Cat	Gend Cat
BS*M	4	BS	F
BS*F	12	BS	F
BA*M	22	BS	F
BA*F	66	BS	M

(e) Aggregated Frequency Data (*n*-way table)

Students Grouped	GPA Num	SATM Num	SATV Num
GPA	1.00	0.39	0.35
SATM	0.39	1.00	0.36
SATV	0.35	0.36	1.00

(f) Association Data (matrix-based)

Figure 1.10 Datatype Examples

35

Table 1.5 Book Organization

Part	Chapter		Datatype
I	**Introduction**		
	1	Introduction	
	2	Examples	
II	**See Data—The Process**		
	3	Interfaces, Environments and Sessions	
	4	Tools and Techniques	
III	**Seeing Data—Objects**		**Datatype**
	5	Seeing Frequency Data	Frequency
	6	Seeing Univariate Data	Magnitude
	7	Seeing Bivariate Data	Magnitude
	8	Seeing Multivariate Data	Magnitude
	www.visualstats.org	Seeing Grouped Data	Grouped
	www.visualstats.org	Seeing Association Data	Association
	9	Seeing Missing Data	Missing
IV	**Theoretical Foundations**		
	www.visualstats.org	Advanced Topics	

each visualize the same set of statistical objects. However, the statistical environment visualization of an object is very different from its statistical analysis visualization.

Statistical environment *visualization.* A statistical environment visualization is a visual representation of the data analysis session, including the steps that are taken during the session and the environment in which those steps take place. The representation consists of images representing the objects of the analysis and the actions of the analyst. A statistical environment visualization emphasizes the relationships between the various objects and actions, rather than the nature of the objects themselves.

Statistical analysis visualization. A *statistical analysis visualization* is a visual representation of a specific statistical object, the representation consisting of images representing detailed aspects of the object. The visualization emphasizes the object's nature and internal structure, not its relationship to other objects.

Datatype. There is a much larger amount of material about statistical analysis visualization than about statistical environment visualization. After all, research focused on statistical environment visualization began only 15 or 20 years ago. We say "only" since research focused on statistical analysis visualization dates back more than 200 years, as we have seen above. Thus, we adopt a second organizing principle to bring order to the large amount of material concerning statistical visualization. That organizing principle is datatype, discussed in Section 1.6.2. The resulting chapter structure is shown in Table 1.5.

Organization of statistical analysis visualization chapters. The chapters on statistical analysis visualization are all example driven, and most of them have the same organization. Following an introduction to the chapter's topic is a section presenting the data that are used throughout the chapter and a section presenting the graphical tools covered in the chapter. These sections are followed by sections covering visual methods for exploring, transforming, and fitting the type of data covered in the chapter. Each of these sections is illustrated by applying the tools to the data. Some of these sections are not included in the book, but are available online at www.visual-stats.org.

1.7.3 Who Our Audience Is—and Isn't

It is our aim to communicate the intrigue of statistical detective work and the satisfaction and excitement of statistical discovery, by emphasizing visual intuition without resorting to mathematical callesthenics. Visual statistics follows its own path. Seldom is there mention of populations, samples, hypothesis tests, and probability levels. We use the "interoccular impact test" because you'll know when it "hits you between the eyes." As we write, we have in mind readers who are:

Novices. This book is written for readers without a strong mathematical or statistical background, those who are afraid of mathematics or who judge their mathematical skills to be inadequate; those who have had negative experiences with statistics or mathematics, and those who have not recently exercised their math or stats skills. Parts I, II, and III are for you.

Practitioners. This book is written for readers who are actively using statistical or data analytic methods and who wish to find a new and complementary path toward information discovery and confirmation. These readers include teachers, data analysts, statistical consultants, and scientific researchers. The entire book is for you, although some topics in Parts III and IV may not be relevant to your needs.

Developers. This book is written for readers who wish to develop new visual statistics methods, including computational and graphical statisticians and their students and co-workers. These readers will be interested in understanding how those working on one visual statistics research and development project solved software design and implementation problems that are similar to those they may be facing in their own project. The entire book is for you.

But this book is not for you if you wish to put it under your pillow and learn statistics by osmosis. As one of the reviewers of our book pointed out, it's not unlike learning to play the piano: The process will take much effort. At times it will be frustrating. Much of the time it may be just plain boring. But in the end it should be gratifying. And we do believe that ultimately, statistics will be fun for you.

1.7.4 Comics

Writing about dynamic graphics is a frustrating and difficult enterprise, as was first, and most eloquently, pointed out by Wainer (1988) in his comments on one of the first complete presentations of these techniques (Becker et al., 1988b). Writing is a static, noninteractive, linear process, whereas dynamic interactive graphics is dynamic, interactive, and nonlinear. This problem can be regarded as an instance of the problem of representing motion via static images (Cutting, 2002). We will do our best to communicate the excitement of cutting-edge dynamic interactive graphics through this ancient medium, but at times we despair. And always, we urge you to abandon the book, at least temporarily, for your computer. Try out what we are writing about, and see if it makes more sense when you see it happening rather than when you just read about it. Then, return to the book for whatever additional light this slower-paced, more reflective medium can shed.

One way to deal with the limitations of the written word is to use videos to demonstrate the techniques. However, after seeing quite a few of these videos over the years, we find that they are too often quite disappointing. Although it is true that videos are an improvement over textual descriptions, they have the drawback that readers cannot work at their own pace. Also, they use a medium different from what is generally used in scientific communication. Furthermore, it takes considerable effort to produce a suitably professional video.

Interactive analysis is generally made up of different steps, which themselves involve several substeps or actions of the user. Traditional, text-only descriptions are difficult to follow because they force the reader to create mental representations of the visualizations and of the effect of actions carried out on the visualizations, without providing the support of actual images.

Of course, books can include figures to illustrate the descriptions, but alas, the figures can be quite complex. The figures also require detailed descriptions. Whereas traditional expositions of statistical results aim for a simple graphic or table that summarizes the results, interactive analysis requires that the reader be shown the actions undertaken at each stage of the analysis. Since we are basically trying to describe narratives of data analysis, one or two figures rarely suffice: We need a chain of figures to effectively illustrate the analysis process. But then we take too many pages. So we must find new ways of transmitting interactive analysis of data.

A medium that is very effective at telling stories via a series of static pictures is comics. As defined by Scott McCloud (1994) comics are "spatially juxtaposed pictorial and other images in deliberate sequence." In comics, each frame captures a visual snapshot of the flow of the events such that the several interconnected frames tell the entire story. The frames are accompanied by text, which points directly to the place that is most meaningful. Using these resources, comics communicate narratives that extend over one or several periods of time, resulting in a powerful instrument that mixes, and in some aspects improves, painting and writing.

Compared with video, comics are a powerful instrument for effective communication of information. For example, Tufte (1997, p. 144) describes two ways of reporting about the dangers of a river: a comic like diagram that mixes text and drawings with a television account of a piece of news. He concludes that the fixed one-dimensional ordering of a film like television narration would be much less eloquent and engaging for conveying the message than the comic like description as portrayed in the diagram. The diagram is superior because the readers can work at their own pace and can pay attention to a number of details that otherwise would pass unnoticed in television. Hence, in this, and possibly in many other similar examples, comics are a better alternative than recorded films or videos for transmitting complex information.

The way that comics mix text and graphics in the same panel is also an advantage with regard to the traditional use of figures in scientific books. As noted by McCloud (1994), there has been a trend in modern times to avoid using text and graphics together, automatically considering a message conveyed in such a way as too simplistic and only appropriate for children. He says (p. 140) "words and pictures together are considered at best a diversion for the masses, at worst a product of crass comercialism." Indeed, applied to our problem of describing statistical graphics, we found this attitude especially regrettable, as mixing words and pictures in the same panel makes the explanations exciting and enjoyable.

Thus, in this book, the reader will find figures that make use of "balloons" pointing to relevant pieces of information (as seen, for example, in Figure 1.6), either giving interpretations or adding more meaning to the graphics. Hence, the reader does not have to look for explanations of the figures in the main part of the text or in the figure captions, but can concentrate on examining the figures themselves to extract their meaning. Also, we have used sequences of pictures to illustrate processes of data analysis that can not be summarized in only one figure (see, for example, Figure 2.7).

We have the impression that we have merely scratched the potential of comics applied to the communication of scientific information. Comics have a rich language that is usually associated with frivolous topics. However, we believe that this association is unfair, and that, to the contrary, comics might strength vigorously our capabilities to convey formal knowledge effectively. We hope that this book provides enough examples to convince you of such potentiality.

1.7.5 Thumb-Powered Dynamic Graphics

A final feature of this book, which is, to the authors' knowledge, an original method of communicating dynamic graphics in print, is the use of what we affectionately call "thumbers"—the small images in the upper-right corner of some of the right-hand pages. These dynamic (though not really interactive) graphics do a moderately good job of communicating the dynamic aspects of a dynamic interactive graphic. And they are very easy to use. They are, however, very timeconsuming to create and format properly, which explains why we don't use them in every chapter.

1.8   Visual Statistics and the Graphical User Interface

Visual statistics posits that a dynamic interactive graphical environment that is mathematically, cognitively, and perceptually appropriate will be fun and relaxing, will encourage visual intuition and playfulness, and will improve and ease the data analysis process. Visual statistics should also make the statistical work of every user, no matter how naive or sophisticated, simpler, faster and more accurate. At their best, statistical visualizations let the data analyst play with the data, forgetting completely about the computer, and freeing up all of the analyst's abilities to focus on understanding the data while the computer slaves over the boring details. For these reasons, visual statistics should widen the audience of users, making statistical data analysis accessible to those who, for example, have no formal training in statistics, are afraid of mathematics, or have had negative prior experiences with statistics.

Of special relevance is the point made by those working in cognitive science that the user's perceptual and cognitive capabilities are of primary importance when sculpting the human–computer interface: The visualizations must pay attention to the visual aesthetics of the moving graphics, making sure that they are constructed and displayed in a perceptually and cognitively accessible way. If these concerns are respected, the resulting system will be one that can ease and improve the data analyst's ability to generate and test hypotheses about information in the data.

Visual statistics has also been heavily influenced by the results of research into the computer science concept of *intelligence augmentation* (IA), a concept introduced by Brooks (1975, 1995). IA is an approach to software engineering that emphasizes the user, the goal being to help the user be smarter. This emphasis is the reverse of that proposed by the better known artificial intelligence (AI) approach to software engineering, which emphasizes making the machine smarter.

1.9   Visual Statistics and the Scientific Method

As we stated in the opening section, seeing data refers to the computer-augmented process that enlists dynamic interactive graphics to help us understand data. Seeing data also refers to the result of that process—to *the seeing* of data. Thus, when a data analyst interacts with a dynamic interactive statistical visualization system in order to gain a better understanding of the data, the data analyst is seeing data. Note that individual steps in the seeing data process may involve graphics that are static, passive, or noninteractive, as well as graphics that are dynamic or interactive, although we suspect that graphics that are dynamic and/or interactive are more effective at helping us see our data. Seeing data also refers to a scientific paradigm by which we gain understanding of our data. It is a rather different paradigm than that used in other areas of the discipline of statistics. And as you proceed throughout the book, and compare it with other statistics books, you will see that we do not take the usual position on how one goes about the scientific enterprise, nor do we have the common view of how statistical data analysis is used to help in that pursuit. We do not often talk of populations and samples, nor of significance tests and probability levels. These topics, which are standard in other statistics books, are not often mentioned here not because we reject their validity and helpfulness. Rather, we do not mention them often because we see them as being familiar and well understood, and because we wish to emphasize the importance of the subjective experience of the expert user as that user brings to the fore all the substantive knowledge that he or she has to understand the data at hand.

We wish to move the statistician's role farther from center stage of the scientific theater, relegating it to a supporting role on stage left or right, making room for the person with expert substantive knowledge to occupy center stage without feeling either upstaged or crowded, thereby enabling the expert to make use of his or her expert knowledge in a way that everyone, statisticians and experts alike, sees as being a statistically acceptable way, using the perhaps radical approach of the visual statistician to augment the expert's substantive knowledge with the traditional statistician's classical approach based on mathematical statistics.

After all, we do agree that statistical data analysis is concerned with generating and evaluating hypotheses about data. For us, generating hypotheses means that we are searching for patterns in the data—trying to "see what the data seem to say." And evaluating hypotheses means that we are seeking an explanation or at least a simple description of what we find—trying to verify what we believe we see.

We believe that one should always strive for the clarity provided by the nontraditional and totally subjective IOI "measure" discussed below. We also believe that the visual data analysis process described below can lead to models with greater usefulness than those derived in the traditional manner. It is our position that the traditional definitions of fit, significance, parsimony, and effect size should be considered, but only in their roles as supporting actors, with the leading role being given to the visual impact as perceived by the expert data analyst. Naturally, the data analyst must judge his or her own degree of expertise and weight the visual impact accordingly.

Visual statistical data analysis supports the scientific enterprise by repeated application of the steps outlined below. Taken together, these steps form a visual statistics paradigm for scientific understanding and decision making. Note that the search for understanding can involve many cycles of exploration and revelation, and that the transformation step may need to be repeated many times.

1.9.1 A Paradigm for Seeing Data

Visual exploration. Exploring data generates hypotheses about patterns in our data. The visualizations and tools of dynamic interactive graphics ease and improve the exploration, helping us to "see what our data seem to say."

Visual transformation. Transforming data to measurements of a different kind can clarify and simplify hypotheses that have already been generated and can reveal patterns that would otherwise be hidden. Note, however, that transformation is not made without implications for the revision step described below.

Visual fitting. Fitting data with a model shows us a model-based view of our data that can be simple and clear. Visual fitting uses visualizations that support envisioning and revisioning, where fitting and evaluation are both classical and visual.

- **Envisioning** involves fitting a model to the data using traditional least squares, maximum likelihood, or chi-square techniques. This process yields estimates of the model's parameters. The traditional techniques are visually augmented by visualizations that show the model-based view of the data.

- **Revisioning** involves modifying the estimates of the model's parameters using visual techniques. We realize that once we envision the data, we may wish to fine tune the envisioned model: Perhaps we have obtained more information or have otherwise improved our understanding. Or perhaps tweaking the parameter estimates will improve interpretability. Revision methods involve visually based direct manipulation techniques, such as sliders or point-moving mouse modes, that enable us to change parameter estimates: Revision may lead to models with greater usefulness, even though they may not fit as well or be as parsimonious as models more traditionally derived.

Visual impact. At all steps of the analysis, whether we are exploring, transforming, or fitting, we evaluate the visual impact of the step using the "interocular impact"[1] —the IOI—to supplement traditional measures of fit (R^2, chi-aquare), parsimony (degrees of freedom), and fit/parsimony trade-off (adjusted R^2, AIC, etc.). If the interoccular impact is such that it really "hits you between the eyes," it is certainly worth further study. It may well be the gem we have been looking for. Although the IOI remains unquantified, the data analyst knows when what the data seem to say has hit between the eyes.

1.9.2 About Statistical Data Analysis: Visual or Otherwise

Visual statistics is a part of a more general activity known as data analysis, the part that uses dynamic interactive graphics to understand data, as the title of the book states. As stated by Tukey in the quote we presented earlier, we are looking into "what the data seem to say," the views we get being "about appearances, not actualities," as Tukey went on to say. The latter phrase cautions us that whatever we see, it is really only what we "seem to see." This precaution should be applied, more often than it is, to all types of data analysis. When we have completed a data analysis, visual or not, we really can't be sure whether our conclusions are right. There are three reasons why we can never be sure:

1. We must use our perceptual and cognitive processes to understand the data, so the process is always subjective—to a degree.

2. The tools we have to understand our data are less than perfect, so the process is always unreliable—to a degree.

3. What we have is just a sample, and it's the population that we wish to see, so the process is always inadequate—to a degree.

1.We do not know the origin of "interocular impact", nor who coined it, but the first author first heard it as a graduate student from his mentor, Prof. Norman Cliff. Nowadays, this term is atributted to Savage (2002).

There is nothing that can be done to remove these strictures, so we must always proceed with caution, realizing that what we are doing is necessarily subjective, incomplete, and inadequate—to a degree. However, it may be that the degree of subjectiveness, incompleteness, and inadequacy is so small that we can confidently report that the data do in fact say what they seem to say—but we never really know.

Our main solace is convergence: If, over many years and many researchers, the data continue to "seem to say" the same thing, we can be more sure that the data "really do say what they seem to say."

Of course, having this particular problem does not make visual statistics unique. Far from it. This problem is faced by classical statistics and its reliance on significance testing to understand data. And the solution—convergence—is the also the same.

Ultimately, in every approach to understanding data, one makes a subjective judgement about the meaning of the result. It is just more obvious that this is the case with the visual approach.

2 Examples

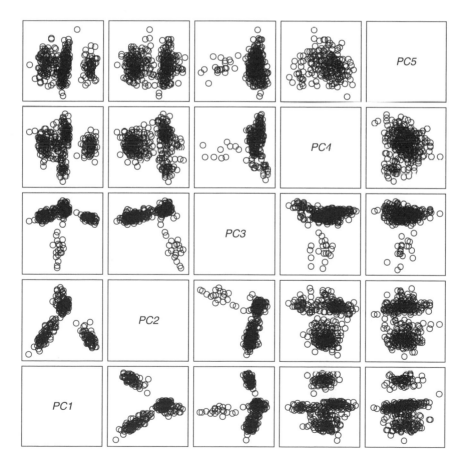

2 Examples

In this chapter we present three examples of visual statistics where dynamic interactive graphics are central in the discovery of structure in data. The first example shows how a spinning cloud of points can reveal structure in data that is very difficult to see using static views of the same point cloud. The second example uses data concerning skin disease to illustrate how one uses dynamic interactive graphics to develop a classification scheme. The third example uses data about sexual behavior and its effects on marriage and divorce.

2.1 Random Numbers

This example shows how a specific dynamic graphic—a plot showing a spinning cloud of points—can reveal structure in data that is very difficult to see using static views of the same point cloud. The example uses numbers generated by Randu, a random number generator widely used during the 1960s and 1970s. Marsaglia (1968) showed that Randu, and other members of the linear congruential family of random number generators, yield numbers that are nonrandom in a special way which can be revealed by dynamic graphics.

Randu is supposed to generate uniformly distributed random numbers between 0 and 1 (inclusive). These numbers should be generated so that all equal-width subintervals within the [0,1] interval are equally likely (uniform), and so that the next number to be generated is unpredictable from the numbers already generated (random).

We use one-, two, and three-dimensional graphics to look for structure or pattern in the numbers. Since the numbers are random, there should be no structure, so seeing structure implies nonrandomness. We will use graphical techniques that are presumably unfamiliar—sequence plots, lag plots, jittered dot plots, spinplots, and so on. We begin by showing how these plots look with normally distributed data, since we assume that such data are more familiar than uniform data.

Normal distribution. A visualization of 3000 univariate normal random numbers is shown in Figure 2.1. The visualization involves four graphics (all are discussed more extensively in Chapter 6). The upper-left graphic is a dot plot. The dots are located vertically according to their generated value, and horizontally according to their frequency within a narrow range of generated values. Since the number of dots within a narrow range of generated values reflects density, the shape of the distribution, as represented by the number of dots, should reflect the population distribution, which is normal. We see that it does.

The upper-right graphic is a shadowgram. It is based on averaging many histograms together in a way that determines the density of the distribution for each pixel in the graph. The graph is then rendered so that density is shown by the darkness of the shade of gray—the higher the density the darker the shade. Again, the plot should look like the parent population (i.e., be normal in shape), which it does.

The lower-left graphic is a lag plot. This plot shows each datum plotted against the datum that was generated previously (a lag of 1). This gives us a two-dimensional

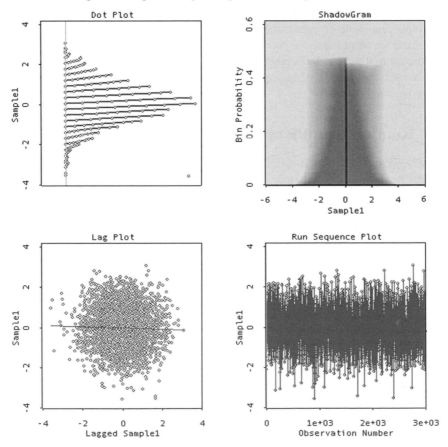

Figure 2.1 Four graphs of 3000 normally distributed random numbers.

plot where the first dimension is the values generated at time *t*-1 and the second dimension is for the values generated at time *t*. For normally distributed values, each dimension should be normal and the two dimensions should be uncorrelated. Also, each dimension should have the same mean and variance (0 and 1 for standard normal values). When plotted against each other, the plot should be bivariate normal, which will look circular in shape, with the density highest in the middle and tapering off normally in all directions. The figure looks as it should.

Since the values are random, they should be unrelated to the order in which they were generated. The sequence plot of values versus generation order should show no structure. This plot is shown in the lower right of Figure 2.1. It has no discernible structure.

Uniform Distribution. We turn now to the data generated by the faulty RANDU random number generator. In Figure 2.2 we see the way that 3000 supposedly random

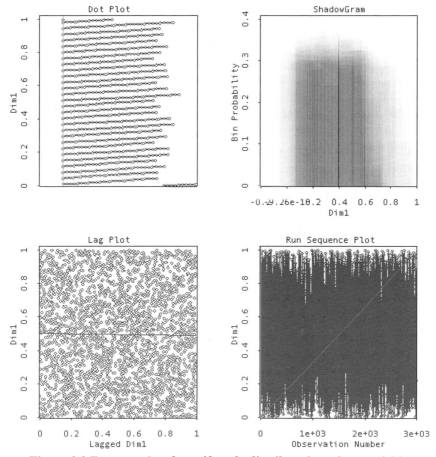

Figure 2.2 Four graphs of a uniformly distributed random variable.

numbers generated by RANDU look with the same graphics as those used for the normally distributed numbers in Figure 2.1. The upper-left graphic in Figure 2.2 is the dot plot for the values generated by RANDU. Since the distribution should be everywhere equally dense, the width of the plot should be the same for all values generated. This appears to be the case.

The upper-right graphic shows an estimate, from our sample, of our sample's population distribution. If our sample is good, the estimate based on it ought to be flat, which it seems to be (the right side drop-off is an artifact of the estimation method). The autocorrelation (correlation of values generated with themselves with a specified lag) should be zero, as it is in the lower-left figure for a lag of 1 (this holds up when we cycle through various lags). Finally, the sequence plot (lower right of Figure 2.2) should show no pattern, which seems to be the case.

So far, so good! The one-dimensional views of the random numbers look as they should if the values generated by the random number generator are indeed random uniform numbers. (Find out more about these and other one-dimensional views of data in Chapter 6.) In addition to the one-dimensional views, we can look at two-dimensional, three-dimensional, and even higher-dimensional views. We do that now.

For a two-dimensional view, we use the 3000 supposedly random uniform numbers to represent 1500 observations of two variables. The main method for assessing the relationship between two variables is the well-known scatterplot, which is shown for our data in Figure 2.3. If the axes of the scatterplot are two random uniform variables, we should obtain a scatterplot with points scattered with equal density everywhere inside a square defined by the interval [0,1] on each axis. There should be no discernible pattern. Looking at Figure 2.3, our numbers, once again, look random.

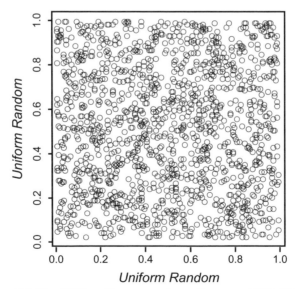

Figure 2.3 The 3000 uniform numbers shown as 1500 2D points.

We can do essentially the same thing in three dimensions that we just did in two dimensions: Arrange our 3000 supposedly uniform random numbers to represent the coordinates of 1000 points in three dimensions. We can then form a spinning three-dimensional scatterplot and look at it for structure. We have done this in Figure 2.4 for two very slightly different rotations of the space.

On the left we see that the space looks like the featureless space it should be if the data are truly random. However, if we watch the cloud of points slowly rotate, or (better yet) if we use the rotation tool to actively rotate the cloud of points, we will eventually see a view of the space that reveals the nonrandom structure lurking within our points. We see the structure in the right space in Figure 2.4. What we see is that the points are arranged in a set of parallel planes within the point cloud. We cannot see this in any univariate or bivariate plot, since the planes are not lined up with the dimensions of the space.

As pointed out by Wainer & Velleman (2001), the effect is like what you see as you drive past a field of corn. Most of the time you just see an apparently random arrangement of cornstalks, but occasionally the cornstalks line up and reveal their nonrandom structure.

Furthermore, if we did not have a dynamic graphic to look at the trivariate space, we wouldn't see the structure either. It took a dynamic graphic, in this case a three dimensional spinnable space, to reveal the structure, and although it wasn't necessary, having the ability to interact with the space made it easier to find the structure.

Finally, note that the parallel plane structure is very well hidden in the space—a very small rotation can obscure the structure completely. For example, the two views in Figure 2.4 differ by only 3°. If you compare the positions of the boxes, you can see how small a difference this is.

If all you see is the left-hand view, you have no idea that there is hidden structure to be revealed with just a 3° rotation. Then, when you see the right-hand figure, the

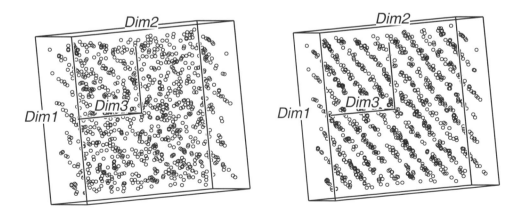

Figure 2.4 Two views of the same 3000 uniform random numbers,
now seen as 1000 points in three dimensions.

structure is revealed, and the IOI test is satisfied. Of course, having now seen the right-hand figure, where the structure has clearly emerged, you can see it beginning to appear in the left-hand view, but you wouldn't see it if you didn't know what to look for.

This is, by the way, a nice example of the IOI test at work. When we see the space rotating, there is a bit of a "glitch" in the rotation; it has already "hit us between the eyes," if only for a moment. Then, if we stop the rotation and back it up somewhat, we see the structure and are amazed.

2.2 Medical Diagnosis

In this section we show dynamic interactive graphics being used to lay the ground-work needed to develop a medical diagnostic tool. We present a more complete version of this example in Chapter 8 where we use dynamic interactive graphics to develop the tool, and where we show how the tool would be used by the diagnostician to assist in making a diagnosis. We use exactly the same clinical and laboratory information that the doctor currently uses to make a diagnosis.

We use a particular set of data to show how a medical diagnostic tool can be developed using the techniques of visual statistics. The data, which are described in detail below, were obtained from patients examined by a dermatologist about problems with their skin. The data include information obtained during an office visit and the results of a laboratory analysis of a biopsy taken from a patient.

When a doctor diagnoses a patient, he or she is assigning a disease category to the patient, using information obtained in the office and from the lab. If a group of possible diagnoses are all quite distinctly different in their symptomology, the diagnosis will be straightforward and will probably be correct. However, if a group of diseases share a number of similar features, the doctor's decision will be more difficult, may require knowledge about a wider array of information, and will be more likely to be incorrect. Of course, if the patient is diagnosed incorrectly , actually having a different condition than the condition diagnosed, the doctor's prescription may be inappropriate. Such misdiagnoses occur because the doctor must weigh very many sources of information and must decide on how much weight each source of information should get.

Frequently, doctors must choose between a diagnosis based on their personal clinical experience and a diagnosis based on laboratory tests. Of course, it is true that the doctor's subjective impressions and experiences are not totally reliable, but then neither are the laboratory tests. Although lab tests can be more accurate, they are not infallible, are often expensive, and can provide diagnoses whose validity is not self-evident. And it would be better if the doctor were able to weigh all of the evidence simultaneously, using both clinical and laboratory results together. But the amount of information that can be relevant can easily exceed even the smartest doctor's abilities, and the problem of how to weigh each of a large number of sources of information is beyond all of us.

Table 2.1 Variables in the Dermatology Dataset

Clinical Variables

Erythema	*Follicular papules*
Scaling	*Oral mucosa involvement*
Definite borders	*Knee and elbow involvement*
Itching	*Scalp involvement*
Koebner phenomenon	*Family history*
Polygonal papules	*Age*

Histopathological Variables

Melanin incontinence	*Pongiform pustule*
Eosinophils in the infiltrate	*Munro microabcess*
PNL infiltrate	*Focal hypergranulosis*
Fibrosis of the papillary dermis	*Disappearance of the granular layer*
Exocytosis	*Vacuolization and damage basal layer*
Acanthosis	*Spongiosis*
Hyperkeratosis	*Saw-tooth appearance of retes*
Parakeratosis	*Follicular horn plug*
Clubbing of the rete ridges	*Perifollicular parakeratosis*
Elongation of the rete ridges	*Inflammatory mononuclear infiltrate*
Thinning of Suprapapillary epidermis	*Bandlike infiltrate*

As seen by statisticians, medical diagnosis is one of many situations in which we classify our new, unclassified data by comparing it with benchmark datasets for which classifications exist, selecting the most similar benchmark and using its classification as the classification for our newly acquired data. Note that the benchmarks may be actual sets of empirical data that have somehow already been classified, or the benchmarks may be theoretical, prototypical datasets that are the idealized exemplars of the classification. It doesn't matter.

Data. The data are observations of 34 variables obtained from 366 dermatology patients. The variables are listed in Table 2.1. Twelve of the variables were measured during the office visit, 22 were measured in laboratory tests performed on a skin biopsy obtained during the office visit. Of the 34 variables, 32 were measured on a scale running from 0 to 3, with 0 indicating absence of the feature and 3 the largest amount of it. Of the remaining two variables, *Family history* is binary and *Age* is an integer specifying the age in years. Eight of the patients failed to provide their age. These patients have been removed from the analysis. The data were collected by Nilsel Ilter of the University of Ankara, Turkey (Guvenir et al., 1998).

Two difficulties must be dealt with before we can visualize these data: (1) The data are discrete—All of the variables except *Age* have four or fewer observation categories; and (2) There are too many variables—humans can not understand 34 variables simultaneously.

If we picture the data directly, we soon see the problems. For example, a matrix of scatterplots of the first three variables is shown in Figure 2.5. A scatterplot matrix has variable names on the diagonal and scatterplots off-diagonal. The scatterplots are formed from the variables named on the diagonal of a plot's row and column. The upper-left triangle of the matrix is a mirror image of the lower right. Scatterplot matrices are discussed in more detail in Chapter 7.

The discrete nature of the variables means that we have only a few points showing for each scatterplot, and that they are arranged in a lattice pattern that cannot be interpreted. For each plot, each visible point actually represents many observations, since the discrete data make the points overlap each other. In essence, the resolution of the data, which is four values per variable, is too low. The fact that there are 34 variables means that the complete version of Figure 2.5 would be a 34×34 matrix of scatterplots, clearly an impossibly large number of plots for people to visualize. Here, the problem is that the dimensionality of the data, which is 34, is way too high. Thus, the data cannot be visualized as they are, because their resolution is too low and their dimensionality is too high.

Principal components. All is not lost! These two problems can be solved by using principal components analysis (PCA). PCA reduces the dimensionality of data that have a large number of interrelated variables, while retaining as much of the data's original information as is possible. This is achieved by transforming to a new set of variables, the *principal components*, which are uncorrelated linear combina-

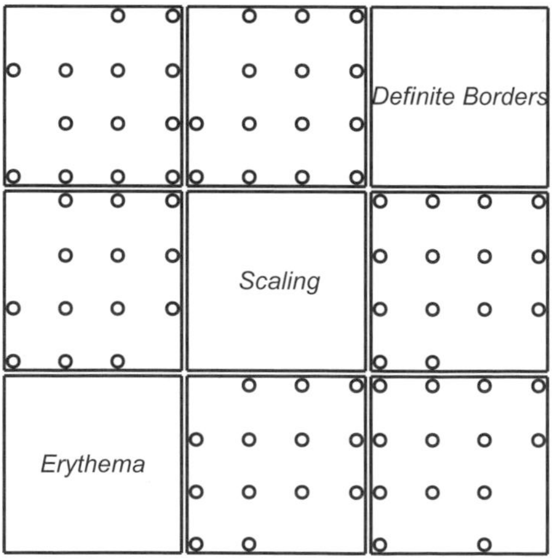

Figure 2.5 Scatterplot matrix for three variables.

tions of the variables. There is no other set of *r* orthogonal linear combinations that fits more variation than is fit by the first *r* principal components (Jollife, 2002).

Principal components have two important advantages for us. First, since only a few components account for most of the information in the original data, we only need to interpret displays based on a few components. Second, the components are continuous, even when the variables are discrete, so overlapping points are no longer a problem.

Figure 2.6 shows a scatterplot matrix of the five largest principal components. These components account for 63% of the variance in the original 34 variables. The general appearance of Figure 2.6 suggests that the observations can be grouped in a number of clusters. However, the total number of clusters as well as their interrelationships are not easily discerned in this figure because of the limited capabilities of

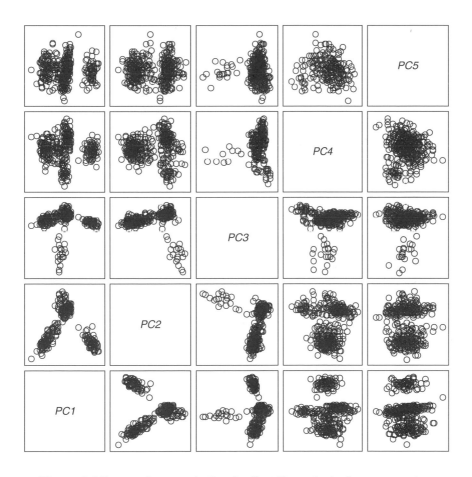

Figure 2.6 Scatterplot matrix for the first five principal components.

scatterplot matrices and because of the small plot sizes. There are ways around both of these limitations, as we discuss in Chapter 8.

Linking. Linking is a powerful dynamic interactive graphics technique that can help us better understand high-dimensional data. This technique works in the following way: When several plots are linked, *selecting* an observation's point in a plot will do more than highlight the observation in the plot we are interacting with—it will also highlight points in other plots with which it is linked, giving us a more complete idea of its value across all the variables. Selecting is done interactively with a pointing device. The point selected, and corresponding points in the other linked plots, are highlighted simultaneously. Thus, we can select a cluster of points in one plot and see if it corresponds to a cluster in any other plot, enabling us to investigate the high-dimensional shape and density of the cluster of points, and permitting us to investigate the structure of the disease space.

Interpretation. Figure 2.7 displays a "comic book"-style account of the process of selecting the groups of skin diseases that were visible in Figure 2.6. Frames of Figure 2.7 are the scatterplots of PC1 versus PC2 to PC5. The frames are to be examined sequentially from left to right and from top to bottom. The last frame is a repetition of the first frame and is intended to show the final result of the different actions carried out on the plots. An explanation of the frames follows.

1. PC1 vs. PC2: This scatterplot shows quite clearly three groups of points, labeled A, B, and C. Two groups are selected, but the third one remains unselected.

2. PC1 vs. PC3: Three actions are displayed in this frame. The groups selected in the preceding frame have been marked with symbols: A has received a diamond ($\Diamond$), and B a cross (+). Also, we can see that dimension PC3 separates group C into two parts: one compact, the other long. We have selected the observations in the long part and called them cluster C_1. At this point we have four clusters: A, B, C_1, and unnamed.

3. PC1 vs. PC4: This plot has the points selected in the preceding frame represented by a square symbol ($\square$). Unassigned observations above and below of squares make two groups. We selected the group with positive values in PC4 and called it C_2, giving us five clusters: A, B, C_1, C_2, and unnamed.

4. PC1 vs. PC5: We assigned the symbol ($\times$) to the group C_2 and selected the top values of PC5. Notice that this selection involved the reclassification of some points that had previously been assigned to the group C_2. The points selected will define group C_3. Notice that there are still some points that keep the original symbol of the points in the plot [a disk ($\bigcirc$)]. We call it group C_4. We now have six clusters: A, B, C_1, C_2, C_3 and C_4.

5. PC1 vs PC2 again: This frame is the same as the first frame in the sequence except that it shows the plot after steps 1 to 4. Note that we ran out of easily seen symbols for marking C_3, so we used gray to identify points. This frame

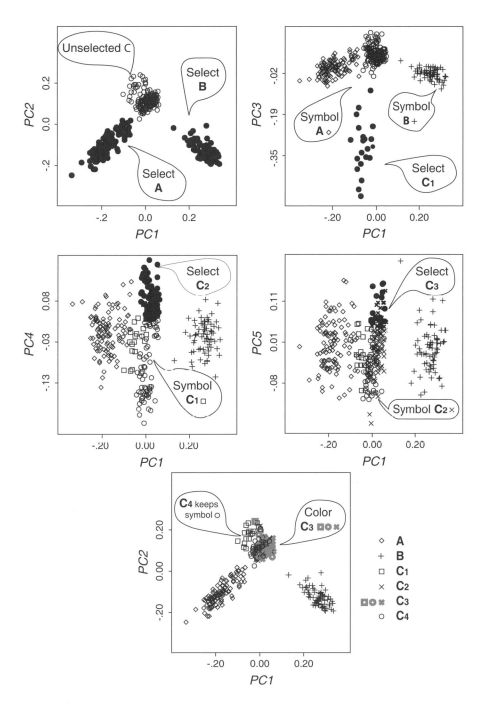

Figure 2.7 Steps in selecting and changing symbols and colors in PCs.

displays very clearly the three big clusters identified at the beginning and also suggests something of the finer structure inside cluster C.

A downside of the plots in Figure 2.7 is that the four clusters identified as C_1 to C_4 are not visualized very clearly. This problem suggests using *focusing,* a technique described in Chapter 4, to remove from the plot the points in the two largest clusters. As the remaining four subclusters are defined basically using PC3 to PC5, it makes sense to use a 3D plot to visualize them. Figure 2.8 displays such a plot after rotating it manually to find a projection clearly displaying the four clusters. This view suggests that some points actually seem to belong to clusters other than those to which they had previously been assigned. Thus, we reassign them, as explained in the balloons.

Validity. We did not mention it earlier, but the data include a diagnosis made by the doctor of each patient. It is interesting to compare our visual classification with the diagnostic classification. Table 2.2 presents a confusion matrix, which is a table that shows the frequency with which members of each of our classes were assigned to each of the diagnostic classes. If our visual classification agrees exactly with the diagnostic classification, the confusion matrix will be diagonal, there being zeros in all off-diagonal cells. To the extent that our visual classification is "confused," there will be nonzero frequencies off the diagonal.

In general, we see that our visual classes correspond closely to the diagnostic classes. All of the patients diagnosed with psoriasis and lichen planus were classified visually into groups A and B. The observations in the cluster labelled C are very well separated with respect to clusters A and B, with only one observation out of place. However, some of the subclusters in C, especially C_2 and C_3 have considerable interchanges between them, suggestion that additional efforts for improving the discrimi-

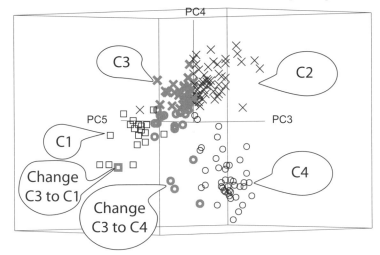

Principal Components

Figure 2.8 Spinplot showing observations in clusters C_1 to C_4

Table 2.2 Confusion Matrix for the Medical Diagnosis Data

	A	B	C_1	C_2	C_3	C_4	S
Psoriasis	111						111
Lichen planus		71					71
Pityriasis rubra pilaris			19		1		20
Pityriasis rosae				39	8	1	48
Seborrheic dermatitis	1			10	48	1	60
Chronic dermatitis						48	48
	112	71	19	49	57	50	358

nation between them are necessary. This problem can result from the difficulties we had to correctly pinpoint the points in frame 4 of Figure 2.7. Of course, there could also be misdiagnoses by the doctor—we can't tell. Nevertheless, 93.8% of the cases are classified correctly which is only 2.4% lower than the classification made originally using a special algorithm called VFI (Guvenir et al., 1998).

2.3 Fidelity and Marriage

Fitting statistical models involves finding explanations or simple descriptions for patterns in data. Typically, there is an outcome or response variable, and one or more predictors or explanatory variables that can be used individually or jointly (as in an interaction) for this purpose. The goal usually is to find the simplest adequate description—the smallest, least complex model—which nonetheless provides a reasonably complete summary of the data. Any statistical model can be cast as a breakdown of the data into two parts: what we have summarized, described, or explained (called the *model*) and the rest, which we do not understand (the residual). That is,

data = model + residual

You can always make the model fit perfectly (by including as many parameters as you have data values), or make the model incredibly simple (just choose an empty model), but the payoff comes when you can find a balance between fit and parsimony.

Traditional statistical methods use various numerical measures of goodness of fit [R^2, χ^2 and parsimony (degrees of freedom)], and often combine these to determine optimal points along a trade-off relation (adjusted R^2, AIC, etc.). We prefer to combine various sources of model information into coherent displays, fitting models visually by direct manipulation. We call this paradigm *visual fitting*.

We illustrate this approach with an example of fitting loglinear models to data about divorce and the occurrence of premarital or extramarital sex. The data, a cross-classified table of frequencies, are shown in Table 2.3. Thornes and Collard (1979) obtained two samples of roughly 500 people, one of married people, the other of those seeking divorce. Each person was asked (a) whether they had sex before marriage, and (b) whether they had extramarital sex during marriage. When these are broken down by gender, we get the $2 \times 2 \times 2 \times 2$ frequency table. A log-linear model

59

Table 2.3 Cross-tabulation of Marital Status Versus Premarital Sex, Extramarital Sex, and Gender

Premarital	Extramarital	Gender	Marital Status	
			Married	Divorced
Yes	Yes	Male	11	28
		Female	4	17
	No	Male	42	60
		Female	25	54
No	Yes	Male	4	17
		Female	4	36
	No	Male	130	68
		Female	322	214

attempts to explain the pattern of frequencies in this table, so-named because it uses a linear model of the logarithm of the frequency.

The question we wish to ask of these data is: How is *Marital Status* related to (how does it depend on) *Gender, Premarital Sex* and *Extramarital Sex*? Before we go any further, ask yourself these questions. What factors influence the likelihood of divorce? Do any have combined (interactive) effects?

We can translate this question into a form that can be used for modeling by a log-linear statistical model as follows. In terms of understanding *Marital Status*, the explanatory variables—*Gender, Premarital Sex,* and *Extramarital Sex*—can be associated in any arbitrary ways; they just describe the sample. Probably, men are more likely to have had premarital sex than women—a *(GP)* association—and maybe also more likely to have had extramarital sex *(GE)*. We don't really care, and lump all associations among *G, P,* and *E* into one term, *(GPE)*.

Empty model. The basic, empty model is symbolized as *(GPE)(M)*, and asserts that *Marital Status* has no association with *Gender*, or with *Pre-* or *Extramarital Sex*.

Saturated Model. At the opposite end of the parsimony spectrum is the saturated model, *(GPEM)*. This model allows marital status to be associated with all of the factors and their combinations, in unknown ways. However, it always fits perfectly, so it cannot possibly tell us anything.

Model-Fitting strategies. What we must do is find a model somewhere in between these two extremes, a model that is parsimonious, yet explanatory. That means that we must search through a variety of models, trying to decide which is best. There are two basic strategies for doing this search:

* **Forward search**. start with the empty model (which will fit poorly), and add terms allowing associations of *M* with *G, P,* and *E* and their combinations, until the model fits well enough.

- **Backward Search**. Start with the saturated model and remove associations of *M* with the others, as long as the model does not fit too poorly.

We use the former method here to illustrate visual fitting. The spreadplot (a kind of multiplot visualization that is introduced in chapter 4) for the initial model, *(GPE)(M)* is shown in Figure 2.9 (on the following two pages). This model fits very poorly, of course ($G^2=107$, df = 7, p < 0.001). The G^2 measure is a badness-of-fit measure. Low values are good, high values are bad. The empty model, reported here, has a very large value of G^2, meaning the fit is very poor, which, of course, it must be, since it has no terms. The hypothesis test, when rejected, as is the case here, indicates the model does not fit the data.

However, the spreadplot is not limited to displaying goodness-of-fit statistics. It also reports in a graphical way other aspects of importance in the model fitting situation. These other elements are indicated in the spreadplot by means of text inside balloons.

The way that the model differs from the data gives us clues about how we can improve our model. We can use mosaic displays to find the specific ways in which the model is different from the data, since mosaics show the residuals (or differences) of the cells with respect to the model. Looking at these differences, we can observe patterns in the deviation that will help us in our search.

Unfortunately, mosaic displays are best viewed in color, and we are forced to use black and white. (We do the best we can, but to be honest, the black-and-white versions shown in Figure 2.9 do not do justice to the mosaic displays. If you can view this online, please do; it will help). So while the color versions of mosaic displays use two different colors for representing positive or negative residuals (and the hue of the color indicates their absolute values), the black-and-white version puts everything in gray. Therefore, we have made special versions of the mosaic displays to use for this book.

Figure 2.10 displays black-and-white mosaic displays for the initial model *(GPE)(M)*. The displays represent positive residuals in black and negative residuals in white. Although this way of coding has the disadvantage of ignoring the actual value of the residuals, we believe that this disadvantage is not very consequential, as, in practice, the analyst pays attention to the patterns of signs of the residuals, such as shown in these black-and-white versions of mosaic displays. Of course, problems can arise with small residuals because they need to be assigned to one color or another. As a solution, we specify cutoff values that we find appropriate for each situation. Values above the absolute values of the cutoff will be displayed in black or white (depending on their sign), and the rest in gray. We used 3 as the cutoff value in Figure 2.10 and as the model fits quite poorly, all the values were above it in absolute values. Thus there is no gray.

There are three mosaic displays in Figure 2.10. Figure 2.10a is the initial display produced by the software. The order of the variables in this display is *PEMG* and was defined simply by the original order of the variables in the datafile. Even though this display looks quite simple and we could possibly interpret it, exploration of the spe-

61

cific values of the residuals suggested that we change the order of the variables in the display to make it more interpretable.

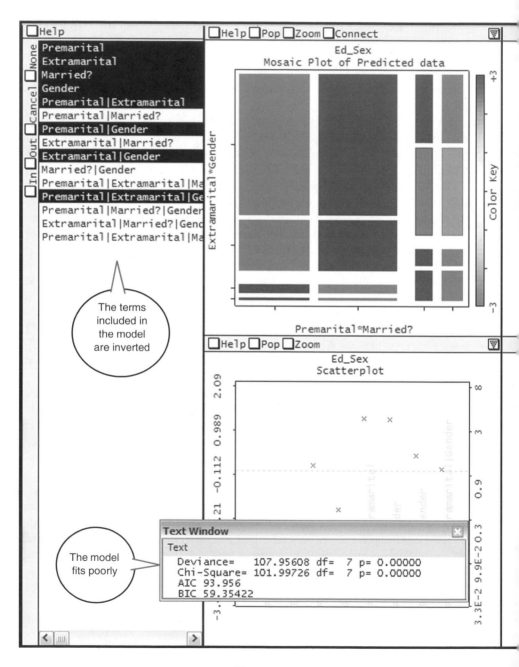

Let's look at the residuals table below Figure 2.10a. The first column of this table is the negative of the second, and similarly for the third and fourth. This is a consequence of the marginal constraints for this model, which make some residuals redundant (Agresti, 1990). Comparing the first and second columns or the third and fourth,

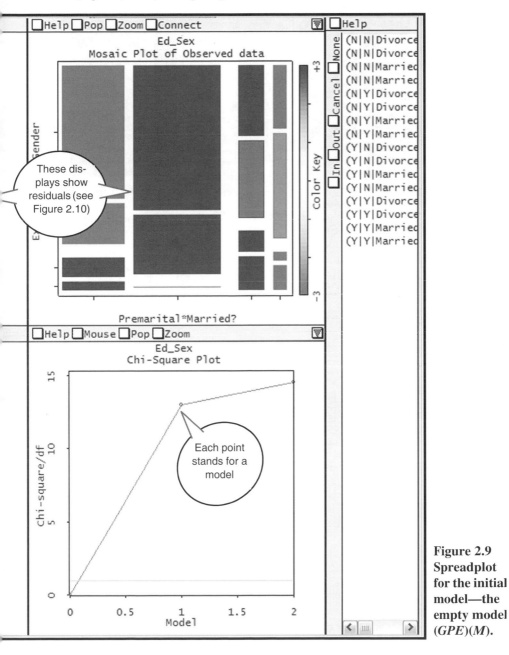

Figure 2.9 Spreadplot for the initial model—the empty model (GPE)(M).

is therefore uninformative. More interesting conclusions may be extracted by comparing the first and third columns in Figure 2.10a (or the second and the fourth).

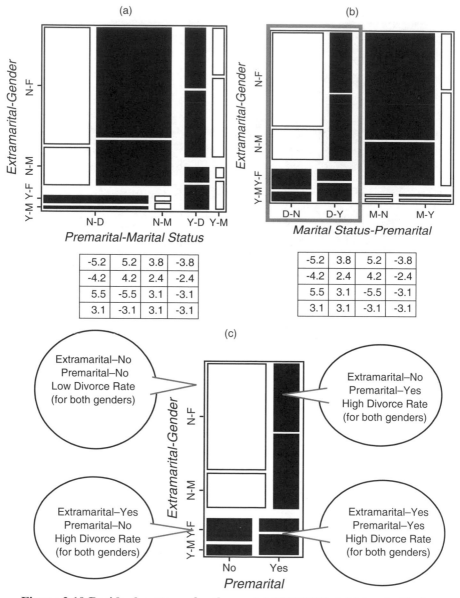

Figure 2.10 Residual patterns for the model *(GPE)(M)*: (a) Mosaic display with the default order of the variables in the software *PEMG;* (b) using order *MEPG*; (c) focused view of rectangular section of (b) (*Marital Status = Divorced*).

Manipulating the mosaic display in Figure 2.10a we transformed it to the mosaic display in Figure 2.10b. Notice that the first and third columns in Figure 2.10a are now the first and the second, and that second and fourth are the third and fourth. This was attained by exchanging the order of the variables *PreMarital* and *Marital Status* so that the order of the variables in the display is *MEPG*. Now we can see that the two initial columns mirror the other two columns (i.e., the divorced part of the display is a mirror of the married part). Hence, we can focus (*focusing* is a technique discussed in Chapter 4) on these two columns without missing any information, as the other part is loaded with exactly the same information. The part we have selected is marked with a rectangle in Figure 2.10b and is displayed separately in Figure 2.10c with balloons that describe the results. As can be seen in this figure, the information can be summarized in only a sentence: people with sexual encounters (*Pre-* or *Extra-*) out of marriage are more likely to divorce than those without them. This statement suggests the model *(GPE)(PM)(EM)*, which asserts a relationship between *Marital Status* and *Pre-* and *Extramarital Sex*, to be tested next.

Figure 2.11a shows the two first columns of the mosaic display for the model *(GPE)(PM)(EM)* with the variables in the same order as in Figure 2.10c. This model still does not fit ($G^2 = 18.15$, df = 5, $p \approx 0.002$) but improves the basic model considerably. In this mosaic display, residuals in absolute value larger than 1 were filled in black or white (depending on the sign of the residual) and the rest were filled in gray. Notice that there are only two black (positive) residuals in the display as well as only two white (negative) residuals. Again, there is evidence of some regularity in the display but as we did previously, we chose to manipulate the mosaic to make it easier to interpret: Thus, we exchanged the order of *Extramarital Sex* and *Gender* variables in

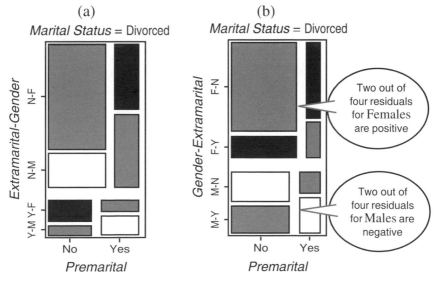

Figure 2.11 Mosaic displays for model *(GPE)(ME)(MP)* focused on *Marital Status* = Divorced. (a) Mosaic display using order *MEPG*; (b) using order *MGPE*.

the plot. The result is shown in Figure 2.11b. The visual effect of this manipulation is that the two positive residuals are located in the upper part of the plot, and the negative residuals are in lower part of the plot. Looking at the labels, we can see that the positive residuals are related with females, and negative residuals are related with males (i.e. females are more prone to divorce than males). This points to a relation between the variables *Gender* and *Marital Status* and suggests the model *(GPE)(PM)(PE)(GM)* as the next one to be tested.

The improvement of model *(GPE)(PM)(PE)(GM)* with respect to the model without the *(GM)* term was modest (G^2 = 13.62, df=4, $p \approx 0.008$). This suggests that we compare models to see if the difference is important. We can do this comparison using the Chi-Square plot shown in the spreadplot of Figure 2.9 by selecting the points corresponding to the two models that we want to compare. The result (ΔG^2 = 4.53 , df=1, $p \approx 0.03$) suggests that the difference is larger than zero, but not by much. Actually, we used our software to put the mosaic displays for both models side by side as well. These are shown as Figure 2.12a and b. In these displays, residuals larger than 1 are shown in black, and residuals smaller than 1 in white (the rest are in gray). A third display, Figure 2.12c, shows the *differences* between the absolute value of residuals of the other two models and can be used to explore which cells of the second model have actually improved the fit out of the first model. So cells in black in this display are those that have larger residuals in the second model than in the first, and cells in

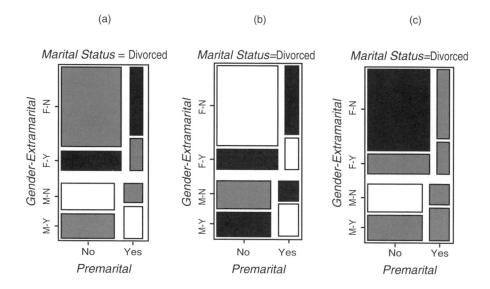

Figure 2.12 Mosaic displays before and after adding the term *(GM)* to the model *(GPE)(ME)(MP)* focused on *Marital Status* = Divorced. (a) Mosaic display for model *(GPE)(ME)(MP)*; (b) mosaic display for model *(GPE)(PM)(PE)(GM*; (c) mosaic display of differences between (a) and (b).

white are those that have smaller residuals (in both cases, the difference must be larger than 1 to use this coding). Cells in gray are for differences not larger than 1.

The visual impression of mosaic displays in Figure 2.12 is that adding the *(GM)* term does not improve the fit at all. On the contrary, the fit of the model seems to worsen, as seven out of eight cells have residuals larger than 1. Looking at the mosaic display of differences in Figure 2.12c we can see that actually only two cells have changed its fit more than 1: males without *Pre-* or *Extramarital* encounters (which have reduced the residual) and females (which have increased the residual). In conclusion, the apparent interaction between *Gender* and *Marital Status* that we observed in Figure 2.11 seems irrelevant at this stage of the analysis of our data.

We will look at Figure 2.12a again to find hints for new terms in the model. The current layout of this plot hinted that females had more divorces than males, but it is possible that a change in the aspect of this display will bring about other suggestions. Figure 2.12a suggests some type of regularities that we can not easily see at this point. Manipulating the order of the variables we arrived at Figure 2.13a, which has the right layout to see a new pattern in the residuals. In this figure, the variables *Pre-* and *Extramarital* are set together on one axis of the plot, and the variables *Marital State* and *Gender* are on the other. In this layout, the two negative residuals (white cells) are in the sides of the plot, and the two positive residuals (black cells) are in the center. If we combine the residuals by *Gender,* as shown by the dashed rectangle in Figure

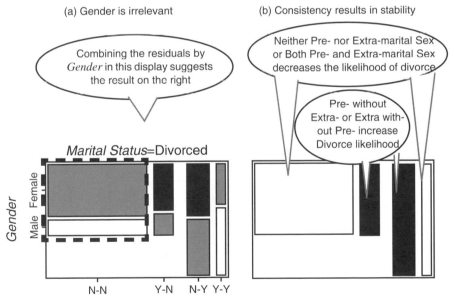

Figure 2.13 Mosaic displays for model *(GPE)(PM)(PE)(GM)*. (a) Using order *PMEG* focused on *Marital Status*= Divorced; (b) Schema of display a when the residuals are combined by *Gender*.

2.13a with those with no *Pre-* and no *Extramarital,* for the four combinations we can visualize the scheme in Figure 2.13b. This pattern suggests a three-way interaction among *Pre-* and *Extramarital* and *Marital Status (PEM).*

The model *(GPE)(PEM),* whic includes the interaction mentioned previously, fits the data well ($G^2 = 5.24$, df = 4, $p \approx 0.26$) and would be accepted as a good model when considering the usual goodness of fit criteria. Yet we can observe in the mosaic display for this model (Figure 2.14) a pattern that suggests that we can improve this model still more. Using a cut-off value of 0.8, the display shows that the females have higher likelihood of divorce than males given this model (constraints on this model make males and females have the same residuals in absolute value). We saw this trend previously, but we rejected it as unimportant because the effect of including this interaction in the model seemed too little. However, this display suggests testing the interaction in combination with the model currently defined.

The model *(GPE)(PEM)(GM)* fits very well ($G^2 = 0.69$, df = 3, $p \approx 0.87$) and even though this model is not very different from the model without the *(GM)* term ($\Delta G^2 = 4.53$, df = 1, $p \approx 0.03$), we regard this model as the most appropriate for our data.

The actions described in the previous paragraphs are very typical of the actions taken during a real data analysis. Normally, the nitty-gritty detail would be left out, but we think it is important to put them in so you can really learn about how to do this type of analysis. The typical cycle of analysis is this: The analyst evaluates the evidence to select the term that best explains the dependent variable. Then the analyst tests that term. If the result is an improvement, the variable or term is left in the equation, but if is not, the analyst may drop it and look at the displays for a new hint.

Thus, interactive dynamic graphics allows you to search for a good model using a process that is not strictly forward or backward, but involves both. The normal noninteractive stepwise approach is an automatic procedure that uses a statistical rule of thumb to add or delete predictors from the model. Often, automated stepwise proce-

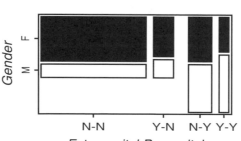

Marital Status = Divorced

Extramarital-Premarital

Figure 2.14 Mosaic displays for model *(GPE)(PEM)* focused on *Marital Status*= Divorced. The order of the variables is *PMEG*.

dures produce poor results because they are limited to simple rules and do not consider all the elements that an analyst would use. Consequently, many statisticians usually prefer to perform the modeling process by hand instead of using the automatic method. Visual fitting has the advantage of providing a concrete, perceptual goal: A good-fitting model will have small residuals and the mosaic will be mostly unshaded. Think of the search for an adequate model as "cleaning the mosaic."

However, the manual process also has a big disadvantage: It makes it difficult to record and describe the actual steps performed by the analyst. Indeed, it often happens that scientific papers or books only report the end model, without discussing the choices preceding the conclusion that such a model was the best. Software does not often help with this endeavor either, as it usually does not provide a way to overlook the models considered and rejected along the process. Fortunately, as we have already discussed in the spreadplot shown in Figure 2.9, we have a display in our software that addresses this problem.

Figure 2.15 shows a display of the fit of the models evaluated in this section. The display shows the χ^2/df value of each of the models considered. The horizontal line in the display stands for the rule of thumb that χ^2/df values below 1 can be considered models that fit well. The display has labels for each of the models we considered. As you can see, this display is an effective way to record and recall the entire set of models that we explored during a modeling session, such as the one we described in this section. And perhaps best of all, dynamic interactive graphics means that all we need to do to return to an earlier model is to click on its representation in the diagram. Then we are back to that model and can proceed from there as we wish.

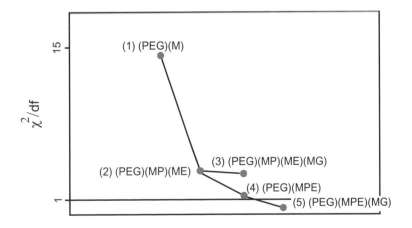

Models

Figure 2.15 Fit of models evaluated during the modeling session.

PART II

SEE DATA—THE PROCESS

3   Interfaces and Environments

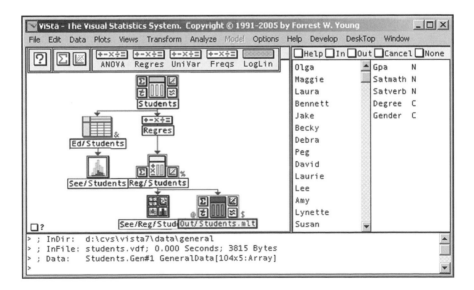

3 Interfaces and Environments

Before the advent of computers, data were analyzed by hand. The analyst would write the data down on a large sheet of paper ruled to represent the rows and columns of a matrix. Then calculations were performed with slide rules, mechanical calculators, and brain power, the sums, products, square roots, and so on, being written down in special cells of the table. This process would continue, sometimes for months (in fact, the first author spent three months rotating a space when he was a graduate student 40 years ago!). Now, of course, things have changed: Obedient machines perform operations in microseconds that once took obedient graduate students months.

Despite the changes, we share much today with the way we analyzed data 40 years ago: Now, as then, data analysis consists of a series of steps, with many of the steps having been worked out long before 40 years ago. For example, now, and long ago, the first step is to prepare the data so that they can be analyzed. Then we get simple summary statistics and basic plots of the data. What comes next depends on the nature of the data, but for any given kind of data the steps now are probably the same as they were back then. For example, if we transform the data and fit a model, it could well be that we use the same transformation and model now as we did back then.

What has changed is the way in which we carry out the steps, even when the content of the steps has not changed. For example, consider the first step, the step where we prepare the data for analysis: Today, we type them into a spreadsheet, whereas yesterday we wrote them down on a piece of paper. When we abstract the prototypical nature of the step, we see that it is the same (recording data in a two-way tabular arrangement) even though the technology (paper and pencil vs. computerized spreadsheet) is markedly different.

Taking a look back at what data analysis was like 40 years ago and comparing that with what it is like now points out that some aspects of data analysis have changed a great deal, and others have not. Looking back we see that the form of data analysis has changed more than the content—the medium has changed more than the message.

What is most important about this retrospective is that it helps us realize that data analysis is form as well as content—medium as well as message. This is a fairly radical idea. Statistics is usually seen as content, not form. But we see, just by looking

75

back, that data analysis takes place in an environment, not in a vacuum. Furthermore, it seems to us that the environment in which data are analyzed has a strong influence on the analyses and visualizations that are performed, on the questions that are asked, and on the answers that are obtained.

Thus, in our view, the environment in which data analysis occurs is of great importance, having a potentially great influence on the conclusions that may be drawn about any given set of data.

At this point, it is very important to remind ourselves of the overall purpose of visual statistics. As we stated on page 1:

> *Visual statistics* replaces algebraic obscurity with geometric clarity by translating mathematical statistics into dynamic interactive graphics. This can simplify, ease, and improve data analysis. When the graphics are mathematically, computationally, perceptually, and cognitively appropriate, they can induce intuitive visual understanding that is simple, instantaneous, and accurate.

We should apply the very same philosophy and technology to the environment in which data analysis takes place, not just to individual visualizations of data and analyses, especially if the environment can have a potentially great influence on the understanding we obtain of our data.

But just what is an *environment*? In common, everyday usage, an environment is the totality of the surrounding objects and conditions. Applying this definition to the digital realm, and realizing that what we see in that digital realm is interfaces to software objects, we can define a computer environment as the set of interfaces that are activated at a particular moment in time, each interface enabling interaction between the human and the computational system's software objects, where a software object is a collection of information and methods for processing that information.

When the computation system is a statistical analysis system, as is the case here, the environment is a collection of interfaces that enable interaction between humans and the system's statistical and statistical-graphics objects, where the objects represent our data and whatever graphics, transformations, analyses, models, and so on, that we have created as part of our effort to understand our data.

Thus, to apply our philosophy and technology to the data analysis environment, we must not only develop a set of statistical objects that enable statistical analysis, but we must also develop an environment with interfaces to these objects. We focus on such interfaces and the environment in this chapter; in the remaining chapters we focus on the statistical objects themselves.

Chapter preview. We present a statistical visualization environment that is firmly connected with the principles and practices of statistics, is built using object-oriented architectures developed within computer science, and that keeps empirical results of cognitive psychology and human–computer interface design in mind. The environment includes statistical objects such as data, transformations, analysis methods and models; a wide variety of dynamic and static graphical objects; and data analysis and visualization tools such as datasheets, workmaps, spreadplots, and the like. Taken together, the interfaces to these objects are a rich visual statistical environment. Notice that this chapter is more closely related with the statistical system developed by us, ViSta, than other chapters. This happens because we have a more intimate knowledge of the internals of this system than we can have of other systems, so although we presume that they work in a similar way, we could only judge it from the externals and could easily be wrong.

3.1   Objects

In this section we very briefly introduce the computer science concept of a software object, the basic building block of a software system. We describe, also very briefly, the major objects involved in our statistical system.

A software object has two aspects:

- *Slots*: **places that can contain information.**
- *Methods*: **algorithms that can process information.**

Objects are organized into hierarchical families: Every object (except the root object) is the child of another object. Every object has a method for creating a new object that is a child of the original object. A child object inherits the information structures and methods from its parent object. Furthermore, the child can have additional information structures and methods that the parent does not have, and the child can have information structures and methods that redefine the parent's structures and methods. Thus, a child can have characteristics and behaviors that are the same as those of the parent, others that differ from those of the parent, and yet others that are unique to the child.

Statistical objects are software objects that implement computationally based statistical analysis methods. In a statistical visualization system these objects fall into three broad categories: analysis objects, graphics objects, and interface objects. The hierarchy of objects is shown in Figure 3.1.

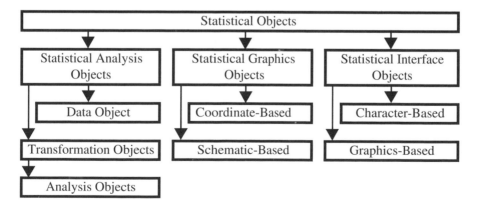

Figure 3.1 Statistical object hierarchy.

In our system, at the root of the statistical object hierarchy is a general statistical object. This is the fundamental object of the hierarchy. It contains all of the information and methods that are meaningful to all the objects in the entire hierarchy, including analysis objects, graphics objects, and interface objects. For example, all objects need to know how to define new objects, delete old ones, copy or rename themselves, and so on.

Inheriting from the root statistical object are each of the three major families of statistical objects: the statistical analysis object, the statistical graphics object, and the statistical interface object. Each of these objects contains all of the methods and slots that are needed by all the objects in its family but are not needed by all the families. For example, the statistical analysis object contains slots for the data array; the name of the data; the datatype; whether there are missing values and, if so, where they are; the activation state of the data; its partitions and elements, and so on.

We do not go into more detail about analysis or graphics objects at this point. We do, however, discuss interface objects in this chapter.

3.2   User Interfaces for Seeing Data

In common, everyday usage, an *interface* "is a device or a system that unrelated entities use to interact." According to this definition, a remote control is an interface between you and a television set, the English language is an interface between two people, and the protocol of behavior enforced in the military is the interface between people of different ranks" (this nice quote is taken from Sun Microsystem's Java Tutorial). The computer scientist's view of an user interface is that it is an aspect of a software system that creates and maintains a mean by which a software system and a user can interact with each other.

Fundamentally, an interface is a set of methods, but interfaces have become of such great importance that software now often has many objects whose sole purpose is to create and operate user interfaces. Such objects are called *user interface objects*, and although an interface is truly a set of methods of such an object, we will think of such objects as if they were objects and name them as such.

There are two types of user interfaces, character-based and graphics-based. With a *character-based interface,* the user interacts with the software system by typing characters on a keyboard, and the system responds with characters displayed on the screen. With a *graphics-based interface*, the user uses a pointer and a keyboard to manipulate windows, icons, and menus that are displayed as pictures by the computer on its monitor.

With a character-based interface the user interacts with an object by using the keyboard. The typed characters are displayed on the screen. The interface may or may not be interactive—if it is, the object's response is displayed on the screen immediately, right after the information typed by the user. Interactivity is useful when you need to know the result of what you just typed before you can proceed. Otherwise, interactivity can be a hindrance. A character-based interface is the best type of interface for entering commands, for solving equations, and for programming.

A character-based interface can be concise and unambiguous, but it is often difficult for a novice to learn and remember, mainly because it is essential to know the correct syntax. Once a character-based interface is learned, it can be very flexible and powerful for advanced users. However, error rates can be high, training is needed, retention may be poor, and error messages may be obscure (Preece et al., 1994).

Graphical user interfaces are based on windows, icons, menus, and pointers—the so-called *wimp interface*. This type of interface has become the standard interface and has become tightly integrated with the operating system. This means that when a user runs a program for the first time, it looks familiar, there being a great deal of transfer from other programs. The familiarity helps reduce the mental effort needed to interact with the program.

The wimp interface provides the developer of a statistical software system with the opportunity to construct specific visual representations of the data, transformations, models, and so on, that form the content of the data analysis. These representations can be tailored to emphasize a particular aspect of the data. We discuss them in Section 3.4.

The phrase "look and feel" has become a very popular phrase in the software design literature during recent years, referring to a very large body of work concerning the visual style ("look") and interactivity style ("feel") supported by the software—in short, the plot's appearance and behavior. In this chapter we consider the look and feel, especially of the dynamic interactive graphical interfaces we use in ViSta for seeing data: the plots and spreadplots. We will see that the look and feel of these dynamic statistical graphics interfaces determine, at least in part, the specific statistical activity most suitable for a given visualization.

3.3 Character-Based Statistical Interface Objects

In this section we discuss interfaces to objects that are either uniquely statistical or are of great importance to the statistical enterprise. Just as is true for interfaces in general, statistical interfaces can be grouped into those that are character-based and those that are graphical. In ViSta, the common uses of a character-based interface to an object are as a command line for interactive processing of commands; as an equation processor for mathematical manipulation of the object's information; as a program editor to create, compile and execute programs that manipulate an object's information; and as a report generator that generates reports about an object's information.

3.3.1 Command Line

The command-line interface in ViSta is shown in Figure 3.2. The interface consists of a character-only command line and a menu system. Note that the first line in the Figure 3.2 begins with a >. This is a command-line prompt. It indicates that the system is waiting for a command. The user then types a "data" command. This command, which is spread out over the next three lines, creates a new set of data. These data are named "normal" (Figure 3.3). They consist of two variables, "Var1" and "Var2". The variable values are created by the random number generator function "normal-rand" which is being asked to generate 300 normally distributed numbers. Since there are two variables, the 300 normal random numbers will be used to form 150 observations. The system's response, in part, is to print a line in the character-based interface identifying the data that have been generated. They are "Normal.Biv#1."

Command-line programs offer a simple way to record the steps taken during a data analysis session. Typing a command produces a piece of output just below it. The expert user can check it and type a new command that results in more output. Programs that do not produce too much output per command are to be preferred because they are easier to understand and may make it simpler to identify which specific piece of information the data analyst was interested in.

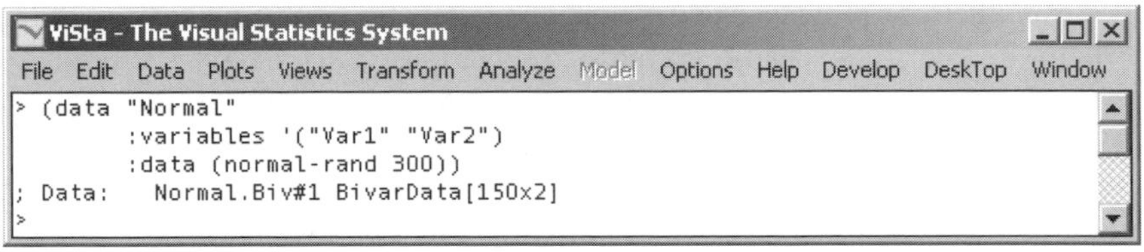

```
> (data "Normal"
        :variables '("Var1" "Var2")
        :data (normal-rand 300))
; Data:    Normal.Biv#1 BivarData[150x2]
>
```

Figure 3.2 Character-based interface: the command line.

3.3.2 Calculator

An interactive character-based interface can be an effective interface for a calculator (Figure 3.3) especially if the syntax of the calculator's language is the syntax for algebra that we all learned when we were children. Although many statistical systems also provide a graphical user interface for the calculator, these do not seem to us to be as useful as the straightforward character-based interface provided in ViSta.

```
? mpg
(16.9 15.5 19.2 18.5 30 27.5 27.2 30.9 20.3 17 21.6 16.2 20.6 20.8 18.6 18.1
17 17.6 16.5 18.2 26.5 21.9 34.1 35.1 27.4 31.5 29.5 28.4 28.8 26.8 33.5 34.2
31.8 37.3 30.5 22 21.5 31.9)
? mean(mpg)
24.760526315789473
? variance(mpg)
42.86731863442389
? normedmpg = (mpg - mean(mpg)) / variance(mpg)
(-0.18 -0.22 -0.13 -0.15 0.12 0.06 0.06 0.14 -0.1 -0.18 -0.07 -0.2 -0.1 -0.09
-0.14 -0.16 -0.18 -0.17 -0.19 -0.15 0.04 -0.07 0.22 0.24 0.06 0.16 0.11 0.08
0.09 0.05 0.2 0.22 0.16 0.29 0.13 -0.06 -0.08 0.17)
```

Figure 3.3 Interactive character-based calculator.

3.3.3 Program Editor

A combination of interactive and noninteractive character-based interfaces can be a very effective interface for writing and editing programs. The immediate feedback of an interactive interface (i.e., the command line) is a very useful way to ensure that you have the correct syntax when writing a program. The interactivity can also be used to support "productivity enhancements" such as command completion and a history mechanism. But once you get the form of the statement right, it is most useful to have a noninteractive interface available (i.e., a text editor) to construct and save the program for later compilation and execution.

3.3.4 Report Generator

A report is a character-based noninteractive interface that displays information about an object in the form of text (Figure 3.4). Reports may include tables, which can be entirely character-based, and pictures, which are graphical, of course. Although such reports are not strictly character-based, they still share most characteristics.

```
ViSta - The Visual Statistics System
SATISTICS SUMMARIZING:
Data: Cars.gen#1
File: D:\CVS\vista7\data\general\cars.vdf
```

VARIABLES (Numeric)	MEAN	StDv	VARIANCE	SKEWNESS	KURTOSIS	N
Weight	2.86	0.71	0.50	0.44	-0.99	38.0
Mpg	24.76	6.55	42.87	0.20	-1.37	38.0
Driveratio	3.09	0.52	0.27	-0.03	-1.34	38.0
Horsepower	101.74	26.44	699.33	0.25	-1.07	38.0
Displacement	177.29	88.88	7899.08	0.92	-0.49	38.0

Figure 3.4 Character-based report.

3.4   Graphics-Based Statistical Interfaces

There are numerous graphical user interface objects that have been designed specifically for statistical data analysis software. We discuss the most important statistical interface objects in this section, including datasheets, variable windows, selectors, desktops, workmaps, plots, and spreadplots.

3.4.1 Datasheets

Certainly the most common interface to a data object is the datasheet, a spreadsheet-like interface similar to the one shown in Figure 3.5. A datasheet is similar to a spreadsheet, having cells organized in a rectangular array of rows and columns. Data can be entered into the cells, and standard editing tasks can be performed. However, unlike a spreadsheet, the datasheet does not permit equations in the cells.

   In many statistical systems the spreadsheet is the most evolved representation. In nearly all statistical systems, many of the abstract things used in data analysis are only realized with the simplest character-based representations: A regression model is usually represented by a few numbers in a table, the numbers being the coefficients of the regression, whereas with sufficient effort, a rich graphical representation can be made that seems to convey much more to the analyst than is conveyed by the numbers alone (Cook and Weisberg, 1999).

▼DataSheet Browser - Br/Cd/Car-Prefs.dsh#1					_□×

| File | Edit | Data | Plots | Views | Transform | Analyze | Model | DataSheet | Help | Window |

Cd/Car-P	About These Data		pgmr2	pgmr3	pgmr4	pgmr5	▲
MulVar[1	Data Header	c	Numeric	Numeric	Numeric	Numeric	
cadillac	List Data	00	8.00	0.00	7.00	9.00	
chevrole		00	0.00	5.00	1.00	2.00	
chevrole	Visualize Data ...	00	4.00	5.00	3.00	3.00	
chevrole	Summarize Data ...	00	6.00	2.00	7.00	4.00	
ford fai	Print Data	00	2.00	2.00	4.00	0.00	
ford mus	Print Source	00	5.00	0.00	7.00	1.00	
ford pinto		0.00	0.00	2.00	1.00	0.00	
honda accord		9.00	5.00	5.00	6.00	8.00	▼

Figure 3.5 A Graphical user interface: the datasheet.

3.4.2 Variable Window

A very simple, yet potentially powerful wimp interface for variables is shown in Figure 3.6. This interface consists of a window containing variable icons, the icons emphasizing characteristics that determine what types of data analysis can be done with the variables. It also supports subsetting, and variable editing, transformation, and visualization functions via the menu items. Having a variety of specialized wimp interfaces like this greatly enriches the overall data analysis environment.

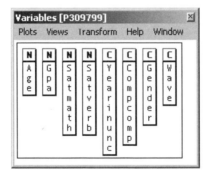

Figure 3.6 Variable window

3.4.3 Desktop

Some statistical analysis systems, particularly DataDesk (Velleman and Velleman, 1985) and ViSta (Young and Smith, 1991; Young, 1994), have developed a graphical user interface based on a modified desktop metaphor, featuring icons representing the statistical objects that a data analyst uses.

The statistical desktop interface is based on the idea of a physical location where data are analyzed, a location where physical and logical tools, datasets, and results would be nearby. ViSta's desktop is shown in Figure 3.7. It combines into a single window four different interfaces. One of these is the character-based command line that we discussed above. The other three are graphical user interfaces that we discuss below. For now, let the short definitions suffice:

- **Workbar.** The workbar consists of the menubar at the top of the window and the buttonbar directly below it. Every statistical data analysis action that the user can take is accessible through the workbar.
- **Workmap.** The workmap is shown directly below the workbar. It consists of connected icons, the icons representing various statistical objects and the connecting lines the flow of work during the session.
- **Selector .** The selector consists of the two lists of names on the right-hand side of the desktop. These lists enable the user to select the specific subset of data to be analyzed.

In Figure 3.7 there is an icon named "Students," which represents the data, and just above it is a toolbar with buttons for help, summary statistics, and visualizations, along with buttons for some commonly used analysis tools. On the right are two lists, one of observations, the other of variables, that provide a simple way of selecting (activating) subsets of the entire dataset. Finally, along the top of the window is a menubar that gives the user access to the full capabilities of the software.

The idea is to display electronic counterparts of the physical and logical objects one would have when doing a data analysis. These objects would include pages the data are recorded on, folders containing those pages, cabinets holding the files, pictures of the data, tables of results, a trash can for discarded results, and so on.

The desktop metaphor involves the concepts of "direct manipulation," "what you see is what you get," and "command consistency." *Direct manipulation* refers to a style of interaction with the desktop objects that mimics the way you would interact with their real-world counterparts. You can "grab" the objects with your pointer and make them do things; you feel as though you are in control. *What you see is what you get* refers to constructing visual objects for the electronic counterparts in a manner such that the visualizations correspond with the user's conceptualization of these objects. For example, a datasheet is visualized to look like a sheet of paper with rows and columns. A plot is visualized to look like a sheet of paper with legends, axes, and so forth. *Consistency* means that wherever you are in the software, a specific action should always have the same effect. For example, a right-click should always produce a pop-up menu, no matter what you right-click on, but the menu should be sensitive to the context defined by what was clicked on.

3.4.4 Workmap

A workmap is a structured desktop that portrays the flow of work during a statistical analysis session. It is a visual dataflow language, creating a diagram that records the flow of steps taken during our search for understanding. This flow of statistical analysis can become quite circuitous and confusing, but it is often necessary if we wish to understand our data thoroughly. We should test different ways that might help us understand the data, look at them from different angles, and chain together exploratory steps until a satisfactory understanding is reached. Some modern computer programs for statistical analysis facilitate this process so that users can more easily analyze their data in many different ways.

A consequence of this approach to data analysis is that the conclusion of an analysis cannot always be summarized by reporting only the result of a single test or procedure. An overview of the process by which one arrived at the concluding test or procedure is often quite useful. It would also be helpful to have an overview that could be used, at each step, to explain the rationale and results of the actions taken at each step, along with alternating actions that were taken or new actions that could be taken. At the very least, we to record the steps taken during the data analysis steps.

Command-line programs make it easy to repeat an analysis step by step. But they are not that good at providing a general overview of the paths examined, including the dead ends that were encountered and the areas that have not beenexplored deeply.

Direct manipulation statistical systems often do not present such a simple way of recording the flow of analysis. Users just point and click and obtain their results, but no trace is left of what steps were taken to obtain the results

The workmap (Young and Lubinsky, 1995; Young and Smith, 1991) represents a method for solving this problem. At first, before starting to analyze data, the workmap is empty, but it grows with each step of the analysis. When a data analysis step is executed, one or more icons representing the statistical objects created during that step are added to the workmap. The icons are connected to previous icons by lines that show the flow of data from one step to the next.

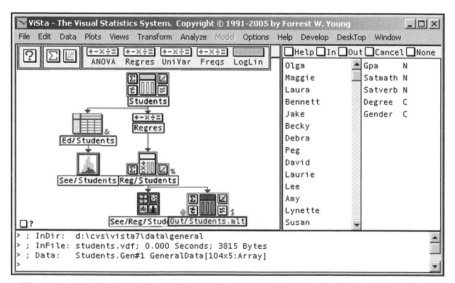

Figure 3.7 Desktop with workmap for regression analysis of student data.

The first few steps of a visual regression analysis of the student data are shown in Figure 3.7 The workmap summarizes these steps:

1.   The data analyst opens the file containing the students data object. The upper-most icon, the one named "Students", appears on the desktop.

2.   The data analyst runs the datasheet editor to fix any problems there might be with the data. In addition to the datasheet window appearing, the datasheet-like icon named "Ed/Students" is added to the map, attached to the data icon with an arrow in order to show the sequence of steps. If changes must be made in the data, a new data icon would be attached to the growing workmap.

3.   The data analyst shows a plot of the data. A plotlike icon named "See/Students" is added to the workmap.

4.   The data analyst uses regression analysis to fit a regression model to the data. Two icons are added to the map. The small horizontal one named "Regres" represents the regression analysis object, and the larger one named "Reg/Students" represents the model object created by the analysis.

5.   The data analyst shows a spreadplot of the model. A spreadplot-like icon named "See/Reg/Students" is added to the workmap.

6.   The data analyst outputs a dataset named "Out/Students.mlt."

Workmaps are dataflow diagrams, a widely cited and used diagram dating back to at least Stevens et al., (1974) although commonly attributed to Gane and Sarson (1979). Dataflow diagrams portray the flow of the original data through various methods, each of which takes in the data as they exist at that moment, processes the data in some fashion, and then produces a new version of the data as the result of the processing. This new version of the data can then become the input to another method, which performs more processing and in turn creates additional new data. As one follows the flow of the data as they move down the hierarchy, we encounter icons that alternate between representing the data and representing a method that processes the data, the hallmark of a dataflow diagram.

The workmap is a dynamic graphic! It is more than just an aid for remembering what has been done. It supports dynamic interaction. The data analyst can interact with the map to revisit previous steps in the analysis, using the workmap as the launching pad for new analyses or new transformations. Subsets of data using different variables or observations can be defined. In short, the full panoply of exploratory data analysis can be accessed dynamically and interactively.

In addition to keeping track of the steps in an ongoing analysis session, and to provide a means for beginning a new analysis path, workmaps can be used to remind oneself or to tell others what the steps of the analysis were that led to the results being considered, and they can also be used to explain the rationale of an analysis to an audience.

Note in the figure that the icons are not all the same. They differ in shape, "decorations," and color (although color is difficult to see in the figure!), characteristics that are used to communicate the role that the icon's action has in the analysis.

Data icons. We show several data icons in Figure 3.8 (they are the larger icons). They are all the same except for the central large portion of the icon. Data icons have six parts, plus the icon's title. The main portion of the icon, which contains the *icon header* and *body*, is surrounded by four *icon buttons* and the *title*. Double-(left)-clicking a specific icon-button produces a default action that is related visually to the appearance of the clicked part of the icon. Right-clicking the same icon button produces a menu of actions that are refinements or generalizations of the default action of the icon button.

Thus, a left-click on the data icon's graphic icon button produces the default spread-plot for the type of data represented by the data icon, and a right-click on the same icon button produces a menu of visualization choices appropriate to the datatype. Structuring the icon in this fashion allows us to provide the user access to more than 80 actions without creating confusion. In essence, the icon functions as six small icons with related but structurally distinct functionality.

The icon header is a red horizontal bar. Double-clicking the header reveals a window with a brief description of the data. Right-clicking the header produces a menu of actions providing information about the data.

The icon body is the largest part of the icon, in the middle. It contains a small schematic representing the datatype of the data. Double-clicking the body of the icon shows an editable datasheet of the data. Right-clicking the body yields a menu for creating, deleting, editing, saving, printing, and exporting the data.

There are four icon buttons surrounding the main portion of the icon, two on the left and two on the right. Each button contains a small schematic representing the action of the button.

The upper-left button contains a summation sign, representing the fact that the button produces summary statistics of the data. Left-clicking the button shows these statistics, and right-clicking produces a menu of statistics available.

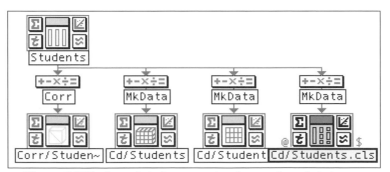

Figure 3.8 Various data icons portraying various datatypes.

The upper-right button contains a schematic of a plot. Left-clicking the button produces the main visualization of the data, and right-clicking the button produces a menu of additional plots and spreadplots.

The lower-left button contains a t to indicate that this is the transformation button. Left- or right-clicking this button reveals a menu of transformations appropriate for the data.

The lower-right button contains the $\approx$ ("approximately equal") symbol, indicating that the button provides access to the analysis menu, a menu of analysis items which make models that are "approximately equal" to the data.

Data Processing icons.   The small icons in Figures 3.7 and 3.8 are statistical data processing icons. They represent a mathematical process that receives data from the icon with the incoming arrow, processes the data, and then sends the processed data out to a newly created icon at the end of the outgoing arrow.

Thus, the workmap is a dataflow diagram with the arrows connecting icons showing the flow of data: The data "flow into" a method, the method performs some sort of mathematical operations that are applied to the data, which then flows out of the method into a new kind of statistical data processing object.

The processing object can be either a transformation or an analysis. If the processing object is a transformation, the new object will be a data object. If the processing object is an analysis, the new object will be a model object. Note that model object icons contain a little drawing of variables as well as mathematical symbols, representing the fact that a model is a mathematical abstraction of data.

Model icons.   The model icons have four parts that are similar to those in the data icons, but there are no little icon buttons for transforming or fitting data, since those operations can be performed by returning to the data and taking the desired action.

Another program that uses icons to represent aspects of the statistical session is *DataDesk* (Velleman and Velleman, 1985). In DataDesk, there are icons at the variable level as well as at the data level, and there are icons for models and graphics. The user's interaction with DataDesk creates these icons sequentially but they are not connected, and thus do not show the "dataflow".

In our experience, the drawback of both programs is that after a time the desktop becomes messy and difficult to use. It is possible, of course, to delete icons (and icon trees in ViSta), but it would also be desirable to be able to create folders on the workmap which themselves contain workmaps. (There are always more things that would be nice to do, but they all take time!)

3.4.5 Selector

The selector, shown in Figure 3.9, is a graphical user interface that enables the user to select subsets of the data by specifying which observations and which variables are used to define the subset. The selector controls the activation of observations and variables (see Section 1.6.1), allowing the user to indicate which observations and variables are to be included in the calculations and visualizations being contemplated.

Figure 3.9 The selector.

The selector interface can take many forms, as is true for any interface, of course. A very common form is to make it part of the datasheet, providing a mechanism whereby the user can indicate whether or not each row and column of the datasheet is active.

Another realization of a selector is shown in Figure 3.9. Here we have a window with two panes, the left one showing observation labels and the right one variable names and types (the complete data are presented in Table 1.2 of the introductory chapter). In this figure two of the observations (Lena and Gina) have been removed from the active subset.

Although the incorporation of the selector's capabilities into the datasheet is a very natural and comfortable addition to the datasheet's capabilities, we tend to prefer the type of selector shown in Figure 3.9 since it is much more economical of screen space, and since we essentially always need to have the selector available but seldom need to see the entire datasheet.

3.5   Plots

A plot consists, in part, of the statistical information the plot is communicating to the user, and in part, of the interface that provides access and interaction between the user and the information. Consider, for example, Figure 3.10. All of the plots shown in the figure are statistically identical scatterplots, but there is more to each of these plots than just the scatterplot. These plots also consist, in part, of interface elements that are designed to help the user better understand the data. These interface elements give the plot a particular "look," and the use of the elements gives the plot a particular "feel,"

Thus, any given plot can have a variety of looks (user interface styles) or feels (user interaction styles).

Let's be clear that these four plots not only present the same statistical information in the same way (i.e., each plot is a scatterplot of the same data) but also provide the user with identical capabilities, options, and features, even though their interfaces look so different. As long as each plot provides the user access to the same menu of capabilities, options and features (such as the one shown), the several plots all present the same information in the same way and provide the same capabilities.

3.5.1 Look of Plots

It is an important point, so we repeat it: Even though the plots in Figure 3.10 all look different, they are statistically identical: They are each a scatterplot of the same pair of variables. But, of course, the plots are not the same. They all look different, and,

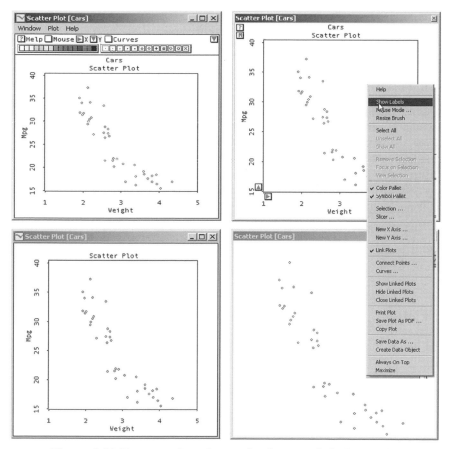

Figure 3.10 Four user interface styles for a statistical graphic.

although we can't see it in the figure, they all feel different when you use them. They all have their own unique look and feel.

Each of the plots shown in Figure 3.10 differs because each has a different visual interface. These interfaces differ in certain fundamental ways: We can look at each of the plots and we can ask: Does the interface involve menus? If so, is there a menubar? Are there hot spots? Are there pop-ups? Does the interface use buttons? If so, is there a button bar? Do we see the plot's background features? If so, which ones? Indeed, these are essential questions about the fundamental components of interface style.

Button-based. The interface can display buttons with tables indicating what the button does (as in the upper-left window) or it can present no such buttons (as in the lower two windows). Although the buttons will nearly always duplicate the action of a menu item (as can be seen by comparing the menu with the buttons in Figure 3.10), the feel of the interaction for the two methods is rather different, as we discuss in the next section.

Dialog-based. The interface can incorporate dialog boxes. Dialogs can more efficiently present complex information than any other interface type. It is possible to put many controls in a dialog box, and the controls can vary from buttons to sliders. Control panels, which are essentially complex dialog boxes, are the most advanced example. And dialog boxes do not take up plot space, since they are in their own separate dialog window.

Menu-Based. Although menus are always a part of the interface, their method of presentation varies in terms of its visibility. The interface can visually cue the fact that menus are available by which the user can interact with the graphic. The most common visual cue is a menubar with menus (as in the upper-left window in Figure 3.10). Another possibility is to show menu hot-spot buttons on a menubar (as in the upper-left window). A third is to show the menu hot spots on the surface of the plot, as is done in the upper-right window of Figure 3.10. On the other hand, there may be no visual cues, as is the case for the bottom two windows.

Plot-based. As can be seen in the four windows in Figure 3.10, the lower right window contrasts with the others in that it does not show what is called the plot's "background" information and features. The information and features include the axes, variable names, legends, tic marks, and so on, of the graphic, which are shown in all but the bottom-right window. A variety of combinations of these plot background information and feature cues can be presented, or not presented, thereby modifying the plot's "look." The "minimalist" view seen in this window dispenses with all of the nonstatistical information and shows only the plot's statistical content (which, of course, is always shown).

We have, then, the possibility of generating a fairly large number of possible looks for any of the dynamic graphics we present in this book. These looks can vary in terms of the way they present (or do not present) menus, the way they use (or don't use) buttons, and the presence (or absence) of various aspects of the plot's background information. The four looks shown in Figure 3.10 are four that we commonly use, as will become evident in later portions of the book. These looks are:

- **Full control.** This look, shown in the upper-left plot, has the most complex interface with the most controls. These controls appear on the menubar and buttonbar (menus, buttons, hot spots and optional pallets).
- **Button-based.** This look, shown in the upper-right plot, shows buttons and hot spots, but hides the menubar and buttonbar.
- **Control-free.** This look, shown in the lower-left plot, shows no controls.
- **Content only.** This look, shown in the lower-right plot, shows only the plot's statistical content.

Other types of looks are possible, given the scheme above, but the ones mentioned here are the ones we have found useful. We do not provide a menu-based look because there is only a single menu and it seems to be a waste of screen space and an addition to screen clutter to have a menubar with only a single menu. Of course, it is possible to pop-up the menu that would be on the menubar, so no functionality is lost and both space and appearance are improved.

The "look" of an interface is not independent from the "feel" of that interface. An interface whose look involves many visible controls suggests an interaction style that involves using the controls, whereas an interface that shows no controls suggests a different interaction style.

3.5.2 Feel of Plots

Just as there are several looks (user interface styles) that are useful for the dynamic interactive graphics used in visual statistics, there are also several feels (user interaction styles) that can be useful. However, whereas it is easy to show the different looks, it is not so easy to show the different feels. Nonetheless, any plot can have several different feels.

Our goal, of course, is to have our plots feel very dynamic and very interactive. To achieve this goal we need to identify the characteristics of interaction style, and we need to learn how variations in graphical interfaces relate to variations in the feel of their dynamism and interactivity.

Plots—indeed, windows in general—have only a limited number of ways of interacting with a user. All of the plots that we discuss in this book have the capability of sensing the motion of the cursor. Of course, the plots can have menus and buttons as well. The menus can be on menubars or can be activated at hot-spots or as pop-up menus, and the buttons can be on button bars or located on the plot itself.

Characteristics of high interaction.   Each way of interacting with a plot, by the cursor, by menus, by dialogs, or by buttons, has its own basic nature that at least in part determines the style of interaction and how the interaction feels. These ways can be distinguished from each other on at least the following four aspects:

- **Immediacy.** When we interact with a plot we want our actions to take effect immediately, rather than having to wait for the effect to take place or having to continue with a sequence of actions before the plot responds.
- **Succinctness.** To make our interactions with a plot as highly interactive as possible we want to be able to do the interaction with the least number of keystrokes or mouse-button clicks. Zero is optimal.
- **Repeatability.** It is highly desirable that the action taken by the user can easily be repeated. This way, the dynamic action will feel even more interactive.
- **Interruptiveness.** The user's action should not interrupt the ongoing behavior of the graphic, and it should not create a discontinuity in the plots' behavior.
- **Flexibility.** When we interact with a plot, we wish to be able, easily and quickly, to change the behavior of the plot that is affected by our actions.

We turn now to a discussion of each of these several ways of interacting with a plot. The discussion is in terms of the characteristics of high-interaction graphics just outlined.

Motion-based (cursor). Motion-based interaction has the highest level of interactivey of the various ways of interacting with a plot. It is immediate, takes only one user action, can be repeated easily and rapidly and is nondisruptive. When you move the cursor, the reaction of the plot is instantaneous, representing a smoothly continuing movement that in no way disrupts ongoing actions of the graphic. When the user moves the cursor, the graphic responds. The action repeats as long as the cursor moves.

A plot that responds to the motion of the cursor as it moves across the plot feels very much more dynamic and interactive than one that does not. When a plot reacts this way, the feel of the plot is very different than when nothing happens when the mouse is swept back and forth (i.e., when the mouse is "selecting"). Thus, plots for which the motion of the cursor causes changes in the plot feel much more interactive and dynamic than any of the other user interaction styles.

The main negative aspect of cursor-based motion-sensitive interaction is its lack of flexibility. To change the behavior of the plot that is influenced by our actions, we must resort to another mode of interaction, such as buttons or menus.

Click-based (button or dialog). Click-based interaction, whether using buttons or dialogs, is nearly as highly interactive as motion-based interaction, but feels rather less interactive. It is immediate, can be initiated with two keystrokes (dialogs may take more), can be repeated with one keystroke, and is totally nondisruptive. It is also flexible since multiple buttons can be placed on the plot or dialog.

When you click a button, nothing more need be done, as the click produces an instantaneous reaction on the part of the plot. Clicking does not interrupt the ongoing behavior of the graphic. Whereas the button or dialog click is a one-shot action, the button or dialog is usually programmed so that the action can be repeated simply by holding the mouse button down while the cursor is on the button. Thus, if anything, click-based repetition is even simpler than motion-based repetition. Finally, button-based interaction is very flexible.

Although a bit of extra effort is needed to initialize a dialog box (i.e., to show it the first time), dialog boxes (including control panels and sliders) have the same level of interactivity as buttons and are a very good way of controlling the dynamic aspects of a plot. While they require steps on the part of the analyst to display them, once visible, the dialog box provides an interface which is not only highly interactive, but is capable of supporting much more complex kinds of interactions than can be supported by a button, since multiple coordinated controls are reasonably easily implemented.

Menu-Based. Menus are considerably less interactive than button-based, dialog-based, or cursor-based interaction. As we will see, menu-based interaction takes several keystrokes to perform, implying that it is not as immediate nor as repeatable as the other types of interaction. Furthermore, the appearance of a menu on the screen completely halts all ongoing behavior, with nothing happening as long as the menus being shown.

The main problem with using menus to control dynamic behavior is that several steps must be taken to accomplish the task: (1) You must either move the cursor to the menu on the menubar or to a hot-spot (this step is skipped for pop-up menus). (2) You must click the menu or hot-spot to make the menu appear. (3) You must move the cursor down the list of menu items to select the item desired. (4) The item selected must be clicked (or unclicked) to cause the action.

So menus are not a good way of controlling dynamic interactive graphics: It takes too many steps to create an action and any ongoing action is stopped completely while the menu is being displayed.

3.5.3 Impact of Plot Look and Feel

Button-based, dialog-based, and especially mouse-motion-based interactions are at the heart of dynamic interactive graphics. They are the most effective at providing an emersive experience that truly gets you into your data. These ways of interacting with the data produce a software "feel" that is remarkably emersive. The potential to provide them as a standard feature of the interface between the user and the data is the defining characteristic of dynamic interactive statistical graphics.

It should also be clear at this point that the look of a dynamic interactive statistical graphic is not totally independent of its feel. The most highly interactive and most dynamic methods of interacting with a graphic require certain types of controls whose presence causes the graphic to have certain types of appearance. That is, buttons, which have perhaps the optimal combination of dynamism, interactivity, and flexibility, are quite likely to be used with a dynamic interactive graphic, creating not only a specific, button-based feel, but also a specific button-based look.

We have spent considerable effort over the course of our research in software design on the question of the look and feel of statistical visualization software and have concluded that different types of statistical visualization tasks demand different types of user interfaces: that is, that one type of look and feel is right for exploring data, another for transforming data, a third for modeling data, and a fourth for communicating results. As we will see in Section 3.6, multiplot visualizations that have the highest degree of interactivity are the most appropriate for exploring data, since the immediacy of the changes in the plots helps us discover structure. On the other hand, visualizations that are less interactive are appropriate for the more contemplative activities, which may require more precision, such as transforming data or fitting models. More will be said on this in Section 3.6.

3.6   Spreadplots

One of the main problems with the visual approach to statistical data analysis is that it is too easy to generate too many plots: We can easily become totally overwhelmed by the shear number and variety of graphics that we can generate. In a sense, we have been too successful in our goal of making it easy for the user: Many, many plots can be generated, so many that it becomes impossible to understand our data.

When we generate too many plots, and if we are restricted to having only one plot per window, we not only have the problem of understanding multiple plots, but we also have the problem of managing multiple windows. In this situation it becomes the case that too much time is spent managing windows, with too little time left for learning about our data.

Several solutions come to mind to improve our task. One solution is to put all of the plots in one big scrollable window, making sure that they don't overlap. Although this may be better than putting every plot in its own window, it still isn't a very good solution: The user now spends too much time scrolling up and down or back and forth trying to locate the right plot. Even simple tasks such as comparing groups can be very time consuming. Another solution is to make sure that the plots all fit on the screen, adjusting size as more are generated. Of course, the plots rapidly become too small to be useful, and just closing them can become a problem.

The spreadplot (Young, et.al., 2003) is our solution to these problems. A spreadplot is a multiview window containing a group of dynamic interactive plots, each of which shows a unique view of the same statistical object. The plots not only interact with the viewer, but also with each other. Since they all provide different views of the same statistical object, they are linked so that changes made in one view are reflected appropriately in the views provided by each of the other plots

Spreadplots can be applied to a wide variety of statistical situations. They can be used to explore data, transform variables, fit models, communicate results, and teach students. Each view in a spreadplot can be any of the dynamic interactive graphics discussed in this book, and can have any of the looks and feels discussed above. When the several views all have the same look and feel, the spreadplot takes on the characteristics of its plots, becoming a multiview dynamic interactive graphic with the look and feel of its plots. However, since there is only one window, the problems associated with managing multiple windows do not exist. Furthermore, a spreadplot can be constructed from a group of plots, which can include any of the dynamic, high-interaction, direct manipulation graphics that we discuss in this book, as well as from more mundane objects such as static graphics, report windows, dialog boxes, or datasheets. As long as it is a graphical object, it can be used.

One of the important features of a spreadplot is that the views can be linked with each other in an extremely general way. The linkage is such that when the user makes changes in one of the plots, the implication of those changes can be shown instantaneously in the other plots. The linkage can be according to features of the data, such as observations, variables, or categories, or the linkage can be according to statistical or mathematical aspects, such as algebraic equations that relate the values displayed in one plot with those displayed in another.

Spreadplots are also very flexible in terms of the style of interaction between the user and the graphic. The interaction can be as simple as just rubbing the cursor around the screen, or as complex as allowing the user to use multiple menus, many buttons, and several dialog boxes simultaneously. Associated with the style of interaction is the visual style of the spreadplot. It can be very simple and elegant, with very few controls evident, or it can be very complex, with literally hundreds of buttons and menu items.

Because of the wide variety of plots we have to choose from, and because of the flexibility of the linkage between plots, the style of interaction with the user, and the visual style of the spreadplot itself, we can design spreadplots to address a very wide variety of specific statistical issues. Thus, some of the spreadplots that we have created are specifically designed for exploring data, others are designed for transforming data, and yet others are designed for modeling data. Of those spreadplots that are designed for exploring data, some are specialized for frequency data, others for relational data, others for certain kinds of magnitude data, and so on.

Finally, there are specialized spreadplots that correspond with graphics that have been proposed previously. The best known example is the scatterplot matrix, a graphic dating back to the 1980s that shows the data analyst an $N \times N$ matrix of N^2 scatterplots, where the N^2 scatterplots are formed by pairing all pairs of numerical variables in a dataset. It is very easy to construct such a graphic using the spreadplot notion; and then it is very easy to generalize the plot matrix notion to create plot matrices whose individual plots can be essentially whatever you like: mosaic plot matrices, quantile plot matrice,and plot matrices where the individual plots are not all the same but are, say, all of the relevant plots when the variables have a mix of datatypes.

Unlike other types of spreadplots, which consist of a wide variety of possible arrangement of rectangular plot cells, a plot matrix has a very proscribed arrangement of square plot cells: If there are N variables, then a plot matrix has N^2 plots. The plots are arranged into a square $N \times N$ matrix, each plot itself being square. The rows and columns of the plot-matrix correspond to variables in the data. We almost always also append additional tall but thin namelists to the left or right side of the matrix of plots. These namelists help identify individual observations in the dataset.

Spreadplots are created by people with expertise in a specific research topic, expertise in statistical data analysis, and expertise in spreadplot construction. Thus, a spreadplot incorporates the expert knowledge and experience of the designers, who must decide which of the many relevant plots are most likely to be useful, how these relevant plots will be presented to the user, and how they will interact with the user and with each other. Spreadplots simplify the data analyst's task when many different plots are relevant to the task at hand, because the expert who designed the spreadplot has made many decisions that the data analyst would otherwise have to make. Spreadplots also simplify the task at hand because window management problems are eased, and because they structure the types of tasks required of the data analyst.

3.6.1 Layout

One problem for visualizing multiple views is that of laying out the plots (Murrell, 1999). Indeed, there are some plots, such as scatterplot matrixes and trellis displays, that are formed just by arranging simpler plots according to certain rules. Scatterplot matrices, for example, arrange scatterplots side by side so that each variable in a dataset is graphed against the other variables, with the graphs being displayed as a row or a column of the matrix. This lets the user rapidly inspect all of the bivariate relationships among the variables, permitting the detection of outliers, nonlinearities, and other features of the data.

The DataDesk program (Velleman, 1997) includes a feature called *corkboards* that helps the user design customized layouts. A corkboard is a window that permits other windows to be pasted into itself. These other windows could provide views of data or models or can all concern some statistical problem that the analyst may find interesting. The plots are empty of data, but methods are provided so that users can carry out their own analyses. This allows the user to develop customized statistical procedures or specialized applications.

JMP also provides a tool called *journal* or *layout* (depending of the version of the software) that can be used for arranging the output of analysis. The user can add pieces of output to journals and layouts, but as they are just images, no longer being live plots with which one can interact, they only serve the purpose of summarizing and tidying up results.

ViSta uses a command in Lisp-Stat that automatically lays out a set of objects (plots, text windows, or dialog boxes) in a two-dimensional matrix of cells with any number of rows or columns. The upper-left schematic in Figure 3.11 shows a two-row by three-column layout matrix.

The number of plots and the number of cells in the layout does not have to be the same. The simplest case is when there is one plot per matrix cell, and the plots are laid out adjacent to each other, one per cell. If there are more plots than cells, then in some cells the plots must be laid out on top of each other, with the last one being shown initially and there being a provision to switch which plot is on top. User interactions with the views can make the hidden plots pop up when required.

If there are fewer plots than layout cells, a cell can be assigned a null value, so that the null cell has no plot in it. A plot next to the empty cell can then be expanded to span to cells. A plot may be expanded to the right, as has been done in the upper-right image in Figure 3.11. Similarly, a plot may be expanded downward, as in the upper-left plot of the lower right image. This behavior is controlled by a a matrix of numbers whose elements specify whether a plot occupies more than a single plotcell within a row or column. Each entry must be a nonnegative integer indicating how many cells are spanned, counting from the current plotcell. Plotcells that have no plot of their own have a span value of zero. There must be two span matrices, one for spanning right, the other for spanning down.

Cells all have the same width and height unless a relative width or height is specified. If a list of relative widths for the columns is provided, the column widths will be made proportional to them. The lower-left image in Figure 3.11 has columns that are approximately proportional to 0.5, 1.5, and 1. Finally, a layout that we have often found useful is the one in Figure 3.11 lower right. Notice that this is a matrix with two rows and four columns, with two lists surrounding four central plots. The lists expand to the height of the spreadplot but are half-wide.

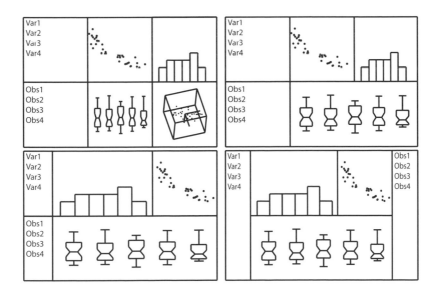

Figure 3.11 Layout of spreadplots.

3.6.2 Coordination

The second problem one encounters when dealing with multiple interacting plots is coordinating their changes. In fact, coordination of multiple views for information visualization is a topic that has received considerable attention from researchers on human–computer interaction. The reason for this interest stems from the observation that developers of software often need to include different sources of data in the interface to help the user make the right decisions.

North and Shneiderman (1997) propose a taxonomy of possible window coordinations based on two concepts:

1. Windows have two components: a view and a collection of information items. Items can be selectable, while views provide support for navigation. Navigation actions are scrolling, slicing, rotating, and so on.

2. The collections of information items in two coordinated windows can be the same (e.g., a bar chart and a tabular list of the same set of data points) or different (the collections are not identical but are interrelated).

Linking corresponds to the situation where two windows have the same underlying collection of information items and the user selects some of the items in a view to see the items in the other view. A situation that is different from what is usually found in statistical applications is when the process of navigating a view makes another window to navigate in synchrony. Yet another situation is when selecting an information item causes other windows to navigate. ViSta, for example, uses the panels of scatterplot matrices as control panels that select variables in other plots, such as boxplots, normal probability plots, and so on, causing the other views to change.

Perhaps the main problem with coordinated views, including the types of linking that are used in dynamical statistical graphics, is that they are very complex to develop—it takes such an extensive amount of high level of programming that most users cannot create them easily by themselves.

There are two types of multiple-plot coordinations that can be implemented by most users. One is the simple point-linking coordination, where the correspondence between points in different plots is the basis for their coordination (Tierney, 1990). The other is observation linking, where the correspondence between observations in different plots is the basis for their coordination (Stuetzle, 1987). However, other types of coordinations are in general not available to be created by data analysts but must be implemented by developers. An associated drawback is that complex coordination schemes implemented by professional programmers are not amenable to further modification.

Note that generic-view coordination architectures are exemplified by Snap (North and Shneiderman, 2000) and by InVision (Pattison and Phillips, 2001), but although they show promise, they have only been used in prototype systems at this time. Another architecture for coordination of multiple views is the Gossip architecture (Young et al., 2003). This architecture is oriented toward coordinating communication between the plots in a spreadplot, so that changes in one plot are reflected in some or all of the other plots. Take, as an example, the spreadplot in Figure 3.12. The different plots and lists that constitute the six views of this spreadplot are coordinated in several ways. For example, the model component list on the left is in charge of computing a log-linear model when the user selects elements of the list. The model based on the selection generates predicted values, residuals, and fit measures that are displayed in the other views. Another type of coordination between the several views is provided by the history view, which is the second plot in the bottom row. It is a line plot that shows the value of the fit measure for each model that the user tried during an analysis session. Selecting a point in this view makes the rest of the plots change so that they show what they were showing when that model was originally fitted to the data.

Spreadplots treat each view as an object that can receive and send messages that control its behavior. Messages can be created by a user by typing commands or indirectly as a consequence of interactions with the user interface (i.e., pointing and clicking, where the clicks causes messages to be sent). A spreadplot manager is in charge of controlling the message traffic. When a plot in a spreadplot experiences some change (e.g., when the user selects a variable in a window that is displaying a list of variables), it sends a message to its spreadplot message manager about the details of the change. The spreadplot message manager then forwards the message to the appropriate objects (or to all objects if it is the first time the message has been sent). Each

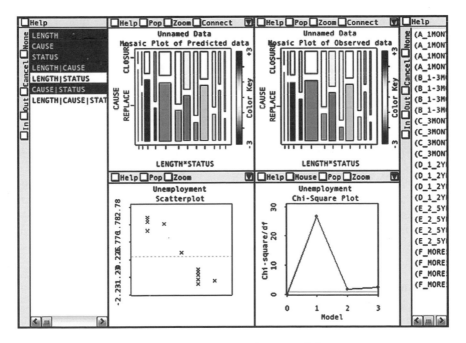

Figure 3.12 Example of a spreadplot

object, having been programmed by the spreadplot developer, then knows how to behave appropriately in response to the message. We say that the spreadplot manager gossips to the other plots about what the plots are doing.

Sometimes, it may be that the message sent by a plot needs to be processed by a statistical object before it is received by the group of objects. That is, the message needs to be processed so that the processed message can be distributed to the group of objects. From the developer's viewpoint, this case is no different than the case where the message itself is distributed to the plots. If processing by a statistical object is needed, the developer must write a method for the statistical object, just as for graphical objects. Then, once the message is processed by the statistical object, the results of the processing will be sent by the statistical object to the manager for distribution to the group of objects.

3.6.3   SpreadPlots

Message Flow. Note the coordination between the plots. Both the slider and the scatterplot matrix react to the user's actions. Specifically, when the user clicks a cell of the scatterplot matrix, a message is sent telling the two large plots to change themselves to show the new focus. Also, when the slider is moved, a message is sent telling the plots to change themselves in light of the change in the transformation's parameter value.

The way in which a spreadplot manages the message flow is shown in Figure 3.13. A *view*, at the bottom of the figure, is, in our example, either a plot or a dialog, but it could be any object with a graphical representation, including datasheets, reports, and so on. Note that a view does not send a message to another view directly. Rather, it sends messages to the spreadplot manager, which then broadcasts the message to all of the objects in the set of objects that form the spreadplot (the cluster of small arrows below the manager icon represents broadcasting). Those objects that need to respond to the message do, and those that do not don't.

These messages cause changes to take place according to clearly defined rules. The scatterplot, for example, evaluates whether the cell in the scatterplot matrix that has been clicked is on the diagonal. If so, a message is sent to the manager that the transformation plot should be shown. If not, a message is sent that a scatterplot should be shown. Also, when the slider is moved, the plots are modified according to the rule instantiated in the equation given above. Additionally, the points in the plots are linked so that the modification of their properties (color, symbol, selection, and so on) propagates to the other plots. See Young, et al. (2003) for more information on spreadplots.

Data flow: Of more interest to us here, for reasons that will become apparent in the last section of this chapter, is the dataflow architecture, a topic not discussed by Young, et al. (2003), but which is depicted in Figure 3.13. The figure has the same general appearance as a workmap, but keep in mind that this is a schematic view of how the various objects involved in creating a spreadplot receive their data.

In Figure 3.13 we see what the actual workmap looks like when a spreadplot is created: It shows a workmap icon attached to a model object. But, what is going on behind the scenes is shown in Figure 3.13. When a spreadplot is created, it extracts an object called a *data space* from the active statistical object. This step isolates the changes that a user may make to the data via the spreadplot from the statistical object, ensuring that the statistical object does not get changed with some sort of inadvertent disastrous side-effects. The data spaces is an object that contains the data to be visualized in some type of canonical form that is designed to optimize the efficiency of the visualization process. Currently, it appears that this canonical form is simply a matrix with rows for observations and columns for dimensions.

Next, the spreadplot manager is instantiated, with it taking in the data space information. It then invisibly instantiates each of the view objects using the extracted data space as the information from which each plot's initial display is constructed. The spreadplot manager then forms the desired layout and displays those plots which it

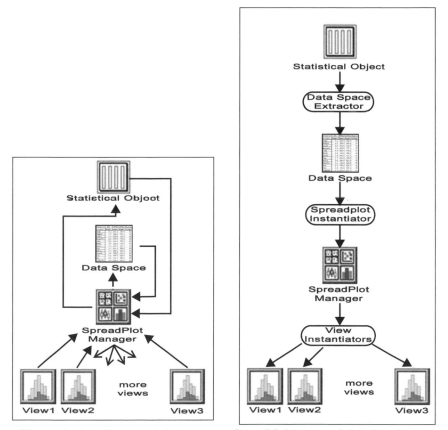

Figure 3.13 Left: spreadplot message flow. Right: spreadplot data flow.

has been instructed to display. The manager then creates the desired message passing structure. Finally, unless told not to, the spreadplot manager shows the spreadplot.

3.6.4 Look of Spreadplots

We have seen in Section 3.5.1 that two plots can present precisely the same statistical information, using identical statistical graphics, and provide identical functionality but look completely different. We also saw in Section 3.5.2 that a pair of plots can differ from each other in terms of their user interaction style even though they present identical information using the same statistical graphics with the same capabilities: It is possible to have two such statistically identical plots, where one plot requires the user to do nothing other than rub the cursor across the screen and watch what happens, whereas the other requires the user to click a button, choose a menu item, move sliders, or manipulate elements of a dialog box or control panel. These different user interaction styles vary in terms of the immediacy of their interactivity.

The same can be said for spreadplots, since, after all, they consist of plots, and especially when all of the plots have the same look and feel, the spreadplot takes on the look and feel of its plots. Thus, when a spreadplot is being designed, the designer must decide for every plot in the spreadplot, what the appropriate look and feel should be. Generally, all the plots will look and feel the same—will support the same kind of user interaction. The spreadplot designer has an additional consideration when creating the look of a spreadplot, and that is how the windowing features of each plot will look when the plots are put together to form the spreadplot. More will be said on that in a moment.

The look of a spreadplot is determined partly by its plots and partly by its container window—the window containing the plots. The container window determines the appearance of each plot's "window decorations"—the title bar, menubar, and borders. The look of everything except the window decorations is determined by the plots and is exactly as it would have been had the plot been an individual plot rather than part of a spreadplot. Nothing new here.

What is new is the part of the spreadplot look that is determined by the container window. The container window determines the appearance of the plotwindow's "decorations," including its title bar, menubar, and borders. The title bar and menubar of each plotwindow can be shown or hidden. If the title bar is shown, it can optionally show or hide its various control icons. The window borders can also be shown or hidden, and if they are shown, can take on a variety of appearances, from a simple line to a full-fledged, depth-cued window border.

In Figure 3.14 we show four spreadplots that are identical except for the window decorations of their plots. The different looks of the four spreadplots result from the use of different container window options at the time the spreadplots were created. These differences are emphasized by making all of the plots "content only." By comparing the four spreadplots, we see that we have a lot of control over their look.

The degree of control means, at one extreme, that the plotwindows can be drawn with no decorations at all. This results in a spreadplot like the upper-left example in

Figure 3.14, and the one shown in Figure 3.15. With this look, we see the background and content of each plot (if the plot is "contentonly," as is the case for Figure 3.14, but not for Figure 3.15, where we see only the content) but nothing to indicate that each plot is in its own separate window. This look is the least cluttered and has the cleanest appearance. We often use this when the entire background and content of the plot windows are being shown, letting the plots themselves act as their own visual frames.

We also commonly use a look that simply draws a line as the window edge, with no title bar, and no window-edge depth cues. This type of look, which is demonstrated in the upper-right section of Figure 3.14 and is used in Figures 3.12, 3.16, and 3.17 clearly demarcates the plots, but does not confuse the user by revealing the unnecessary information that each plot is in a separate window and then all of the windows

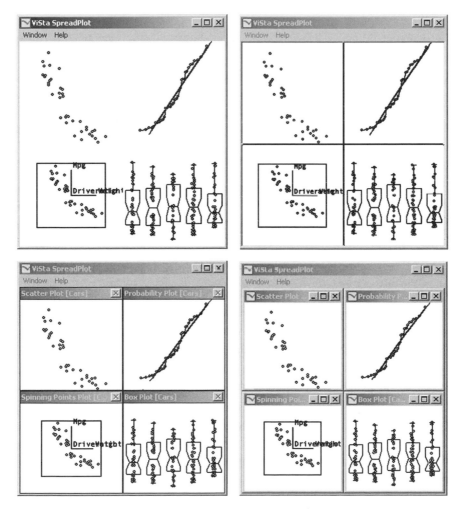

Figure 3.14 Four container-controlled looks for a spreadplot.

are placed inside another window. This look works very well in plotmatrices and augmented plotmatrices such as Figure 3.16.

We occasionally use the look shown in the lower-left panel of Figure 3.14. It is especially handy for students and others learners as well as for making presentations at meetings.

Finally, we can make each plot window look just like a window with depth-cued borders, a title bar with a title, and a full set of controls for minimizing, iconizing, maximizing, and closing the plot window (lower-right example in Figure 3.14). We never do this, because we want all of the individual plot-windows to act together and do not want the user to be able to control each one separately. Note that we also never show each plot's menubar, since each plot has only one menu. Having a menubar with one menu wastes screen space and adds visual clutter. Rather, we provide the pop-up version of the menu instead.

A final point is that we always present the container window's menubar. If you look at any of the figures showing a spreadplot, you will see a menubar with HELP and WINDOW menus. The help window provides help items for both the spreadplot as a whole and for the individual plots. The window menu provides items for the individual plots and for other aspects of the system. It would also be possible to include each plotwindow's menu on this menubar, which sounds like an excellent idea that we overlooked implementing!

3.6.5 Feel of Spreadplots

The feel of a spreadplot is determined by the feel of its plots. Generally, we require that all plots have the same feel, so that the user has a consistent experience no matter which part of a spreadplot is being used. Thus, we have the same types of spreadplot feel as we do types of plot feel:

- **Motion-based (cursor).** Spreadplots whose plots provide motion-based interactivity feel much more highly interactive than any of the other styles.
- **Button click-based.** Spreadplots whose plots provide button-based interactivity feel quite dynamic, although less than motion-based.
- **Dialog click-based.** Spreadplots whose plots provide dialog-based interactivity have the same level of interactivity as those that are cursor based but require extra keystrokes to make them visible. Click-based interactivity is more flexible than motion-based interactivity.
- **Menu-based.** We do not provide spreadplots with menu-based interaction by itself, although pop-up menus are available in all plots as a means of gaining access to the full capabilities of the plot.

3.6.6 Look and Feel of Statistical Data Analysis

The extensive experience the authors have had with dynamic graphics leads us to speculate that different looks and feels are appropriate for different types of statistical analysis activities. When we explore data, the visualizations that have the highest degree of interactivity seem to be the most appropriate: The immediacy of the changes in the plots seems to helps us discover structure. What is important in this situation is the overall structure of the data, with the importance of the fine detail receding into the background. So, for exploratory data analysis, visual impact should be stressed and precision relegated to a back seat.

On the other hand, when we are testing hypotheses, fitting models or transforming variables, then visualizations that are less interactive may be more appropriate. Such visualizations promote inspection and contemplation, encourage reflection and attention to detail, and emphasize concern for precision and careful fine tuning. In the next several subsections we explore the relationship between the feel of the interface and the statistical activity to which we are interfacing.

Look and feel for exploring data. The control-free, motion-based look and feel seems to provide the best interface for exploring data. The control-free look, shown in Figure 3.15, removes all buttons, buttonbars, menus, and menubars, as well as all window decorations, thereby creating a clean, uncluttered appearance. However, the plot backgrounds, which provide needed reference information to help us understand the data during our exploration, are still part of the look. Note that we also don't need the added flexibility of the buttons and menus, since motion-based interaction is generally all that is needed for exploration. Of course, if we do need to access additional features, we can do so via pop-up menus.

The spreadplot shown in Figure 3.15 is the bivariate numerical spreadplot discussed in Chapter 7, a spreadplot designed to explore bivariate magnitude data through the use of very high interaction techniques. The cursor, located in the upper-left plot, indicates that the spreadplot is in brushing mode since it consists of a brush with an attached dashed rectangle. Being in brushing mode means (discussed below) means that when the user moves the cursor back-and-forth across the spreadplot, the points inside the rectangle are highlighted, and if labeling is on, the labels associated with the highlighted points are displayed, as we see in Figure 3.15.

Thus, the scatterplot has the feel of a very-highly interactive dynamic graphic. Indeed, as the user moves the cursor across the various plots in the spreadplot, the specific subset of graphical elements that are highlighted and labeled changes, always corresponding to the specific set of elements of a plot that fall within the brush's rectangle. The change happens immediately, drawing our attention to the structure of the data.

But keep in mind that if the various plots in the scatterplot are linked, it isn't only the plot elements within the brush's rectangle that are highlighted and labeled. The "linked-to" elements in the other plots of the spreadplot are also highlighted and labeled, although labels are off for all five small plots because label overlap in these small plots is so bad that the plot is less useful with labels than without.

Of course, the nature of the highlighting depends on the characteristics of the plot, especially on the nature of its graphical elements. We see this in the figure: The plots of this spreadplot have several different types of graphical elements. Therefore, they have several different types of highlighting:

1. For the parallel-axes plot (the tall and narrow plot second from the right), highlighting means that its graphical elements, which are line segments, are drawn when highlighted but are not drawn (shown as unconnected points) when not highlighted. The highlighted line segments can be labeled.

2. For the namelist (the tall narrow list of labels at the right), its graphical elements (the names) are shown in reverse video when highlighted, in regular video when not highlighted.

3. For the histogram (bottom row, second from left) its graphical elements (tiles of the bars) are drawn, when highlighted, as solid rectangles, whereas they are

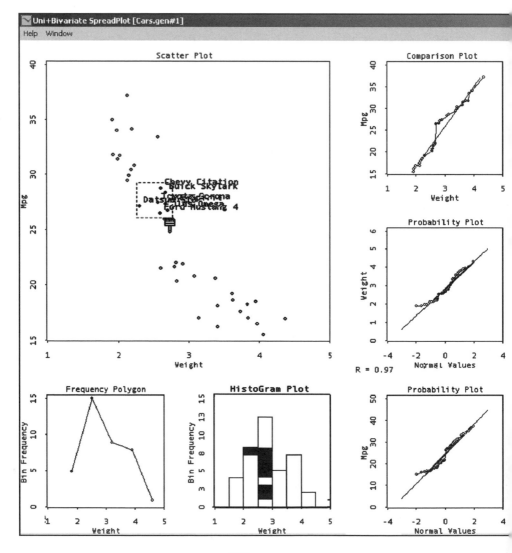

drawn as empty borderless rectangles normally.

4. For the comparative plot (top of the column of three small plots) and probability plots (the other two in the column), the graphical elements are points, and they are highlighted by being drawn as solid dots rather than as empty dots.

You have to use your imagination to envision what the user sees as the cursor is rubbed around the screen. Since the spreadplot is in brushing mode, the elements of the plot being brushed, as well as linked elements in other plots, will be flashing on and off as the brush moves about, leading the eye to structure that may be present. Brushing is a very-high interaction activity, and is conducive to discovering structure.

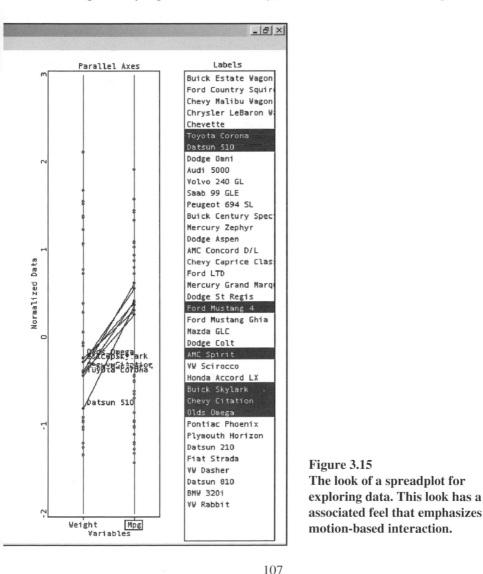

Figure 3.15
The look of a spreadplot for exploring data. This look has a associated feel that emphasizes motion-based interaction.

Look and feel for transforming data. Now consider the spreadplot shown in Figure 3.16, a spreadplot designed to support transformation of variables to bilinearity. In this spreadplot, what we wish to do is to make all of the bivariate plots shown in the scatterplot matrix look as linear as possible, for it can be shown that variables that are independent, normally distributed have bivariate relationships that are linear. The content-only click-based look and feel seem to be most appropriate for transforming data. All we are really interested in is the shape of the bivariate relationships, not really even needing the reference material provided by the plot backgrounds.

As discussed in Chapter 7, this task requires study of the individual bivariate plots and manipulation of a slider to transform the variables shown in a plot so that their

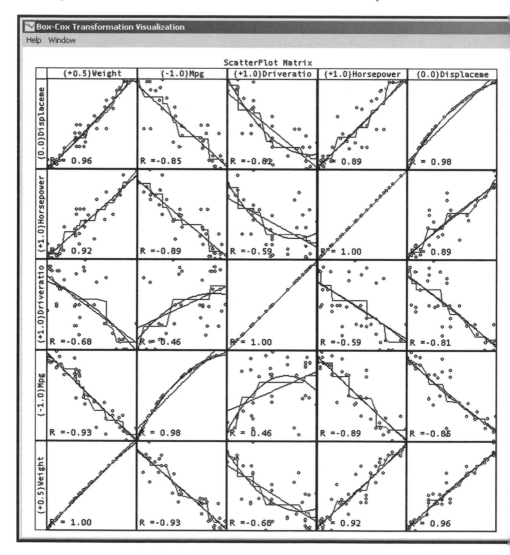

relationship looks linear. Other than variable names, we do not need any of the reference material provided on the plot backgrounds. We really need to focus on the shape of the bivariate relationships, Thus, the content-only look is most appropriate.

The goal of this data transformation situation is to linearize the relationship between all pairs of variables. Thus, the spreadplot's scatterplot matrix has a special *plot selecting mode*: When we click on a scatterplot matrix that is in this mode, the small plot we click on is selected. This is different than anything that we have seen up to this point far. Our click does not select points in the small plot. Rather, it selects the entire plot.

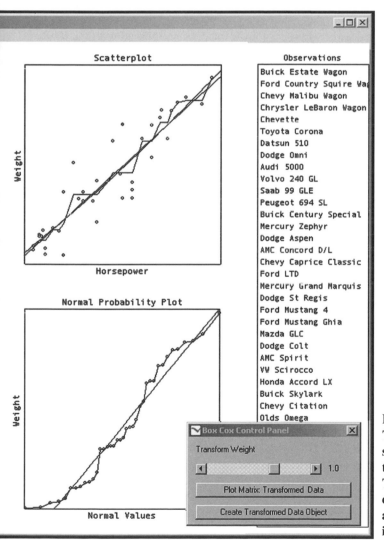

Figure 3.16
The look of a spreadplot for transforming data. The spreadplot emphasizes menu and dialog-based interaction.

The scatterplot matrix has been programmed so that we can focus in on any one of its small plot cells to make the bivariate relationship shown in the plot cell more linear. We do that by clicking on the desired plot cell, usually the one that looks least linear. The click on the plot cell causes a large version of the scatterplot to be shown at the lower right and causes the corresponding comparison plot (QQplot) to be shown above it. Thus, the mouse is not in the high-interaction motion-based state used for brushing. It is in the lower-interaction click-based state that is more appropriate for plot selecting.

However, the slider is highly interactive, lending a high-interaction feel to the spreadplot: When the slider is moved, it uses the equations proposed by Box and Cox (1964) to transform the data. The equations are applied dynamically as the slider is moved, making the bivariate relationship selected either more or less linear. As the user moves the slider back and forth, the transformation is modified with the result displayed, changing dynamically and smoothly in real time, enabling the user to find the position for the slider that produces the most linearity.

Notice that as the slider is moved back and forth by the analyst, the large scatterplot is not the only plot being affected. The large comparison plot is also affected. Furthermore, since the transformation changes the linearity of the relationship between the two variables by changing the two variables themselves, all of the small scatterplots in the y-variable's row and the x-variable's column of the scatterplot-matrix are affected. Thus, when the data analyst moves the slider, both large plots change, and an entire row of small plots, plus an entire column of small plots, also change, all smoothly and in real time.

Look and feel for fitting models. The button-controlled, click-based look and feel appears to be the for fitting models. Our main interest is in details of the fit and in interpreting the model. There are usually a fairly wide variety of actions that we might want to take to evaluate the fit and model, and having them "just one click away" makes for the richest experience. Thus, having click-based access to tools for interpreting the model provides the needed flexibility. Of course, we still want to be able to identify points in the usual point-brushing, high-interaction way. As an example, in Figure 3.17 we present the visualization spreadplot for multiple regression. Notice that its' look is that of a full-control spreadplot, with a control panel for high-interaction changes. Let's examine how the regression spreadplot works.

First, brushing works in the usual way, and it is combined with linking and labeling in the usual way as well. Thus, we can simply move the cursor across the various plots, and whenever an activatable plot element is within the confines of our brush rectangle, the element is activated, highlighted and labeled.

Note also that there are actually a minimal number of buttons on the plots, each one having a Help, a Mouse, a Pop and a Zoom button. Clearly, the Help button provides help about the plot. The Mouse button switches between the two standard mouse modes of brushing and selecting. The Pop button pops the plot out of the spreadplot so that it can be dealt with individually, and the Zoom button makes the plot larger.

The control panel, which is where most of the data analyst's actions will take place, is divided into three parts: Variables, Options, and Demonstrations. The Variables section controls which variables are shown in the plot cells. As shown, the *Mpg* variable, which is the response variable, is being shown. If we were to click on another variable in the control panel, all four plots in the spreadplot would change to provide information about the newly chosen variable. In the Options section we see that Regression Line and Residual Lines are both selected. When these were clicked on, they caused the regression and residual lines that you see in the regression and fit plot to appear. If they were clicked on again, these sets of lines would disappear. In the Demonstrations section we have access to two regression demonstrations, one for Influence Points and the other for Restricted Range.

3.7   Environments for Seeing Data

Now that we have defined what we mean by *objects* and *interfaces,* we can define what we mean by an environment: An *environment* is defined as the union of all of the object interfaces that are active at any given moment. An environment is also defined, albeit indirectly, by the objects that are active at any given moment, since the interfaces that define the environment directly, must be interfaces to objects. The environment changes over time as the selection of active interfaces changes. Thus:

   Environment. An environment is the entire collection of interfaces and objects active at a given moment in time.

Let us consider the changes that can occur in the environment as the data analyst proceeds though a data analysis. Many of the actions that can be taken by a data analyst either activate or deactivate an interface object, thereby changing the environment. These actions modify the environment by creating new interface objects (the analyst opens a new datasheet or displays an additional spreadplot) or by activating or deactivating an existing interface object (the analyst closes a plot, or reopens a closed but not deleted selector window).

While not all of the actions that a data analyst takes change the environment, it is important to understand that as a data analyst searches through the data for understanding, the environment adapts itself to the nature of the ongoing analysis. When the steps are a focused progression toward that understanding, the environment converges on a structure that reveals that understanding.

Wilkinson (1999) uses the term *Analytics* to refer to the interactions that occur between the data analyst and the analysis system during this search for understanding. Analytics involve the use of methods that produce new statistical objects when they are applied to existing statistical objects. In our terms, we call this a *step* in the data analysis. Each step of a data analysis involves using the methods of a statistical object to process the information on a statistical object, where the processing creates a new statistical object.

When one looks at the diagram that results from such methods, one sees that it is hierarchical. For example, when we reconsider the workmap of the visualized regression analysis shown in Figure 3.7 from this point of view, we notice that it is hierarchical. It begins with the acquisition of data by the system, which is visualized on the workmap by the topmost icon in the hierarchy (named "Student"). Then the analyst takes the series of steps outlined in the discussion of Figure 3.7). As we follow these

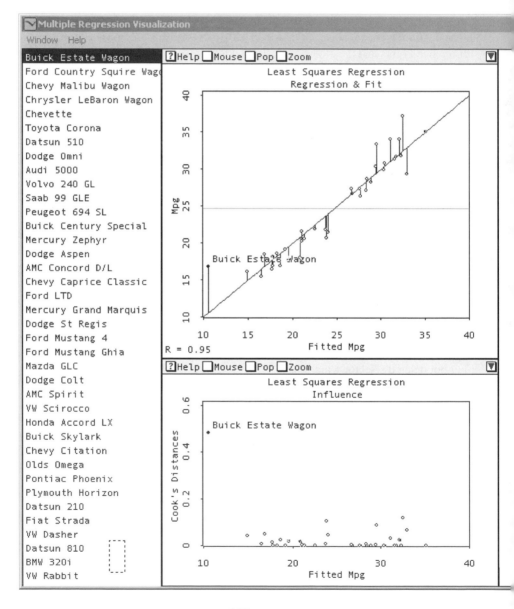

steps and as we watch the workmap grow, we see the hierarchical nature of the diagram emerge. This is also true for the workmap shown in Figure 3.8, which reveals the structure of the steps taken to read in the data and then to calculate the correlations and frequency datasets that we used to demonstrate datatypes (the datasets are shown in Table 1.2). In all cases, we can represent the steps taken during a data analysis session by a tree diagram, such as the workmaps shown in this chapter.

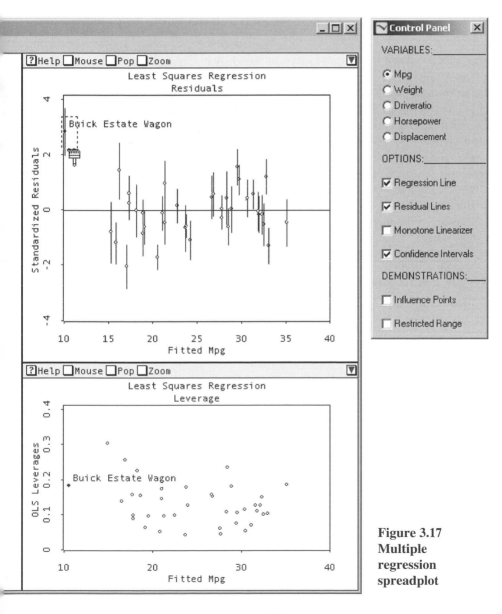

Figure 3.17 Multiple regression spreadplot

3.8 Sessions and Projects

As we have seen, a data analysis environment is the entire collection of interfaces and objects that are active at any given moment. The environment changes and evolves over time as the user steps through the data analysis process. Every time that the analyst takes another step in the analysis, the environment changes.

Session. A session is a time sequence of environments. Since an environment is a collection of interfaces and objects, a session is also an evolving time sequence of changing interfaces and objects.

A session is created by the data analyst as the analyst pursues better understanding of the data, one step at a time. Thus, a session is a time sequence of environments, and since an environment is a collection of interfaces and objects, we can see that a session is an evolving sequence of interfaces and objects, each step in the sequence differing from the preceding one by a change in an interface. A session begins when the analyst opens a new and initially empty data analysis environment, and it ends when the environment is closed.

Naturally, the steps of a session vary in complexity, some of them being very simple, some very complex. Although these steps are portrayed on the workmap, their sequential order is not. Despite this limitation, the workmap is entirely sufficient for recording and reproducing the data analyses it portrays, although the original order of the steps is lost.

Project. A data analysis project consists of a sequence of data analysis sessions that follow a unifying investigational thread.

Each of the project's sessions occupies a chunk of time: A project is a sequence of sequences.

3.9 The Next Reality

Since, as we saw above, an interface makes interaction possible between the software and the user, and since the environment is defined by the interface methods that are active at the moment, it follows that the environment determines the richness and the capability of the relationship that the user has with the software.

When one thinks of an environment, one probably thinks of a three-dimensional space containing visible objects that appear to be solid and space-filling. Almost certainly, one is probably not thinking of a 3D computerized environment, but of the natural environment.

3.9.1 The Fantasy

When one turns one's attention to a computer-created 3D environment, one notes that the 3D analog of the 2D desktop is the 3D workspace, a concept that has no impact outside the research laboratory (and, to be honest, not much inside the research laboratory either). One reason for this has been the expense of the equipment necessary to create a 3D workspace, although this reason no longer applies. Another reason is the complexity of controlling such a workspace, both in terms of the pointing device and of the cognitive and perceptual processes needed.

One possible way to deal with these problems is to create a virtual reality representation of the workspace, since with a virtual reality interface the user employs navigation and orientation skills that are familiar and natural. The goal of virtual reality is to give the user the illusion of immersion in the environment. Not only are the abstract statistical objects that the data analyst needs to use represented in the workspace by electronic analogs, so is the data analyst himself or herself.

As we stated in Chapter 1, we anticipate the development of a virtual environment for data analysis. This environment would contain visual representations of the data and of the mathematical, statistical, and graphical procedures relevant to the data. It would also contain representations of an expert consultant's suggestions of what to look for in the data and how to look for it; representations of the actual steps taken by the data analyst while exploring; and representations of the actions taken by the data analyst to confirm or disconfirm the hypotheses.

If we did have a virtual data visualization environment, it would have to contain objects representing not only the data but also the tools we use to understand the data, as well as transformations, models, reports, graphs, datasheets, and so on, that are part of the analysis. Such an object system is a parallel universe, if you will, to the virtual universe—a parallel universe that exists inside our software structures and methods. If we were to go about constructing such a virtual data analysis environment, the environment, the objects it contains, and the way they are represented in the environment would result from interface methods to their underlying program objects. For such a virtual environment, for every program object that the data analyst might wish to use, there must be an interface method to that program object which constructs an appropriate virtual object. This virtual environment would contain virtual objects representing the "things" used by the data analyst: the variables, datasets, datasheets, graphs, reports, transformation methods, analysis methods, statistical models, and so on.

And, lest we forget, the data analyst interacts with the data and other statistical and graphical objects in various ways. Usually, the interaction involves some type of tool that creates something new about the data. The tool could be a data editor, which would create a revised version of the data; or, the tool could be a report generator, which would create a new report about the data; or the tool might be a graphical method, which creates a plot of the data or an entire collection of interacting plots. It is of vital importance that any virtual data analysis environment that we create let these actions occur in a "natural" way.

3.9.2 The Reality

Advances in dynamic interactive 3D graphics have made it possible to implement, at least on an experimental basis, virtual reality environments for data analysis—VR data environments, if you will. These experimental environments are capable of creating an illusion of 3D reality that is much more compelling than has heretofore been possible. As a consequence, it is now possible for data analysts to interact with representations of their data, and of their graphics, tables, and other data analysis objects, that appear to be solid and space-filling.

A specific kind of image, called an *anaglyph image*, makes the 3D illusion possible. An anaglyph image is produced using two views of an image taken from slightly different viewpoints. These viewpoints correspond to the position of the two eyes. The two images are rendered in two contrasting colors (red and blue, for example). Then, using glasses with color filters, the right eye sees only the image intended for the right eye, and the left eye sees only the left image, thereby simulating normal vision. The result is 3D imagery that vividly portrays the illusion of depth. Anaglyphs are the most commonly employed technology for simulating stereoscopy.

Visuals-Pxpl. The first known attempt to achieve a VR-data visualization environment was that of Young and Rheingans (1991), whose work was reported with an accompanying video supplement. They developed Visuals-Pxpl, an implementation of Young's six-dimensional Visuals graphics system on a special-purpose, massively parallel, one-of-a-kind graphics computer called Pixel-Planes developed at the University of North Carolina by Fuchs and his co-workers (Fuchs et al., 1989).

The article published about Visuals-Pxpl focused on the construction of a six-dimensional data space, on its real-time projection onto the user's 3D virtual space, and on the real-time control and modification of the 6D to 3D projection. As described, the system was not immersive, but presented a full-color 2D image of the virtual space on the screen of a relatively small display.

An immersive display system was created in the mid-1990s, although the developments were never reported. This system was a reimplementation of Visuals-Pxpl that displayed the virtual space on a head-mounted stereoscopic display system developed by the University of North Carolina's computer science department. The display featured two small screens, one mounted in front of each eye, with the system being programmed to calculate two stereoscopic views. The system then presented the left-eye view on the left-eye screen and the right-eye view on the right-eye screen. The head-mount also featured a triaxial LED system that was tracked by sensors in the room so that the location, position, and viewdirection of the user was known at all times. Given this information, the system could, in real time, update the displays to show the user the appropriate images of the data space in which he or she was wandering.

VRGobi. The CAVE system developed by Carolina Cruz-Neira and colleagues (1992) was applied to data visualization (Symanzik et al., 1997). The CAVE is a projection-based system that uses 3D computer graphics, position tracking, and auditory feedback to immerse users in a 3D environment. The CAVE consists of stereographic images rear-projected onto three sidewalls and front-projected onto the floor (therefore it is a four-projections system versus the PlatoCAVE and the MiniCAVE discussed later, which were based in one projection only). The illusion of 3D is created using special glasses. The position of the user is tracked via a magnetic-based tracker, a cyberglove, and a handheld wand. This tracking allows the system to receive input from the user in order to create the appropriate 3D images and to be manipulated via gestures.

The software program XGobi was adapted to run on the C2 system (a second, larger CAVE-like environment), and christened VRGobi. The main difference between the two systems is that "with XGobi the user interface is rather like a desktop with pages of paper whilst the C2 environment is more like having the whole room at our disposal for the data analysis" (Symanzik et al., 1997 p. 44). This change meant that a number of user interface elements had to be modified to adapt to the new conditions. The authors concluded that the CAVE is "remarkably different from the display devices that are commonly available for data analysis" having "huge potential for data analysis" (Symanzik et al., 1996, p. 7).

PlatoCAVE. The PlatoCAVE system (Wegman and Symanzik, 2002) was an immersive display system installed in a room approximately 20 feet on each side. The system consisted of a Stereographics CRT projection system driven by a Silicon Graphics workstation. The images that were projected on a wall of the PlatoCAVE were approximately 15 feet (3 meters) in diagonal measurement. The effect was described as quite stunning, as the images would be seen as standing out of the wall, with the system being capable of producing a three-dimensional full-color image. However, its high price limited its use to the research laboratory.

MiniCAVE. A second system, the MiniCAVE, replicated the features of the PlatoCAVE using less costly hardware. Thus, using personal computers instead of workstations and changing the type of projector (LCD instead of CRT), they were able to build a virtual reality system for a tenth of the cost of the previous system. Although the new system was, in essence, a downgraded version of the first system, it incorporated speech recognition as an input mechanism (Wegman et al., 1999). Speech recognition was included because the desktop metaphor, with its human–computer interface based on windows, menus, icons, and pointing devices (i.e., the WIMP interface) did not seem to be well suited for the 3D workplace. Furthermore, physical control devices such as gloves or wands were awkward, to use while speech seemed, at least in principle, a more natural way to interact with the visualization.

3DVDM. 3DVDM is a 3D visualization system that is focused especially on the visualization of large databases. This system also uses the CAVE environment and can manage thousands of observations smoothly (Nagel and Granum, 2004).

But—we must return to reality.

3.9.3 Reality Check

Once the novelty of these systems has worn off, it seems natural to question if their added cost is associated with improved data analysis. This question was addressed in a paper that compared the performance of a group of 15 statisticians using VRGobi and using XGobi (Nelson et al., 1999). The results were encouraging, suggesting that VRGobi was better for visualization tasks that required subjects to identify clusters. However, the users found it more difficult to interact with VRGobi than with XGobi, a result that is confounded with the fact that the subjects were also more familiar with desktop-based interfaces than with virtual reality interfaces.

Although we believe that there is great potential in these systems, it always seems to be the case that virtual reality never quite lives up to its potential. So although there certainly are some very good demonstration systems, the number of successful systems for VR-data analysis seems to remain at zero. The jump from the laboratory to the real world is certainly not trivial.

Reality is a hard act to follow.

4

Tools and Techniques

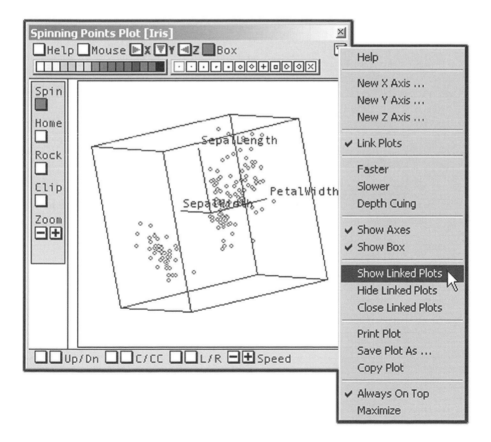

4

Tools and Techniques

Statistical visualization involves many activities, most of which are familiar from other computation situations. However, some of the activities will be unfamiliar because they are unique to the dynamic interactive graphics involved in statistical visualization. We must do ordinary data management activities, such as reading in the data and selecting the right variables or observations. But we must also do activities that are unique to statistical visualization: activities such as generating and using dynamic plots to explore our data, coordinating the plethora of plots that we obtain when we have been exploring our data for awhile, and keeping track of the overall process of data exploration in a way that lets us both retrace our steps and go back to an earlier stage of exploration to branch off in a new direction.

The techniques for carrying out the familiar activities are well known and do not need to be discussed here. However, the techniques for the activities that are unique to statistical visualization are not well known, and we do need to discuss them. It is these techniques that are the focus of this chapter.

Dynamic interactive statistical graphics provide the potential user with a wide variety of controls and windows, some of which will be familiar from other types of computer programs, and some of which will be novel, since they are unique to statistical graphics. In this chapter we focus on those controls and windows that are unique to statistical graphics, although to some extent we must also review the familiar techniques in order to explain those that are novel.

The material is organized into three sections. The first is on the controls used in statistical graphics windows, the second is on windows that present a single dynamic plot of data, and the third focuses on windows that present, organize, and coordinate multiple dynamic plots of data.

On the asthetics of dynamic plots. At first glance, the graphics constructed by a dynamic interactive graphics software system may well look flat and colorless, like nothing so much as a bunch of dots or lines, sometimes colored rectangles, all looking like graphics where aesthetic issues are secondary.

New users in particular may get the wrong impression, thinking that the technical quality of these plots is inferior to the presentation graphics that our proverbial new user is accustomed to seeing. However, as has been pointed out by many (e.g., Tufte, 1983), superfluous ornamentation is not needed by the truly beautiful!

In fact, the aesthetic in dynamic interactive graphics concentrates on creating graphics that move smoothly and rapidly, even for large data, rather than ones that use fancy renderings that may improve the appearance of a graphic, but add little substance and cannot happen in real time. We go light on the window decorations so that the dynamism can be fast. We reduce the lines and characters that have to be drawn for each observation, so that we can increase the number of observations that can be dynamically graphed.

There is a parallelism between static graphics, presentation graphics, and dynamic graphics, on the one hand, and certain kinds of automobiles, on the other. As implemented in the large well known statistical systems, static graphics, are like huge, powerful, but lumbering trucks ("juggernaut lorries," as the English would say, or "18-wheelers" to Americans); they can muscle their way through any set of data and prepare a picture of it, no matter how big the data. Dynamic graphics, on the other hand, are like a small nimble sports car. They don't have the power of static graphics to process millions of records of data, but they have the nimbleness needed for negotiating the tricky curves of exploratory data analysis. Dynamic graphics is also unlike presentation graphics software, which allows you to add all the tail fins and hood ornaments that you please to make you results as eye-catching as possible, even if it is just superficiality that one sees. Each type of graphic has its place and usefulness, and of course, its own special ways in which it can be misused.

Of course, a dynamic interactive graphic is much more powerful than a static graphic. In fact, the initial view provided by a dynamic graphic is often the very same view that the corresponding static graphic would provide. Then, the power of the dynamic graphic comes into play: It uses this initial view as a place from which exploration of the data can be launched.

Static graphics are only able to show a specific view of the data. Dynamic interactive graphics are prepared to respond to actions from the user to show (or hide) information that can help to reveal things that otherwise would remain unseen.

This dramatic increase in the power of dynamic interactive graphics over static graphics does not come without a cost: Dynamic interactive graphics are much more complicated to use. You don't just look at the picture, you interact with it. So there must be a well-designed user interface which has controls that enable the instantaneous interaction that is required to really dig your data.

Chapter outline. We discuss the types of controls that have been implemented to control graphics, especially dynamic interactive graphics, and the techniques used to carry out the control functions of these tools. We then turn to the look and feel of plots and how this relates to the tools and techniques. We then delve into the specifics of how these tools and techniques are used in statistical graphics, and how the look and feel affects this use. We focus the discussion on the three basic statistical graphics: datasheets, plots, and spreadplots (multiplot windows).

4.1   Types of Controls

The interface of our dynamic, highly interactive statistical graphics system is based on the now common model of a user who interacts with the system via a mouse or other functionally similar point-and-click device. With this device the user finds a particular spot on the screen and clicks on it. We assume that the user has a two-button mouse and that the meaning of a click with either button can be modified by holding down a particular key, such as the shift key or control key, often referred to as the *meta key*. Thus, in essence, the user can select a spot and click on it in four ways: normal click, right click, metaclick and metaright click.

The interface, then, must have elements to click on, and the elements must be as unobtrusive as possible, without being so hidden as to prevent the user from finding them and using them. Some of the elements that are used by various statistical visualization systems are discussed below. The discussion is coordinated with Figure 4.1, which shows a ViSta window and menu for a spinnable point cloud. The user interface for the spinnable point cloud is one of ViSta's most complex. It includes several "standard" controls (including the Help, Mouse, X, and Y buttons, which are used by a wide variety of graphics). There are also "specialized" controls that are unique to the spinplot. These are grouped on the left and at the bottom of the plot. The figure also shows a menu, and, of course, the point cloud itself. The elements on this graphic are discussed later.

4.1.1   Buttons

Small buttons can be added to plots, but too many of them make for a messy appearance. Since they are activated by just a single click, they are convenient for actions that occur immediately. Also, since buttons can be made to change their state (change color, for example) they are particularly appropriate as on/off toggles, such as standardizing/unstandardizing data, or turning a feature such as labeling on or off.

There are various ways in which buttons can be used, and the spin plot window exemplifies all of them. These are as follows:

- **Simple buttons.** In Figure 4.1 the Help button is a simple button which, when clicked on, produces a help window for the spinnable plot. The Mouse, X, Y, and Z buttons are also simple buttons that either change a system state or produce a specialized menu or dialog box. (If there are more than four variables, a dialog box is shown to select a variable for the axis. If there are four variables, the axis is switched to the undisplayed one without dialog a box being shown. A pop-up menu would be more convenient than a dialog box.)
- **Toggle buttons.** In Figure 4.1 we see that the Box button is in the "on" state (it is gray rather than white), and we see the "box," which is the wire-frame cube drawn around the point cloud. By clicking repeatedly we turn the button on

and off, and the box around the point-cloud appears and disappears. The Spin button works the same way, toggling spinning of the point cloud, its axes, axis labels (which are showing), point labels (which are not showing), and the surrounding box, on and off.

- **Incrementer button pairs.** In Figure 4.1 there are five pairs of buttons that increment/decrement an argument value that controls some aspect of the display. These are the zoom button pair, which zooms the point cloud in and out, and the Up/Dn, C/CC, and L/R buttons, which control rotation vertically (up/down), circularly (clockwise/counter clockwise) and horizontally (left/right). All of these button pairs are affected by the speed button pair, which controls how fast the other buttons work. By using the right button and the meta key, these button pairs have interestingly modified effects.

- **Menu hot-spot buttons.** These are buttons on the screen that are demarcated by a special symbol indicating that what the button does is cause a menu to pop-up at the button's location. In the ViSta interface the symbol is a red triangle on the button. Thus, in Figure 4.1 the hot spot (which is mostly covered by the menu and cursor) has a down-pointing triangle, indicating that a menu can be pulled down (popped-up) by clicking on the hotspot. The menu is shown.

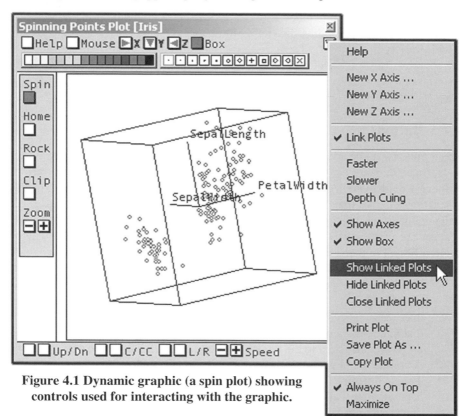

Figure 4.1 Dynamic graphic (a spin plot) showing controls used for interacting with the graphic.

In addition to buttons, which are good for immediate-interaction events, which don't take up too much screen space, but which can make the screen look cluttered if there are too many, there are a number of other controls.

4.1.2 Palettes

Palettes are essentially groups of tiny buttons that are used for selecting a value for a characteristic of the plot that has multiple alternative values, where the effect of the choice is immediately visible when a selection is made. In Figure 4.1 we see two palettes, one for point color (represented in the book by grays, but on the screen by colors), the other for point symbol. If a set of points in the point cloud is selected and one of the color palette buttons is clicked, the color of the points selected changes to the clicked-on color. The same is also true for point symbol. Note that a palette takes up a fairly large amount of screen space. For this reason each palette can be toggled on and off by an item of the menu.

4.1.3 Menus and Menu Items

As you know, menus consist of menu items, where each item is an option that is available to the user for modifying the actions that take place in the window. Because it takes more clicking and movement of the mouse to produce a menu and choose an item than it takes to click on a button, the items should control options whose effect is not necessarily immediately visible.

Menus and buttons both have strengths and weaknesses. On the one hand, menus, unlike buttons, do not fill up space on the screen and do not cause visual clutter. On the other hand, menus are hidden most of the time, so it is quite likely that some options or capabilities will remain unused just because the users do not see their menu items.

As you can see in Figure 4.1, some of the buttons are also represented by menu items (the X button is the same as the New X-Axis menu item, for example). This may not be the best, but it seems like common practice to provide many redundant ways to access the same features, apparently to avoid the risk of the feature being overlooked.

4.1.4 Dialog Boxes

Dialog boxes pop up generally after a menu item is selected or a button is pressed. Dialog boxes are one of the most complex controls, being a resource that is used when many options are available for a given situation. They can include forms, several types of buttons, and so on. Modal dialogs force the user to use and dismiss the dialog before continuing with his or her interaction with the plot. They are not recommended because they prevent exploration of different values or options. Modeless dialogs (see Figure 4.2) can remain on the screen while the plot is used to explore the data dynamically, and they can be clicked on to change their options, many of which

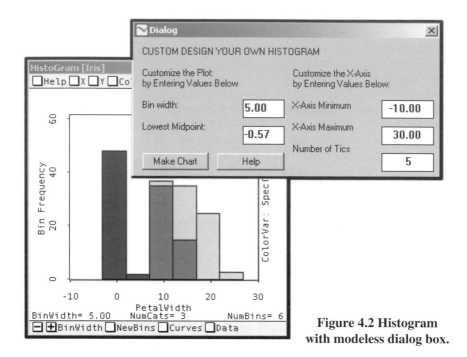

**Figure 4.2 Histogram
with modeless dialog box.**

will have an immediately visible effect on the point cloud. The downside is that they take up a lot of screen space, and they are sometimes very difficult to program so that they work in all situations. The dialog in Figure 4.2 lets the user set the parameters of the histogram so that they can search for a histogram with a desired appearance. The dialog has immediate effects on the histogram.

4.1.5 Sliders

Sliders enable the user to input an argument value that falls in a specified range and to increment or decrement the argument value dynamically. The argument's value is used by the software system in some way specific to the particular situation, with the resulting effect being shown as a change in the graphical display. Thus, operating the slider enables the user to see how the values of the slider determine the nature of the graphics being displayed, with, if desired, the graphics changing smoothly and continuously as the argument's value is scrolled throughout its range. A control panel with two sliders is shown in Figure 4.3.

Sliders are related to scroll bars, a standard user interface element in all current computer interfaces. Scroll bars are described in the human–computer interface guidelines of many operating systems and are familiar to all of us. Data visualization sliders are sometimes simply scroll bars with an associated display of the current value of the argument as it is being changed by the movement of the slider. Eick (1994) has proposed several improvements of the slider in relation with data analysis.

4.1.6 Control Panels

Control panels, which are the most complex user control device, are modeless dialog boxes designed to control a dynamic graphics process. An example of a scatterplot's guidelines control panel is shown in Figure 4.3. This panel can add several guidelines to the scatterplot. Each guideline results from an analysis of the two variables of the scatterplot, with the analysis producing the guideline. The analyses include principal component analysis (for display of the principal axis guideline and residuals); linear, polynomial, or monotonic regression (for showing regression guidelines); and smoother fitting (for displaying the smooth guidelines).

4.1.7 The Plot Itself

A current trend that is popular in some programs is to make the plot itself the control interface. When this is done, clicking with the mouse on parts of the plot pops-up dialog boxes or menus that can be used to modify the clicked-on part of the plot. This technique is interesting but is limited in that some options cannot be related to specific parts of the plot, and it is not always clear to the user which parts of the plot will have actions represented by dialogs or menus, and which parts will not.

4.1.8 Hyperlinking

Hyperlinking has been used in statistical visualization programs to connect plots or analysis output to other, presumably related plots or output, or to other information

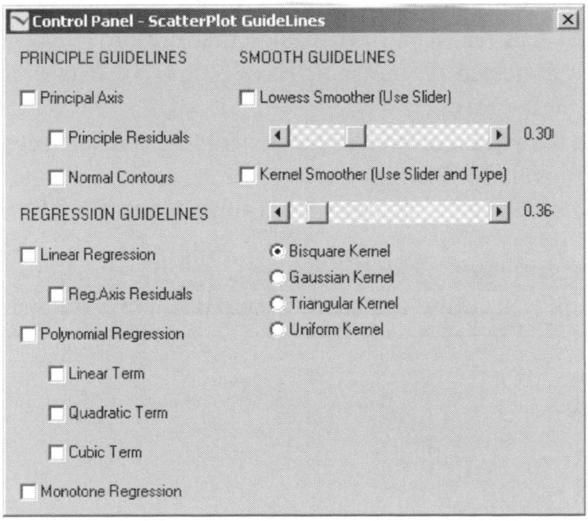

Figure 4.3 Control panel with two sliders.

127

altogether. This technique can be used to avoid excessive amounts of information being displayed on the screen. At first only a part of the output is shown, with the user being able to request more information by clicking on hyperlinks. This is an excellent way of dealing with the information overload problems we have about above.

4.2 Datasheets

Datasheets are the basic tools for representing and manipulating tables of data. The datasheet columns represent variables and the rows represent observations (Figure 4.4). The first column of the table commonly takes the role of showing the label of each observation, and the first row often shows variable names, with the next rows showing variable information such as measurement type and role in the study.

Datasheets are essentially simple spreadsheets, the main difference being that the cells of a datasheet always contain data, never the formulas that are such a powerful feature of spreadsheets. Although datasheets can be seen as being weak or watered-down versions of spreadsheets, datasheets are usually one small component of a statistical system that has far superior calculation abilities. As such, datasheets do not need to have equations in their cells because the statistical environment can instantiate the formulas and perform the computations defined by such equations.

There are many implementations of datasheets, particular when one realizes they are simple spreadsheets. Of the vast number of spreadsheets, some are commercial and some noncommercial, and many are huge programs loaded with features that are irrelevant to data analysis. Statistical analysis systems usually have simpler datasheets that are more focused to our needs.

Of all the customary features, we find that the following list includes those most useful in data analysis:

Editing. Quick editing of values in spreadsheets is a convenient capability. It is not unusual that real data have values that are obviously incorrect and that need to be corrected on the spot. Easy editing of data is a very convenient facility in these cases.

Searching. Searching for a value or label using the spreadsheet is important when the dataset is relatively large.

Permuting. Often the data are arranged in some relatively arbitrary fashion: They

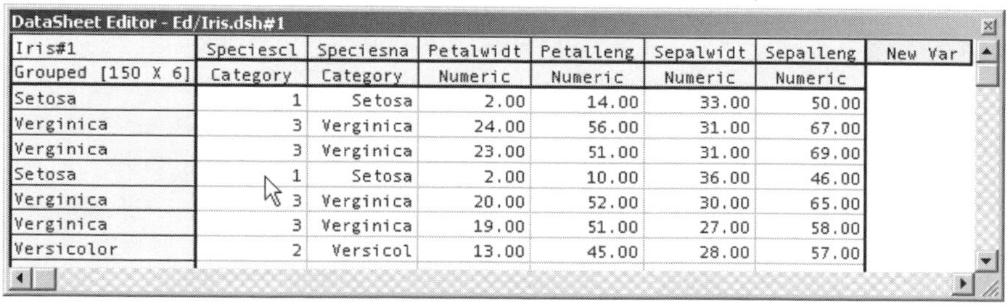

DataSheet Editor - Ed/Iris.dsh#1								
Iris#1	Speciescl	Speciesna	Petalwidt	Petalleng	Sepalwidt	Sepalleng	New Var	
Grouped [150 X 6]	Category	Category	Numeric	Numeric	Numeric	Numeric		
Setosa	1	Setosa	2.00	14.00	33.00	50.00		
Verginica	3	Verginica	24.00	56.00	31.00	67.00		
Verginica	3	Verginica	23.00	51.00	31.00	69.00		
Setosa	1	Setosa	2.00	10.00	36.00	46.00		
Verginica	3	Verginica	20.00	52.00	30.00	65.00		
Verginica	3	Verginica	19.00	51.00	27.00	58.00		
Versicolor	2	Versicol	13.00	45.00	28.00	57.00		

Figure 4.4 Datasheet.

128

may be arranged according to their observation order or according to their alphabetical position. It may be useful to rearrange the order in some other fashion to support, for example, comparisons between observations or variables. Many datasheets let you select a row or column and drag it to a new position.

Sorting. Datasheets also often allow you to sort the rows or columns into an order that corresponds to the values in one of the rows or columns. This can be quite powerful. For example, when the data for the Challenger spacecraft were reordered according to temperature rather than observation order, it was immediately obvious that temperature was the problem.

Subsetting. Also, many datasheets permit the selection of variables or observations so that the actions taken by the analyst use only the rows and columns of the datasheet that are selected. Since this can be done in an immediate feedback way, it increases the power of the overall statistical system.

Retyping. It has been proposed that good data analysis does not assume types of data, but instead, infers it from the analyses that have been done (Velleman and Wilkinson, 1993). Our preference is to determine datatype by changing variable types to see if there is any effect on the visualizations and analyses that have been performed. If not, the weaker measurement assumptions suffice.

Linking. The observations and variables can be linked with the various graphs that are produced, providing a way of labeling points and dimensions of plots. The techniques that can be used with graphs, such as selecting and setting colors or symbols, can be used here as well.

4.3   Plots

Before getting into how one interacts with the plots to activate their dynamic features, we present a way of conceptualizing the basic nature of the wide variety of plots that are available to the data analyst. This conceptual scheme organizes the plots according to the nature of the object used to represent the data, a characteristic that has some effect on the methods we can use to interact with the plot.

The various graphics that have been proposed all have a specific type of object that is used to visually represent the feature of the data that are being graphed. For example, the scatterplot uses point symbols to represent the data's observations.

We refer to the basic "plot thing" that is used by the plot to represent your data, the plot's glyph (a symbol that imparts information nonverbally), a word that has been used elsewhere with essentially the same meaning. We note that glyphs can be organized according to their dimensionality. Note that we are referring here to the dimensionality of the glyph, not the dimensionality of the plot. This leads us to the following way of thinking about plots:

Points. Many plots use points as the plot's glyph. This is a zero-dimensional glyph which, of course, cannot be seen. So we use tiny point-symbols as the glyph representing the point. An example of a plot of this type is the very familiar scatterplot, where each observation is represented conceptually by a point and is represented physically by a tiny point symbol.

Lines. Some plots use lines as the plot's glyph. Lines are one-dimensional objects that are drawn as such on the screen. Examples that we discuss in this book include the parallel-coordinates plot, the run-sequence plot, and the lag plot.

Areas. Still other plots use two-dimensional geometric figures such as rectangles or wedges as the plot's glyph, where the area represents the observed data— pie charts, bar graphs, and mosaic plots are familiar examples of such graphics.

Schematics. A final type of plot are those that use schematics to represent data. Well-known examples include boxplots and frequency polygons, to mention just two. (Although it may be stretching the definition, we could call these plots nondimensional, since the dimensionality of the schematic is irrelevant.)

The chapter is written as though all plots are point-based. This is, of course, not true. But since the chapter focuses on point-based plots, we should keep in mind the other families of plots as we proceed through the chapter, asking ourselves whether the particular way of interacting with a dynamic graphic depends on the glyph type of plot under consideration.

When we ask ourselves that question, it seems to us that for the most part, you can substitute *line* for *point* throughout the chapter; however, there are very few line-based plots, and many point-based plots. If dynamic interactive time-series graphics were developed further, line-based plots would be more common. The best example of a line-based plot is the *parallel-coordinates plot* (also known as a *profile plot*). In this plot the line is the fundamental glyph, and there is no consideration of points. On the other hand, the run-sequence plot and the lag plot use lines as the plot's glyph, but unlike parallel-coordinate plots, they are connected-point plots, where the glyph is a line that is point-based and therefore acts totally like a point-based plot.

It also seems to us that quite often you cannot substitute *area* for *point* in this chapter. The difficulty is that in an area-based plot, such as a mosaic plot, the area, which is the building block of the plot, represents several observations, whereas for a scatterplot, for example, the point, which is the plot's glyph, represents a single observation.

We discuss a wide variety of different ways of interacting with dynamic plots. These ways include

- **Activating plot objects:** selecting and brushing, the two major ways in which the window's objects can be activated
- **Manipulating plot objects:** labeling, linking, focusing, and morphing, the major actions that can be taken on the objects in essentially all plots
- **Manipulating plot dimensions:** permuting, changing scale, and changing aspect ratio, all of which have to do with the dimensions of the plot.

We take these up in turn in the coming sections.

4.3.1 Activating Plot Objects

The objects in the plot—usually the points—can be either activated or not activated. Activation is usually denoted by using a visual effect, such as changing color, brightness, or shape. The objects activated can then become the focus of whatever future actions the analyst decides to take.

Activation is often a prerequisite for taking other actions, but it can also be an operation of interest in its own right. The important aspect of activation is that the activated objects are made to stand out from the rest of the objects in some way. Then the activated objects can be used for identifying interesting features of the data.

For example, Figure 4.5 shows a group of activated points in a scatterplot. The visual effect of activation is obtained by changing the color or symbol of the activated points with the color of the points. Other techniques for emphasizing activated points are changing color, size, brightness, or reversing the color and white parts of symbols.

Mouse modes. Activation is usually done by using the mouse for either selecting or brushing, as described below. In ViSta the mouse has two default modes: selecting and brushing. When the mouse is placed in selecting mode, it will do the selecting method described below. When it is in brushing mode, it will do the "brushing" method described below. ViSta also allows the user to type commands that activate plot objects and supports programmable mouse modes, which effect the actions taken by the mouse. Thus, in ViSta's regression module, a mouse click can cause the system to limit the regression to those values of the X-axis variable that are larger, say, than the value of X at the point that was clicked, to see what the effect of limitation of

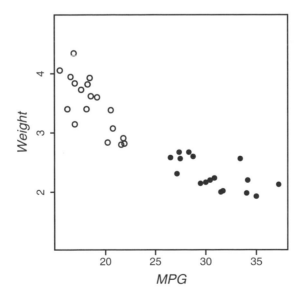

Figure 4.5 Selected points in a scatterplot.

131

range would have on the regression. Of course, there must be a way to de-select objects. Generally, activating nothing (i.e., clicking on an area not containing an object) produces this effect.

There are two ways of activating an object, by selecting it or by brushing it. We discuss these next.

Selecting . Selecting is a method of interacting with dynamic graphics which causes objects in the graphic to become active. Selection is usually accomplished using the mouse. The standard mouse selection techniques are:

- **Individual selection**: accomplished by single-clicking an object on the screen.
- **Multiple selection**: sually, holding down the meta key (shift or control) while objects are single-clicked.
- **Area election**: dragging the mouse to create an area of the screen for which contained objects are selected. The area can be rectangular or free-form (using a lasso-like tool).
- **Multiple area selection**: using the meta key and dragging the mouse to select objects contained within several areas.

Brushing. Brushing is a direct manipulation tool that is similar to selecting. In this technique, an active area with the shape of a rectangle is defined. Moving the mouse causes the active area to be moved, with objects contained within the active area being activated. Activation is denoted using the same type of visual effects used for selections (i.e., changing color, brightness, shape, etc.). Brushing is usually discussed in connection with linking and can be used, for example, to explore a bivariate plot conditioned by a third variable. Brushing is discussed in several sources (Becker and Cleveland, 1987; Becker et al., 1988; Cleveland, 1994b).

While selecting and brushing are quite similar generally in the effect they produce on the elements on the screen, brushing imposes a bigger workload on the computer. For example, in a scatterplot, in order to get a really smooth effect, it is necessary to refresh the plot each time that a new point enters or goes off the rectangle area. Selecting is a discrete process in which refreshing is necessary only after the user releases the button of the mouse. Consequently, slow computer environments will be able to implement selecting much more easily than brushing.

4.3.2 Manipulating Plot Objects

Labeling. Placing labels on a plot is a complicated problem for static plots. Numerous heuristic (Noma, 1987) and computational (Christensen et al., 1992, 1995) strategies have been used to produce labels that do not overlap. This is a critical issue for static and printed plots when they need to display all the labels (e.g., in maps).

This problem is solved in dynamic graphics by not showing the labels unless the user carries out actions to reveal them. If the user chooses only one label at a time, overlapping cannot happen. However, when the user wants to view simultaneously labels that correspond to objects that are close in a space, it is possible that some of them will be covered by others. Figure 4.6 shows a scatterplot where three points have been selected. The Buick Estate Wagon can be read even though it goes out of the frame of the plot. However, the other two are very difficult to read. In this case, the user can reposition the mouse very slightly to try to obtain a different selection, but this can be difficult or impossible in many situations.

Some of the dynamic solutions that have been used for this problem are:

- Small menus or pop-up windows that turn up for overlapping or close points.
- Sequential listing of labels (e.g., each click shows a new label).
- Linking to an external list of labels that shows the observations selected.

The problem remains that if the points are not exactly overlapping and if learning the exact values is of importance, it is still difficult to distinguish one from another. The techniques described in the section on focusing and excluding can be used to enlarge the area of interest and to explore it with less possibility of overlapping labels.

Linking. It is usually the case that a single display cannot show you everything that is interesting about a set of data. Even for the simplest data, we will often see different features of interest in different displays, if only because most graphics are designed to help us explore one (or at most a few) features. Thus, it is usually necessary to examine several different graphics in order to get the full picture of the problem at hand. (In Chapter 6 we give an extended example of this, looking into a single variable and finding interesting views in several different graphics.)

The problem is that very quickly there are too many plots to remember, let alone to understand. Linking is one of the important steps in solving this problem. When a plot is linked with other plots, a change in one plot can be made to cause changes in the

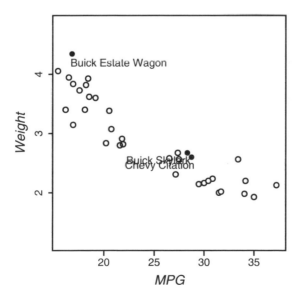

Figure 4.6 Displaying labels in a scatterplot.

other plots. The linkage is through the data shared by the plots, with the data's observations being the usual basis for the linkage.

Generally, the units of information that are linked are the observations in the dataset. When observations are linked, then actions carried out on the representation of a observation in one plot are propagated to the representation of the observation in whatever other views are currently displayed.

So, for example, changing the state of a point, its color, symbol, or other aspect in a plot is mirrored in other plots, data listings, or other views. Figure 4.7 shows two scatterplots of four variables for the same observations. When the plots are linked, then if the observation Buick Estate Wagon is selected in the left plot, the same observation is selected automatically in the right plot.

Note that the representations of the observations do not have to be the same in each of the plots. Note also that more than two plots can be linked. Thus, if we linked a histogram (which uses tiles to represent the observations) to the scatterplots in Figure 4.7, then when an observation is selected in any one of the three plots, a tile of the histogram and a point in each scatterplot is highlighted.

Note that linking is a very general concept. Not only can we have multiple plots linked simultaneously, and not only can we have different representations involved in the linkages, but we are also not restricted to a one-to-one relationship. Linkages can be a one-to-many relationship between many plots using many representations. Finally, the linked objects do not have to be observations. They may be variables, or matrices, or whatever is represented by the plot as an object with changeable features.

The main aspect of linked plots is that when a change is made to some aspect of the data as they are represented in one plot, the change can be portrayed in some fashion in additional plots. In a high-interaction, highly dynamic system the changes happen instantaneously in real time. In such a system, the combination of linking and brush-

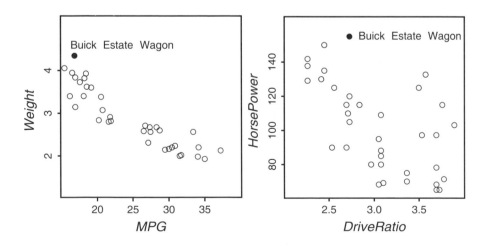

Figure 4.7 Two linked scatterplots.

ing is particularly powerful. With this combination, when one of a set of linked plots is brushed, the appropriate objects are highlighted in the other plots. Linking was first described by Stuetzle (1987) and is implemented in Lisp-Stat (Tierney, 1990), ViSta, DataDesk (Velleman, 1997), Manet (Unwin et al., 1996), and other programs.

Linking is an example of the general problem of coordinating multiple views in a computer system, as discussed by North and Shneiderman (1997), who present a taxonomy of all possible multiple-view coordination techniques. Focusing and excluding

Figure 4.8 shows an example of focusing on a part of a plot. The cloud of points in the scatterplot on the left looks like it could be split into two separate parts. To look into this, the points with lower *MPG* are selected and a regression line is calculated and shown. The line has a very different slope than the regression line for all of the points, which is also shown in the figure. Focusing on just the points selected produces the plot on the right, where we can study this part of the plot in more detail.

Focusing refers to the capability to look at just a subset of the data, focusing on a group of points or other elements of interest to get a better understanding of their structure. *Excluding* is just the opposite: removing a group of points or other elements which for some reason seem to distract from seeing the structure that may exist in the rest.

There are two versions of focusing. In once, the scale of the plot is not adjusted automatically after focusing. This lets the user easily compare the focus-on with the focus-off view, but the focus-on view may be very small or located at the edge of the plot. In the other, the scale of the plot changes when the focus is changed so that the plot is scaled and located to take up the entire viewing area. This makes focus-on/focus-off comparison difficult, but permits full inspection of the structure in either case.

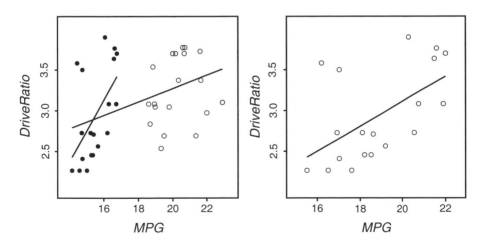

Figure 4.8 Focusing on a part of a plot.

Notice that focusing in statistical graphics does not work like zooming in other computer programs. Zooming enlarges the points in the selected area, whereas focusing leaves the points the same size but moves them farther apart.

Changing colors. Using the capability to change the color of points is a way to emphasize any structure that you think you may see in your data. It is also a way to search through the data to see if the structure is really there.

For most plots, when you first see the plot, all of the points are the same color. Then, while you are exploring your plot, if you see something interesting (a cluster of points or an outlier, for example), you can emphasize it by changing colors of relevant points. This gives you a picture of the data that makes the structure easier to see.

If the plot in which you changed colors is linked to other plots, the colors of the points in the linked plots will also change, allowing you to see if the structure is also revealed in the other plots. This is the way to gather more evidence about the structure that you think may be in the data.

Furthermore, when the points represent the observations in the data (as they almost always do), the color information can be saved as a new categorical variable, using color names as the categories of the variable. These variables can be used subsequently for other analysis (e.g. cross-tabulations, discriminant analysis) to help clarify their meaning.

Color palettes are a very effective way of using color to search for structure, since they make color easy to access and since they have immediate consequences. If a program does not have color palettes, then menus providing access to color should be used. Dialog boxes should be avoided for changing color because they usually obscure the points that are being colored.

Perhaps the main awkwardness with symbols and colors arises when a region of a space is too densely packed with points, making it difficult to identify individual points. Interactive capabilities can be very helpful in dealing with this problem, as shown in Figure 4.9. In this figure there are two plots. In the upper plot the points have been identified with six different symbols. However, only three groups can be clearly perceived. By focusing on this part of the plot we get the view shown in the lower part of Figure 4.9, where it is quite a bit easier to see the various symbols.

Changing point symbols. Note that you can also change the symbols used to represent points, and that everything that was stated above about point color also applies to point-symbols. However, whereas point color is very effective at communicating structure, point symbols are not, because the symbols may superpose and form uninterpretable blobs, making them not very effective at communicating information to the viewer. A possible solution is to use what Cleveland calls *texture symbols*, symbols specially designed to do well in case of overlapping (Cleveland, 1994a).

Changing point labels. Finally, point labels can be a very effective way to identify structure in your data, but in some programs (including ViSta) it is so awkward and clumsy to change the labels that what would otherwise be an effective way of communicating structure becomes ineffective.

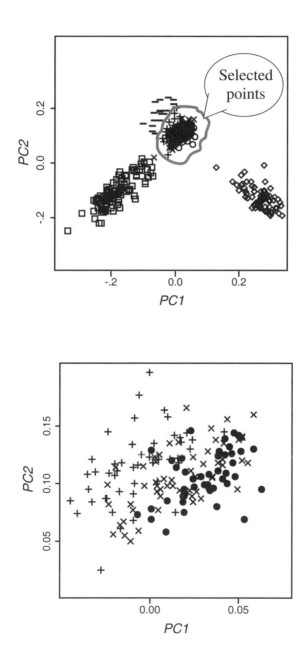

Figure 4.9 Focusing on part of a plot to see overlapping points.

4.3.3 Manipulating Plot Dimensions

In this section we discuss techniques for manipulating the dimensions of the plot, such as re-ordering the variables or dimensions or some other feature; changing the absolute or relative size of the axes, etc.

Reordering. There are several examples in the data visualization literature which point out that *reordering* the elements in a plot can determine what we see in a plot. For example, Tufte (1997) explains that the ordering of the information in the plots that were used to decide whether to launch the space shuttle *Challenger* was the cause of not seeing that problems lay ahead: The tables and charts concerning the possibility of failure in cold weather that were prepared by NASA's engineers were sorted by time instead of temperature. When they were resorted by temperature, it was clear that temperature was of great importance.

There are guidelines which suggest how to order information in several situations. For example, it is recommended that bar charts show the bars ordered by the frequency of the categories (Cleveland, 1994a).

Recently, Friendly and Kwan (2003) introduced a principle that is intended to rationally order the presentation of information in complex and high-dimensional plots. They suggest that visual communication can be facilitated by ordering according to *principal effects*, as they call it, with the definition of principal effects depending on the type of data. In essence, the principal effect is, mathematically, the most important information contained in the data. More precisely, they propose defining principal effects as being the appropriate one of the following four definitions: It is the *main-effects* of *n*-way quantitative data, the *associations among factors* for *n*-way frequency data, the *correlations* for multivariate data, or the *group mean differences* for MANOVA data. In all of these cases, the order of presentation can be optimized by using the values of the singular values to define the order.

Having a well-defined mathematical basis for determining how to order the presentation of information in displays of data is clearly an important first step toward produce compelling visualizations. We can look on such a definition as providing us a rational and (at least mathematically) optimal ordering to start with. The displays, of course, should enable and encourage the analyst to seek new orderings that can be more revealing than those produced by default.

Examples of dynamic reordering include the implementation of mosaic plots in Manet (Hofman, 2003). In these plots, categories of a variable can be ordered by dragging their representations in the plot. This technique would be useful to accomplish the goal mentioned in the preceding paragraph: helping the analysis improve the order shown by default.

Changing scale. The scale of a plot determines the size of the plot. Scatterplots usually choose their initial scale so that all its points will fit in the space available. This is an appropriate default, but it will be inadequate on some occasions, such as, for example, when there is one outlier observation that is very far from the rest of the points, leaving little space for the other points. In this case, focusing can be used to remove the isolated point from the view by modifying the scale of the plot.

Control of the scale is necessary to allow for comparisons between plots. As an example, Figure 4.10 shows four histograms. The histograms in the top row show the information about a variable (miles per gallon) for two groups of cars (U.S. and non-U.S. cars). If we wish to compare these two histograms, we need to note that the scales on the two plots are not the same, a fact that makes the comparison difficult. Notice, for example, that the scale of the frequencies for U.S. cars is higher than for non-US cars, so the height of the bars cannot be compared directly. On the other hand, the lower row of histograms in Figure 4.10 has the same scale on both axes, making comparison easier.

The trellis display (Becker et al., 1996) puts several plots side by side and sets a common scale for them, thereby making these comparisons easier to carry out. However, while the trellis display requires contrasts that are specified in advance, it is often the case that comparisons occur to the data analyst while exploring the data. Thus, we need to be able to set common scales in a simple way. Dialog boxes are used in many programs, but they can be very inconvenient, as they require the user to do

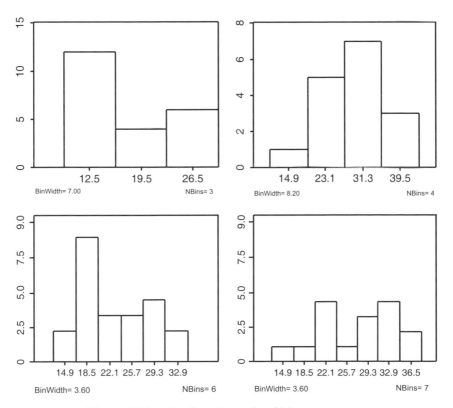

Figure 4.10 Adjusting the scale of histograms.

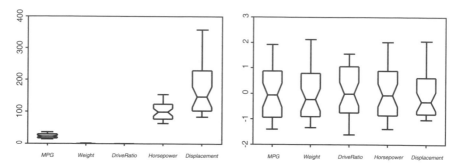

Figure 4.11 Multiple parallel boxplots: Unstandardized and standardized.

too many things (looking up the values of the scales, opening the dialog boxes, setting the values, etc.). DataDesk has a very elegant way of setting common scales. Dragging a window and dropping it on another puts the same scale in both plots, but discovering this feature may be unlikely.

Another problematic situation related to scales is when a plot displays several variables simultaneously. The multiple parallel boxplot, discussed later in the book, is an example. ViSta includes a button in these plots that standardizes the variables in order to facilitate comparisons among them. Using this button, the user can switch between the unstandardized version and the standardized version of the boxplot, as shown in Figure 4.11. If two plots are scaled in the same way but one of them changes, the correspondence may be lost. Graphical linking as defined by Wilhelm (1999) refers to the capability of linking graphical features of plots, including scales, to avoid losing this correspondence.

Changing the aspect ratio. The aspect ratio $a = h/w$ refers to the height-to-width ratio of a plot. Plots with an aspect ratio of 1 have the same width and height. Values higher than 1 are plots that are taller than wider and values lower than 1 are plots that are wider than taller. Cook and Weisberg (1994) point out that many statistical programs always use an aspect ratio of 1, which is not always the best. They show an example very similar to the one in Figure 4.12 and Figure 4.13. The two fig-

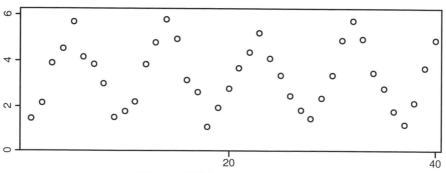

Figure 4.12 Aspect ratio of 1/4.

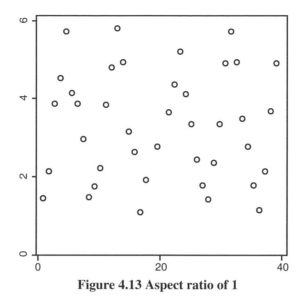

Figure 4.13 Aspect ratio of 1

ures present the same data but use different aspect ratios. While Figure 4.13 reveals nothing, the shape of the relation is quite evident in Figure 4.12.

Even though some rules have been proposed for determining proper aspect ratios (Cleveland, 1994b), interactive modification of them is an important feature of dynamic plots. When the plots are part of stand-alone windows, this property is automatically a result of the properties of the windows—you stretch or shrink the window and the aspect ratio changes. However, these capabilities do not result in sufficiently accurate control of the aspect ratio, and a way of maintaining an exact ratio of 1 is required. Lisp-Stat has commands that can fix the aspect ratio to 1, so that changing the window size only changes the size of the plot, not its aspect ratio.

4.3.4 Adding Graphical Elements

Another way of adding information to a plot is by using graphical elements such as lines or circles, much as in drawing programs. After all, statistical graphics are a type of graphics, and consequently, it seems natural to draw on them for incorporating remarks, pinpointing elements, and providing explanations. These nonstatistical annotations can be very useful, but we do not describe them here because they have been widely described. However, there are similar features that are specifically statistical and are described later in the book, which we preview here.

One statistical use of adding lines is simply to connect the points of a point plot in a specific order, such as in the order they appear in the dataset. This turns a scatterplot into a time-series plot, for example. Many statistical software programs provide such a capability.

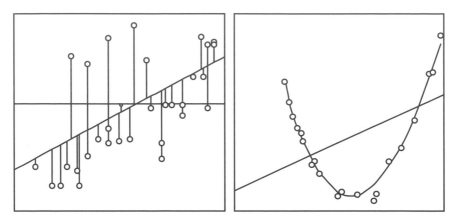

Figure 4.14 Statistical uses for adding drawings.

Two somewhat more advanced examples are shown in Figure 4.14. On the left we have fit a line to points in a point cloud. We are seeing not only the best-fitting line but the residual lines as well. In the second example (on the right) we have two variables that are very strongly, although not perfectly related by an apparently polynomial function. We have fit a straight line to them, and we have also fit a lowess function to them. Note that the lowess function is not a connect-the-dots function, although it looks like that in much of the plot. But be sure you notice the bottom of the function where there is some scatter around the curve.

Other type of drawing figures can be added. Circles, for example, can be drawn proportionally to a third variable and added to an scatterplot, producing a *bubble* plot. Bubble plots show the relative magnitude of one variable in relation to two other variables.

Some programs offer drawing tools to add ovals, lines, polygons, and so on. These tools are useful for customizing displays to emphasize features found while exploring the data. For example, ovals can be used to surround a cluster of points, text can be added to provide an explanation for an outlier, or arrows can be used to illustrate the relation of one part of the part with another. The goal of this practice can be to make presentations to an audience but it can also be a way of recording information about the analysis for the analyst to use later. JMP is the program that is best in these capabilities.

PART III

SEEING DATA—OBJECTS

5

Seeing Frequency Data

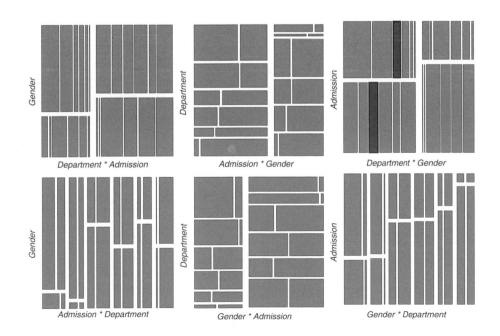

5

Seeing Frequency Data

Data are a set of facts suited to the rules of arithmetic and organized for processing and analysis. Although the facts must be suited to the rules of arithmetic, they do not necessarily have to be numerical. The apparent contradiction is resolved when you realize that even nonnumerical facts can be categorized and counted. For example, if we have a group of people, we can count how many are males and how many are females, how many are Christians or Jews or Moslems, or how many are employed or unemployed, and so on. As counting involves the rules of arithmetic, these facts are suited to the rules of arithmetic.

However, facts such as a person's gender, religion, or employment status are not data. They are just facts. To be data they must be organized. We emphasize this because there are two ways in which facts like these can be organized: as categorical variables or as a frequency table. When organized as variables, the fundamental unit of observation is a category, whereas when organized as a table, the basic observation is a count. We use the term *frequency data* for tabulations of counts, reserving the name *categorical data* for datasets made up exclusively of categorical variables. Frequency data have been tabulated, existing in an aggregated form. Categorical data, on the other hand, have not been aggregated or tabulated, existing as unaggregated raw data.

Statistical methods for numerical data (correlation, regression, tests for mean differences, etc.) were developed early in the twentieth century. Advances in techniques for analysis of categorical variables, more common in the social and biomedical sciences, had to wait several decades before being developed to the same degree of generality. If techniques for analysis of categorical data were slow to develop, the invention of methods for visualization of categorical data had to wait even longer. Thus, while frequency data analysis techniques came to a mature state only recently, the methods for its visualization are even younger, as it was not uncommon that books on frequency data analysis (except for correspondence analysis and related techniques) did not portray graphical representations. Instead, activities such as exploring frequency data were carried out simply by examining the data in contingency tables. Graphical displays for categorical data were rarely utilized.

Yet, progress in visualization of categorical data continued, and the number of visualization techniques applicable to categorical data eventually increased sufficiently to justify a book. This book, by Friendly (2000), described a wide range of graphical methods for fitting and graphing discrete distributions, two-way and *n*-way contingency tables, correspondence analysis data, logistic regression, and log-linear and logit models; it also provided computational procedures and macros to perform these analysis. As a result, the book made a large variety of methods for visually exploring categorical data both available and accessible.

However, Friendly's book focused exclusively on static displays; no mention, apart from a note in the introduction, was made of dynamic displays. This absence has an easy explanation: If visualization of categorical data was a relatively new subject, applications of dynamic concepts to this field have been even scarcer. This chapter, intended to complement Friendly's book, is limited to dynamic extensions of these techniques.

As well, for reasons of space, we restrict our attention to dynamic techniques applicable to frequency data. We also exclude methods that can be seen as simple extensions of those discussed in other parts of this book, such as correspondence analysis (which can be displayed with linked scatterplots or biplots) and logistic regression (which can use dynamic plots similar to those discussed for regression analysis).

The plan of this chapter is the following. First, we introduce the examples used to illustrate the techniques described in this chapter. Second, we explain techniques for interactive manipulation of frequency data arranged in tables. Third, we introduce interactive capabilities for multivariate frequency data graphics. Finally, we deal with log-linear models and how to specify them using an interactive graphical interface.

5.1 Data

In this chapter we will two datasets to exemplify the concepts we introduce. These datasets are introduced in the next two subsections.

5.1.1 Automobile Efficiency:

Table presents data about the country of origin (six categories) and the number of cylinders (four categories) of 38 automobiles sold in the United States in 1978–1979. The data are organized as variables. Notice that the metric of the variable *Cylinders* can be considered to be either numerical, ordinal, or categorical. We will take advantage of its special status in our analysis. The country of origin variable is clearly categorical. Table 5.2 presents the same data, but this time organized as a cross-tabulation of the *Country* and *Cylinders* variables.

5.1.2 Berkeley Admissions Data

Table 5.3 shows data on applicants to graduate school at the University of California at Berkeley for the six largest departments in 1973 classified by *Admission* and *Gen-*

Table 5.1 Automobile Data Organized as
a Categorical Dataset

Cars	Cylinders	Country
Buick Estate Wagon	8	U.S.
Ford Country Squire Wagon	8	U.S.
Chevy Malibu Wagon	8	U.S.
Chrysler LeBaron Wagon	8	U.S.
Chevette	4	U.S.
Toyota Corona	4	Japan
Datsun 510	4	Japan
Dodge Omni	4	U.S.
Audi 5000	5	Germany
Volvo 240 GL	6	Sweden
Saab 99 GLE	4	Sweden
Peugeot 694 SL	6	France
Buick Century Special	6	U.S.
Mercury Zephyr	6	U.S.
Dodge Aspen	6	U.S.
AMC Concord D/L	6	U.S.
Chevy Caprice Classic	8	U.S.
Ford LTD	8	U.S.
Mercury Grand Marquis	8	U.S.
Dodge St Regis	8	U.S.
Ford Mustang 4	4	U.S.
Ford Mustang Ghia	6	U.S.
Mazda GLC	4	Japan
Dodge Colt	4	Japan
AMC Spirit	4	U.S.
VW Scirocco	4	Germany
Honda Accord LX	4	Japan
Buick Skylark	4	U.S.
Chevy Citation	6	U.S.
Olds Omega	6	U.S.
Pontiac Phoenix	4	U.S.
Plymouth Horizon	4	U.S.
Datsun 210	4	Japan
Fiat Strada	4	Italy
VW Dasher	4	Germany
Datsun 810	6	Japan

Table 5.2 Automobile Data Organized as a
Frequency Table

Country	Number of Cylinders			
	5	8	6	4
Sweden	0	0	1	1
France	0	0	1	0
U.S.	0	8	7	7
Italy	0	0	0	1
Japan	0	0	1	6
Germany	1	0	0	4

der (Bickel et al., 1975). This dataset has been used as an example of the application of log-linear models in several places (Agresti, 1990; Friendly, 2000; Rindskopf, 1990; Valero-Mora et al., 2004). This dataset has three variables: *Gender* of applicants (male/female), *Department* (A to F), and *Admission* (yes/no). For such data we might wish to study whether there is an association between *Admission* and *Gender*. Are male (or female) applicants more likely to be admitted? The presence of an association might be considered as evidence of gender bias in admission practices.

Table 5.3   Berkeley Admissions Dataset

		Gender			
		Male		Female	
		Admission			
		Yes	No	Yes	No
	A	512	313	89	19
	B	353	207	17	8
Department	C	120	205	202	391
	D	138	279	131	244
	E	53	138	94	299
	F	22	351	24	317

5.1.3 Tables of Frequency data

An important advantage of frequency data over numerical data is that the former can often be completely displayed in a table so that readers can reach their own conclusions or perform their own analysis. The compactness of frequency data also allows us to explore the data in a way that is very difficult to carry out with numerical data: by looking at the data directly, without resorting to computing any type of statistical, graphical, or numerica, summary. This is possible because tables of frequency data for two or three variables often contain only a handful of figures, which can be examined directly without much effort.

Unfortunately, as explained in many introductory statistics books, absolute values of frequency data can be deceiving without consideration of the marginals of the variables (sum of rows or columns in cross-tabulations of two variables). Not taking this element into account would lead naively to considering a value to be too high or too low in a contingency table, when in fact it is the contrary. Absolute frequencies have to be adjusted in many cases to be interpreted correctly.

A simple solution, which is often also discussed in introductory statistics books, is transforming the absolute values to percentages or proportions. These relative quantities are more amenable to correct interpretations, and in fact, many nonsophisticated users are convinced that tables of percentages and counts are the only frequency data analysis techniques they need.

As a result, large statistical packages have traditionally attempted to provide modules specialized on computing tables of frequency data. For example, SPSS and SAS both have components focused on displaying counts and percentages of categorical variables that can lay out the data in several ways (nesting, stacking, multiple responses) and formats (setting borders, titles, alignments, etc.). Indeed, many users of these packages feel that the capabilities featured by these modules fulfill their needs completely with respect to analysis of categorical data.

Interactive manipulation of tables has emerged recently as a feasible approach to the exploration of frequency data for those users only interested in results such as proportions or percentages. The classic statistical packages worked using commands, but recent updates have promoted interactive versions that can be manipulated using direct manipulation techniques. Also, some spreadsheets incorporate characteristics that are very similar to those portrayed by statistical software and that can be utilized for obtaining similar results. In this section we review methods and techniques that allow ua to carry out simple manipulation of frequency data interactively.

The section is organized in two subsections: techniques for working with categories, and techniques for working with variables. in the former we examine reordering, joining, and excluding the categories, and in the latter we discuss including/excluding, relabeling, computing percentages, setting the positions, and partitioning the variables.

When regarded as appropriate, as in other parts of this book, we mention the statistical systems/packages that implement the features discussed here, but we also discuss features not yet implemented in any system.

5.1.4 Working at the Categories Level

We use the cross-tabulation of the number of *Cylinders* and *Countries* in Table 5.2 as the example for this subsection. A cursory examination reveals several problems with this table. First, the categories in the variable *Cylinders* are not sorted. Second, this table presents many "holes" (values equal to zero) that are caused, on the one hand, by countries that manufacture few models of cars (such as France and Italy), and on the other hand, by countries that have a bias regarding the number of cylinders they use (Germany and Japan focus almost exclusively on four-cylinder cars and eight-cylinder cars are built only in United States). Third, the category of five-cylinders cars has only one representative (from Germany), creating a column almost completely empty.

In this section we show interactive actions useful for manipulating the categories of variables to solve problems such as those displayed in Table 5.2. The actions are illustrated in Figure 5.1, which portrays a number of steps culminating in a table more apt for analysis and interpretation than the Table 5.2. The steps are explained one by one in the following paragraphs.

Reordering. This operation can simplify and enhance, sometimes dramatically, the perception of associations in tables of frequency data. For example, it is very apparent that categories of the variable *Cylinders* in Table 5.2 might be sorted accord-

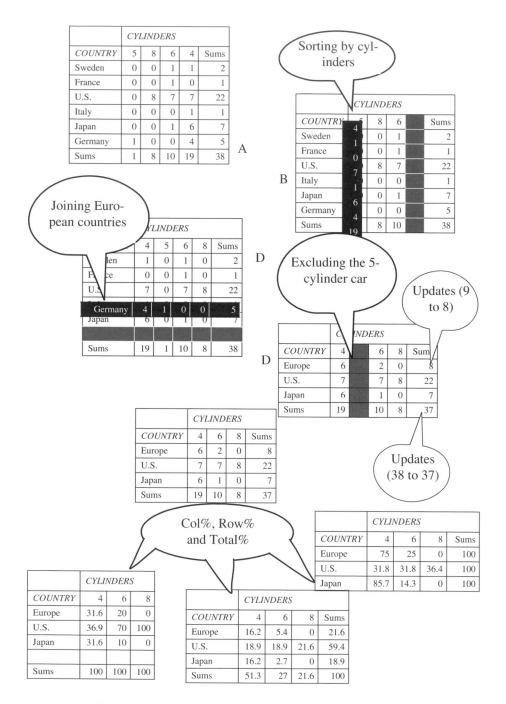

Figure 5.1 Manipulating interactively the data in Table 5.2.

ing to their value. This action is performed interactively in panel B of Figure 5.1. Another reason for reordering the categories is to put together rows or columns that need to be compared (say, cars from United States with cars from Germany). Of course, tables could also be sorted according to numerical criteria such as the value of the frequencies or the principal effects.

Joining categories. When the table has cells with values equal to zero, a possible solution is to combine some of the categories. These combinations should, however, take into account substantive reasons that justify the new categories. Let us see how we could do this in our case. As can be observed in Table 5.2, there are countries that manufacture only one or two vehicles. Since these cars are all european, we could create a new category combining all of them together (panel C of Figure 5.1).

Excluding categories. If combining a category with others is not feasible, we may prefer to exclude the category from the analysis. The only car with five cylinders in Table 5.2 is an example of this situation. Since putting this car's category in another category does not seem appropriate, we choose to exclude it from the table. After doing this, the marginal sums and totals must be updated (panel D).

Interactively *reordering* categories is an operation usually available in spreadsheets but not in statistical packages. *Joining categories*, in turn, can not be performed as represented in Figure 5.1 by any program that we know, but the operations of selection and replacing in DataDesk can be a decent alternative for such tasks. Manet and Mondrian can perform this operation directly on the plots (see Section 5.2.2). Finally, *excluding categories* can be also carried out in DataDesk by using the selection operation.

5.1.5 Working at the Variables Level

Manipulations of tables at the variables level involves including or excluding variables in the table, combining variables (also called relabeling), computation of relative quantities such as percentages or proportions, and partitioning the table.

Including/excluding variables. The Berkeley data provide an example where considering the cross-tabulation of only two variables, *Gender* and *Admission,* demonstrates the difficulties that are associated with getting meaningful answers to the question of whether there is gender bias in the admission process. As shown in Table 5.4, looking at this cross-tabulation suggests that Females are rejected more often than Males. However, this table ignores the information we have concerning the various departments. As is shown in the last panel of the Figure 5.2, when the variable *Department* is included in the table we can see that females have higher admission probabilities in several departments. Indeed, had we ignored the division by *Department* and focused on only two variables, we would have been grossly misled by Table 5.4.

On the other hand, sometimes we may wish to exclude a variable from tables if we find that it is independent of the other variable(s) being considered,. In this case,

Table 5.4 Percentage of Admissions by Gender in the Berkeley Dataset

Gender	*Admission*	
	Yes	No
Male	69.65	30.35
Female	55.48	44.52

excluding a variable simplifies the presentation without biasing it, since the variable excluded is independent of the others.

Including/excluding variables interactively in tables is available in some spreadsheets. Dragging the variable in or out of the table will recalculate and show a new table. ViSta can add new variables to a table, but it cannot exclude them, so, to create a subtable, it is necessary to re-create the entire table. DataDesk does not allow including or excluding variables, but you can replace one for another.

Relabeling. This technique consists of combining two, or possibly more, variables in a new one (Bishop et al., 1988). The new variable has a number of categories equal to the product of the number of categories in the combined variables. Bishop et al. (1988) present as an example a study of survival in breast cancer patients. This example had five variables: *Degree of chronic inflammatory reaction* (minimal or moderate-severe), *Nuclear grade* (malignant appearance, benign appearance), *Survival* (yes, no), *Age* (Under 50, 50–69, 70 or older), and *Centre of diagnosis* (Tokyo, Boston, Glamorgan). The goal of the study was to evaluate the effect of *Centre of Diagnosis* and *Age* on *Survival*. As the other two variables, *Nuclear grade* and *Degree of chronic inflammatory reaction*, were regarded as a description of the disease state and were interrelated, they were always to be present together in any model or analysis. Hence, the authors relabeled these both to a new one called *Histology,* with the four categories: malignant appearance–minimal inflammation, malignant appearance–greater inflammation, benign appearance–minimal inflammation, and benign appearance–greater inflammation.

This manipulation is not implemented directly in any software package, but DataDesk and ViSta implement a transformation that combines two categorical variables straightforwardly. The variables created in this manner can be used for making tables with the variables relabeled.

Computing percentages. Comparison of rows and/or columns of tables are facilitated if they show relative quantities such as percentages or proportions instead of absolute values. Row percentages, for example, allow us to compare the profiles of the rows of the tables and see whether they are similar across columns. Absolute values would make these comparisons much more difficult because of the differences in magnitude of the marginals of the tables.

However, using percentages does not remove all the possible questions. First, we have the choice of computing different—row, column, or total—percentages for each cell of the table, and all of them give a valid but somewhat different picture of the data (see the last part of Figure 5.1 for possible percentages for the car data). Hence, exploring the data becomes more effortful than when we consider only one figure for each cell. Second, ignoring absolute frequencies when interpreting the percentages is, as a rule, meaningless. For example, if the absolute frequencies are small for some rows or columns, high variations in percentages matter less than when the frequencies are large. In summary, percentages can be misleading if viewed out of context.

The solution to the problem above employed by some statistical packages is to display several values in each cell of the table. Thus, each cell might include the absolute frequency and one or several percentages. This solution, however, overloads the tables and makes them difficult to read, so interactive programs let you hide or exhibit the figures using a menu or a control button (DataDesk or JMP, for example, use this approach). In this way, questions arising at any moment can be investigated by displaying figures in the cells that had been kept hidden until then. Once they have been evaluated by the user, these figures can be hidden. Hence, interactive control over what is or is not shown in a table helps users evaluate tables of frequencies more easily.

Arranging the variables in the table. Figure 5.2 gives an example of changing the layout for the data about admissions in Berkeley. Table 5.3 has been repeated in the upper part of Figure 5.2 to help the explanation. The main question to answer with these data is if there were different rates of admission for males and females in six departments. *Admission* is, therefore, the dependent variable, and *Gender* and *Department* are the independent variables. Using the rule of putting the dependent variables in the columns and the independents in the rows, Table 5.3 should be modified so that the variable *Gender* is in the rows. Percentages of admission for males and females in each department could be used to explore such question. There are two possible versions of the table. Putting *Gender* first and *Department* second is not satisfactory because it makes difficult to carry out the comparisons of interest. The other layout, with *Department* first and *Gender* second, puts together the values for males and females in the departments, so it is easy to see that the differences in percentage of admission by department are quite similar except in department A, where females have an advantage over males of about 20%.

Figure 5.2 suggests a way to change the position of the variables in the table interactively. The method uses *drag and drop* of the names of the variables in the table into their new positions. This method of manipulating tables of frequencies is available, for example, in a very famous commercial spreadsheet. By using this method, the user can easily explore different configurations of the tables, facilitating the process of searching a layout that responds better to the problem at hand.

Partitioning the table. Complex tables can be simplified by choosing a category of a variable and showing the frequencies of the rest of variables conditioned to this category. In the context of maximum likelihood estimation, Bishop et al. (1988) suggest that this strategy is particularly useful when the variable used for partitioning is a dichotomy variable. SPSS calls this feature *layers*, and it provides the capability of

	Gender			
	Male		Female	
	Admission			
	Yes	No	Yes	No
A	512	313	89	19
B	353	207	17	8
C	120	205	202	391
D	138	279	131	244
E	53	138	94	299
F	22	351	24	317

Department / Gender

Drag and Drop *Gender*

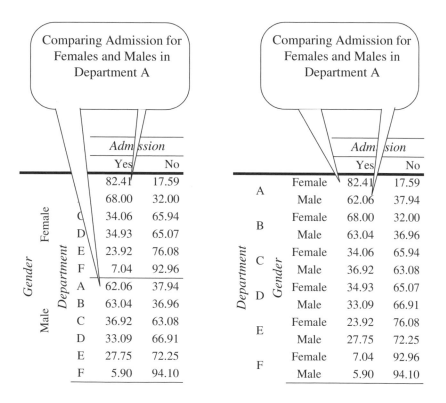

Comparing Admission for Females and Males in Department A

	Admission	
	Yes	No
	82.41	17.59
	68.00	32.00
C	34.06	65.94
D	34.93	65.07
E	23.92	76.08
F	7.04	92.96
A	62.06	37.94
B	63.04	36.96
C	36.92	63.08
D	33.09	66.91
E	27.75	72.25
F	5.90	94.10

Comparing Admission for Females and Males in Department A

		Admission	
		Yes	No
A	Female	82.41	17.59
	Male	62.06	37.94
B	Female	68.00	32.00
	Male	63.04	36.96
C	Female	34.06	65.94
	Male	36.92	63.08
D	Female	34.93	65.07
	Male	33.09	66.91
E	Female	23.92	76.08
	Male	27.75	72.25
F	Female	7.04	92.96
	Male	5.90	94.10

Figure 5.2 Arranging the variables in the table.

combining several variables in each layer. This strategy allows us to reduce the complexity of large tables by focusing sequentially on one category at a time.

5.2   Frequency Plots

We mentioned earlier that until recently the graphics for categorical data were not as well developed as their counterparts for numerical data. Also, the coverage of statistical software of the graphical methods for categorical data were not as systematic as for quantitative data. Finally, even in the case that methods for visualizing categorical data were available and had been implemented in statistical packages, it still happened that users of statistics lack the habit of using them on an everyday basis as often as they might.

However, recent advances in graphical methods for categorical data have paved the way for changing the situation above, as new visualization displays have been developed with the aim of being comparable in scope to those available for quantitative data (Friendly, 2000b). Of these, mosaic displays (Friendly, 1994, 1995; Hartigan and Kleiner, 1981, 1984) have probably created more interest, as they possess a property not easily found in other displays: they extend naturally to multidimensional tables. This contrasts with previous methods, which were limited to one- or two-way views of categorical data.

An example of the power of mosaic displays was given by Hartigan and Kleiner (1984), who presented a mosaic display of a $3 \times 6 \times 7 \times 12$ table (1512 cells!). Indeed, this display required some study to be understood, but much less than would be necessary to see any pattern directly in the table of frequency data.

In this section we focus exclusively on dynamic interactive versions of mosaic displays. Readers interested in a complete account of static displays for categorical data can refer to (Friendly, 2000b) for a discussion of several techniques specific to different situations. We do not examine other graphics for frequency data, though, as the main focus in this book is on the dynamic interactive extensions to the graphics, not the displays themselves. Hence, we restrict ourselves to a display of more interest with the hope that lessons taught here can be applied to other situations if needed.

This section is divided in two parts. First, we describe the basic features of mosaic displays, to indicate what they show and how to read them. Then we describe the particular interactive features that are useful for dynamic graphics. We use as an example the Berkeley data, introduced in Section 5.1.2.

5.2.1   Mosaic Displays

Mosaic displays are graphical methods for visualizing n-way contingency tables and for visualizing models of associations among its variables (Friendly, 1999). The frequencies in a contingency table are portrayed as a collection of reticular tiles whose areas are proportional to the cell frequencies. Additionally, the areas of the rectangles can be shaded or colored to portray quantities of interest, such as residuals from a log-linear model (discussed in Section 5.3).

A mosaic plot can be understood easily as an application of conditional probabilities. For a two-way table, with cell frequencies n_{ij} and cell probabilities $p_{ij} = n_{ij}/n_{++}$, a unit square is first divided into rectangles whose width is proportional to the marginal frequencies n_{i+}, and hence to the marginal probabilities $p_{ij} = n_{i+}/n_{++}$. Each such rectangle is then subdivided horizontally in proportion to the conditional probabilities of the second variable given the first, $p_{j|i} = n_{ij}/n_{i+}$. Hence, the area of each tile is proportional to the cell frequency and probability,

$$p_{ij} = p_i \times p_{j|i} = \frac{n_{i+}}{n_{++}} \times \frac{n_{ij}}{n_{i+}}$$

The steps above are exemplified for the three variables of Berkeley data in Figure 5.3. There are three mosaic plots in this figure. The first step splits the entire rectangle into two areas proportional to the categories of *Gender*, so we can see that there are more males than females in the data. The second step divides the previous tiles (male/female) according to the number of applicants in each department. These new tiles would align vertically in the two columns if the proportion of males and females was the same in all departments. However, the plot reveals that there are more male than female applicants in departments A and B, while departments C to F have relatively fewer male than female applicants. Finally, the third mosaic plot displays *Admission* given the other two variables, *Gender* and *Department*. There is much information in this last display. As an example, the two tiles for the males in department A (marked with a thicker border) show that about two-thirds of males were admitted and one-third rejected at this particular department. This contrasts with the results for females at this department (upper-right corner), which have larger proportions of admission (80%).

Spacing of the tiles of the mosaic provide an important aid for interpretation when there are more than two variables in an axis. This can be observed in Figure 5.3, where the separation between the categories of *Gender* is larger than for the categories of *Admission*, making it easier to see the building blocks of the mosaic display.

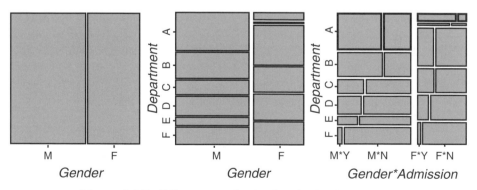

Figure 5.3 Building a mosaic plot for the Berkeley data.

A guide for interpreting mosaic displays is how the tiles align at the various splits. Nonaligned tiles mean that there is interaction between the variables, and aligned tiles, that the variables are independent. For example, we know that there is an interaction between *Gender* and *Department* in the second display in Figure 5.3 because the horizontal lines do not align. The lack of alignment of vertical lines in the third display of Figure 5.3 reveals the interaction between *Admission* and *Department* given the categories of *Gender*. Finally, we can compare the profile of these vertical lines for both genders to see if there is interaction between *Gender* and *Admission*. The lines have similar profiles across *Gender* except for department A, suggesting that there is no interaction for the rest of the departments. In other words, fitting a three-way interaction model to the data of all the departments apart from department A would be unnecessary in this case.

Static implementations of mosaic displays are available in some commercial programs (JMP, SAS, SPlus). As well, now there is noncommercial software that surpasses in many respects the commercial software. For example, MOSAICS (Friendly, 1992), which is available as a web page on the Internet, computes mosaic displays, log-linear models, and has interactive capabilities such as querying the mosaic bars to see the frequency or the residual values of the cells. The free R implementation of the S language provides the vcd package containing mosaic plots. Also, two programs developed at the University of Augsburgh, Manet (Hofman, 2003; Unwin et al., 1996) and Mondrian (Theus, 2003), are available as free downloads. Finally, the program ViSta (Valero-Mora et al., 2003, 2004) has interactive mosaic plots and a module for log-linear analysis.

Notably, the free software features a large number of dynamic interactive characteristics. These characteristics are reviewed in the next subsection.

5.2.2 Dynamic Mosaic Displays

In the following part of this section we discuss dynamic features that are, or could be, implemented in software for mosaic displays. Notice that none of the programs mentioned in Section 5.2.1 has all the characteristics discussed below, and also, that the specific details of how a user interacts with the program in each case may vary importantly. Hence, this section should not be considered as a review of existing software but as a summary of the features that all the programs offer at the moment.

Some of the dynamic manipulations of mosaic displays discussed here are very similar to those that can be performed on tables of frequency data (Section 5.1.3). Therefore, in some cases, we use the same headings that we used in that section.

Including/excluding variables. Exploration of data often starts with a number of variables to be displayed that can be reduced or increased depending on the questions asked. This action could be carried out by dragging the variables in and out of the display's window, using the same method already described for frequency tables in Section 5.1.5.

ViSta is the only program that provides the choice of drawing the mosaic displays interactively, by adding to the display sequentially variables selected from a list. However, the operation of excluding variables interactively is not supported.

Arranging the variables in the display. As shown in Figure 5.4, mosaic displays have a different look depending on the position of the variables. So, for the variables in the Berkeley data, we can draw six different displays. As different positions emphasize different aspects of the data, it seems important to provide ways of changing them.

Manet and Mondrian both have a vertical list of the variables that can be altered by dragging up or down one variable at a time. The list so modified indicates a new order of the variables in the display that replaces the one used as default by the software. As an application of this technique, let's focus on the mosaic displays [2,1] and [1,2] in Figure 5.4. Both displays have *Admission* as the last variable, but one can be obtained from the other simply by exchanging the other two variables. Although both displays allow us to learn the lesson of bias in department A, we find [1,2] easier to interpret, but this can be a subjective opinion.

Reading the labels and the values. The mosaic display is very successful inconveying global impressions of the data, but sometimes we wish to see the exact labels and values. However, as displays with too many details may look ugly, interactive versions of the mosaic can hide them until the user interrogates the plot.

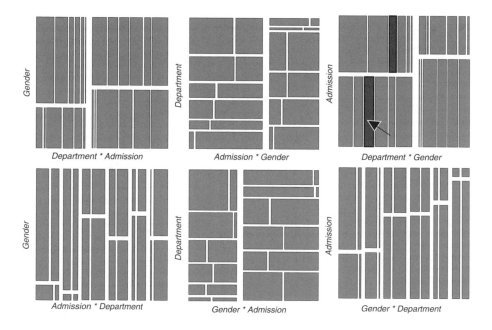

Figure 5.4 Mosaic displays for the Berkeley data.

Labels of the categories may be placed at the axis of the mosaic, but in the opinion of some people, they ruin the aesthetics of the display. Thus, they argue that mosaic displays can be deciphered without resorting to putting labels of the categories, simply using interactive interrogation of the display. This approach is used in Manet and in Mondrian. ViSta and MOSAICS both include labels for the categories in a display, but it is also possible to interact with the plot to visualize the labels if, for reasons of space, they do not fit.

The area of the bars of a mosaic display portray the relative value of the cells of the frequency table. However, if relative values are not enough and it becomes relevant to check the *absolute value* of a cell, all the programs discussed here can show the exact value of the cells by interacting with the bars of the mosaic display. Additionally, as the bars can use colors to represent other values, such as residuals of log-linear models, interrogating the bars displays the exact value of a cell's residual.

Connecting columns or rows. In the presence of interaction, columns and rows of mosaic displays are not aligned. This creates a problem when reading a mosaic, as it can be difficult to match the rectangles that correspond to the same categories. ViSta addresses this problem via a feature called *Connect*, an example of which is shown in the third plot of the first row of Figure 5.4. The arrow denotes that the user has selected a rectangle in the display with the *Connect columns* option activated. The result of activating this option is that the other rectangle corresponding to the same column as the one selected has also been selected.

Relabeling. Standard mosaic plots are built, as described above, by splitting the square alternately in the horizontal and vertical directions. An example of a standard mosaic plot is Figure 5.5 left. This display has *Department* and *Gender* on the horizontal axis and *Admission* on the vertical axis. This plot is the same as the one shown

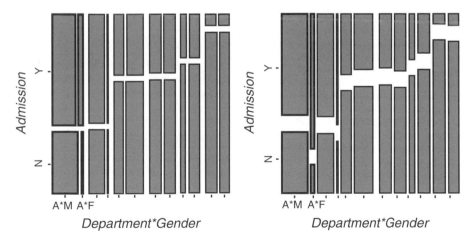

Figure 5.5 Changing the direction of the split

at [2,6] in Figure 5.4. At this case, the critical comparison of *Admission* for Males and Females within Department A is carried out looking at the difference of widths between the bars marked with a thicker border in Figure 5.5 left. The bars on the left are for Males, and the bars on the right are for Females. As the empty spaces between the two bars for *Admission*=Yes and the two bars for *Admission*=No are not aligned, the bars indicate interaction of *Gender* and *Admission* within Department A. Interpretation of this interaction is effectuated by examining the difference of width of the bars for Males or Females. As an example, it can be appreciated that the bar for Females admitted is wider than the bar for Females non admitted, while the reverse is true for Males. This points to the aforementioned advantage of Females over Males in Department A.

Figure 5.5 right portrays the same variables and in the same order as Figure 5.5 left, but after combining the variables *Department* and *Gender* in a variable, and then using the resulting variable for making the display. We saw this operation applied to frequency tables on Page 154 with the name of **Relabeling**. The advantage of this version of the mosaic display is that the critical comparisons of Males and Females is now done along the vertical direction.

As an example, the comparison of the levels of *Admission*=No for Males and Females turns out to be the difference of height between the two bars with a thicker border in Figure 5.5 right. This comparison portrays the differences already mentioned very clearly.

Reordering categories. Categorical variables are usually unordered, but we are free to arrange the categories in ways that best reflect the nature of the data. Friendly (Friendly, 1994) uses the information about residuals from a model to reorder the categories, so that the positive residuals fall in one corner of the mosaic, and the negative residuals in the other. This reordering makes more apparent the over-representation and the under-representation of some cells with respect to the model analyzed compared to the unordered version.

We saw examples of reordering tables of frequency data in Section : *Reordering*.

MANET and Mondrian feature manual reordering of the categories of mosaic plots via a linked bar chart. Dragging a bar in the bar chart using the mouse causes the automatic change of the linked mosaic display.

Joining/excluding categories. We discuss this feature applied to frequency tables in the sections named Joining categories and Excluding categories on Page 153. Mosaic plots seem a natural place for including features of this type, but, unfortunately, any of the programs considered here support this characteristic.

Activating elements. Activation of elements is normally a step that is performed before a subsequent action performed on the element selected. However, if the element selected is emphasized using any type of effect (color, thickness, etc.), the activation can be used for analysis too.

Activation in mosaic plots is implemented in MANET and Mondrian. Clicking on each of the rectangles activates the whole rectangle, which shows up in a different color than the one used by default. Activation is of special importance if used in combination with linked views.

Linking views. We saw linking in Section 4.3.1 and mentioned that it was a general concept that included several aspects. One aspect of linking applied to mosaic displays was mentioned when we discussed that reordering the categories in a bar chart automatically reorders the category in a linked mosaic display.

Another aspect of linking relevant to mosaic displays, or graphics of categorical data in general, is of linked activation. This feature refers to selecting elements in a graph so that the same entities (observations or groups) in other (linked) graphs are also activated. The archetypical example of linked views is linked scatterplots, but the concept can be applied to categorical displays.

Linked activation in mosaic displays differs from the scatterplot situation because its source is aggregated data. Therefore, activating a region in a mosaic display activates the entire class of values to which this region refers, such as when selecting a rectangle in a mosaic display activates the corresponding elements in other plots that this rectangle summarizes. This is not a problem if we are working with frequency (i.e., aggregated) data, but if we have a dataset with mixed variables, we are moving from one-to-one relationships to one-to-many relationships. The existence of one-to-

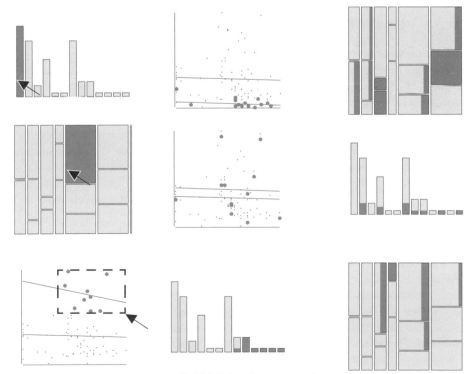

Figure 5.6 Linking frequency data.

163

many relationships is the point where categorical displays differ from the standard situation.

Figure 5.6 is a representation of the various scenarios for linked activation when frequency data are involved. The figure has three different plots of a hypothetical dataset composed of five variables: two numerical (displayed in an scatterplot) and three categorical (the first one displayed in a bar chart and the other two in a mosaic display). The scatterplot shows regression lines both for all the points and for the points selected . Let's suppose that the user interacts with the first plot in each row. The situations and a short account of the results follow:

1. The user selects a bar in a bar chart. This is equivalent to selecting all the observations in a given category. This results in selecting a group of points in the scatterplots and parts of the mosaic display.

2. The user selects a cell in the mosaic display. The effect is to select a group of points in the scatterplot and (partially) several bars in the bar chart.

3. The user selects a group of points in the scatterplot. This produces the partial selection of bars in the bar chart and in the mosaic plot.

Manet and Mondrian are the only programs that implement linked activation in mosaic plots currently.

5.3   Visual Fitting of Log-Linear Models

In the previous sections we have shown how to use tables and graphics to describe frequency datasets and to get a first impression of important features of the data such as individual values, patterns, or associations among the variables. As an example, the percentages in the last table of Figure 5.2 could be used to describe the possible bias in the Berkeley data and how one of the departments differed notably from the rest. As it is usually good practice to spend some time looking at the data with these basic tools in the initial phases of data exploration, we believe that the dynamic interactive extensions introduced above have a considerable interest for those performing statistical analysis with frequency data. However, if we want to test whether our descriptions actually fit the data, we need to go an step further and specify models that can be compared to our data.

Although there are several alternatives available with this same purpose, log-linear models provide the most comprehensive scheme to describe and understand the association among two or more categorical variables. For this reason, log-linear models have gained wide acceptance in recent decades, and every major statistical package now includes capabilities for computing them. Also, there are several excellent textbooks that address this subject (Agresti, 1990; Andersen, 1996; Ato and Lopez, 1996; ; Christensen, 1990), all at intermediate or advanced levels. Indeed, log-linear models for testing hypotheses for frequency data have a status similar to that of classical techniques such as regression or ANOVA for testing hypotheses for numerical data.

Unfortunately, despite the recognition currently enjoyed by log-linear models, dynamic interactive software for computing them is not as advanced as for numerical data. Thus, many of the features that have been implemented successfully for numerical data, discussed in the first part of this book, are absent from the programs for computing log-linear models.

However, the freeware programs mentioned in the introduction to Mosaic displays (Manet/Mondrian, MOSAICS, and ViSta), feature dynamic interaction applied to log-linear models. Thus, all these programs incorporate interactive tools for specifying, computing, and then displaying diagnostics of log-linear models. Furthermore, as these diagnostics may suggest modifications to the model, the programs also make it easy to modify the model and see the consequences in the diagnostics, using what can be called *interactive stepwise procedures*. In summary, these programs illustrate ways that dynamic interactive features can also be applied to the modeling of frequency data.

In the rest of this section, we review the features of the log-linear model module in ViSta. This module, named LoginViSta, works as a plug-in (i.e, it can be modified externally without modifying the internals of ViSta. The main advantage of Login-ViSta over programs mentioned previously lies in its flexibility for specifying *all* types of log-linear models. Another unique feature of LoginViSta is a spreadplot that integrates a number of graphic and numerical diagnostics for models, some of which are very innovative. The section is organized according to the elements in the Login-ViSta spreadplot and the way they work together to supply an interactive dynamic environment for log-linear models.

5.3.1 Log-Linear Spreadplot

A log-linear model can be understood as a linear model (regression or ANOVA) for the logarithm of the frequencies or counts of the frequency data table:

$$\log m = X\beta$$

where m is a column vector of fitted frequencies, X is the *model matrix* (sometimes called a *design matrix*) and β is a column vector that contains the parameters. As the predictors are all categorical, they have to be coded according to one of the habitual methods. For example, for a 2×2 table, the saturated model (which includes all the main effects and interactions for a dataset) with dummy coding (the one used throughout this section) can be represented as

$$\begin{pmatrix} \log m_{11} \\ \log m_{12} \\ \log m_{21} \\ \log m_{22} \end{pmatrix} = \begin{bmatrix} 1 & 1 & 1 & 1 \\ 1 & 1 & 0 & 0 \\ 1 & 0 & 1 & 0 \\ 1 & 0 & 0 & 0 \end{bmatrix} \begin{pmatrix} \mu \\ \lambda_1^A \\ \lambda_1^B \\ \lambda_{11}^{AB} \end{pmatrix} \tag{5.1}$$

The formulation of log-linear models in the form of linear model puts them into the framework of generalized linear models (GLM's) (McCullagh and Nelder, 1989), a framework that also includes models such as the logistic regression, Poisson regression, and the standard model with normally distributed errors. This formulation has the advantages that any log-linear model may be expressed in a compact form and that it is easy to express variations with respect to basic models.

The design matrix can be used to set a large variety of models. Of these the most important type are those called *hierarchical models*, which are specified such that when a term is included in the model, all the lower-order effects are also included. *Nonhierarchical models* do not follow the rule of including all the lower-order effects of the terms included in the model. Nonhierarchical models are more flexible than hierarchical models, but those, in turn, are easier to interpret, as in many applications, to include a higher-order interaction without including the lower interactions, is not meaningful (Agresti, 1990; Vermunt, 1997). However, there are occasions where nonhierarchical models can be interpreted in a reasonable manner, often providing alternative explanations to those provided by hierarchical models (Rindskopf, 1990).

Finally, if one of the variables in the study can be considered as dependent and the rest as independents, the model can be formulated as a *logit model*. Logit models are traditionally regarded as different from log-linear models, but it can be proved that there is an equivalent log-linear model for each logit model. Logit models are simpler to analyze than log-linear models because they assume that the main effects and the interactions of the independent variables are already included in the model. This has the advantage that specifying, testing, plotting, and interpreting the results of logit models is considerably easier than doing the same for equivalent log-linear models. All these different models can be specified via the design matrix.

Log-linear models are specified in ViSta using the spreadplot shown in Figure 5.7. We discuss the panels of the spreadplot in the following sections, from left to right, up to down, as follows:

- *Model builder window*. Specification of log-linear models, hierarchical and not hierarchical, is carried out using this window. This panel is discussed in Section 5.3.2.
- *Predicted and observed mosaic plots*. Mosaic plots are explained in Section 5.2 and this panes is discussed in Section 5.3.3.
- *Past models window*. This window gives an overview of the process of modeling. This window can be used to rewind the analysis to any prior model. This pane is discussed in Section 5.3.4.
- *Parameters plot*. Parameters are an important, if cumbersome, element to consider for interpreting a model. This plot provides a visual help for such task. This panel is discussed in Section 5.3.5.

5.3.2 Specifying Log-Linear Models and the Model Builder Window

One of the key points of software programs for fitting log-linear models is how easily they let the user specify the model. Although some programs allow writing the model matrix manually, the process is too time consuming and error-prone to be practical, especially when there are more than two variables to consider and several models to test. Hence, command-oriented software programs for log-linear analysis define special notations that alleviate the task considerably, as they write the model matrix from succinct high-level descriptions of the model. So the saturated model for a dataset with three variables, with names *A*, *B*, and *C* can be indicated simply as [*ABC*]. These command-oriented programs give control over several elements of the model matrix, empowering the user to test many possible variations from the basic models. Examples of programs with this capabilities are GLIM, LEM (Vermunt, 1997), and the procedure CATMOD in the SAS system. The LOGLINEAR procedure in SPSS can also be used to construct contrasts for testing nonstandard models. Of course, S, R, and S-Plus also have superb capabilities for log-linear modelling.

At this time, however, the programs that use graphical user interfaces do not provide all the capabilities of those based on commands. The number and type of models that can be fitted taking advantage of interactive techniques is typically very low compared with the large number of possible log-linear models discussed in the literature. In particular, hierarchical models are sometimes well covered, but nonhierarchical are usually no covered. The problem seems to stem from the complexity of designing an interactive dynamic visual matrix language for specifying the model.

In the rest of this subsection we show how to specify log-linear models using Login-ViSta, using as examples some models that could be considered for the Berkeley dataset introduced in Section 5.1.2.

Models for the Berkeley dataset. The Berkeley dataset has three variables—*Admission* (A), *Gender* (G), and *Department* (D)—with a total of $2 \times 2 \times 6$ cells. The saturated model for this example, denoted [*AGD*], is

$$\log m_{ijk} = \mu + \lambda_i^A + \lambda_j^G + \lambda_k^D + \lambda_{ij}^{AG} + \lambda_{ik}^{AD} + \lambda_{jk}^{GD} + \lambda_{ijk}^{ADG} \tag{5.2}$$

This model fits perfectly but is the least parsimonious because the number of parameters equals the number of cells in the table. A more interesting model would be the one excluding the *Gender* × *Admission* interaction. This model, in case of fit, would dismiss the gender-bias admission hypothesis. The model is

$$\log m_{ijk} = \mu + \lambda_i^A + \lambda_j^G + \lambda_k^D + \lambda_{ik}^{AD} + \lambda_{jk}^{GD} \tag{5.3}$$

This model is still a hierarchical model, as all the lower interactions of the terms in the model are included. This model is denoted as [*AD*][*GD*]. The parameters of this model can be interpreted in the following way:

- λ_i^A, λ_j^G, and λ_k^D are related to the differences in one-way probabilities. Thus, the parameter for *Gender* is related to the differences in the number of male and female applicants.
- λ_{ik}^{AD} describes the association between *Admission* and *Department*. These parameters reflect the degree to which departments differ in their admission rates.
- λ_{jk}^{GD} refers to the different rates of males and females applicants across the departments.

167

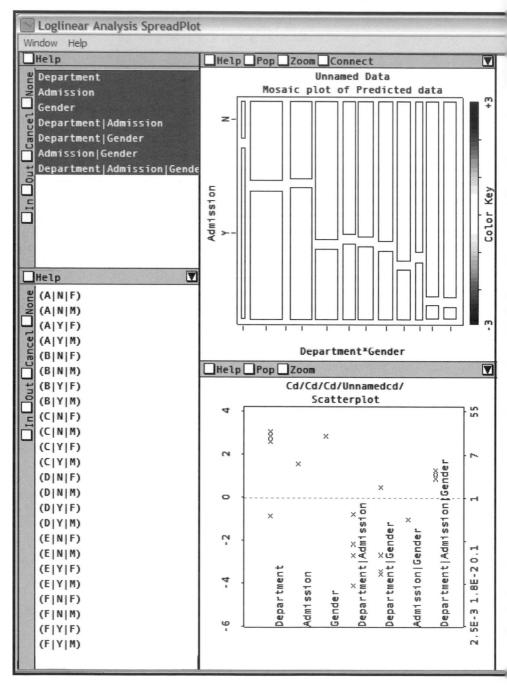

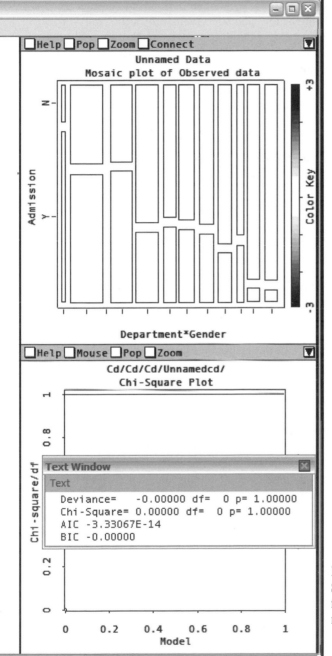

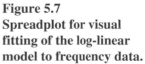

Figure 5.7
Spreadplot for visual fitting of the log-linear model to frequency data.

If model (5.3) does not fit, a model that includes the *Gender* × *Admission* interaction should be tested. This model would simply add the λ_{ij}^{AG} term to the previous model. However, this model could be found excessive, as it states that gender bias occurs in a generalized way in all departments. Another option is to consider the non-standard model in which the interaction *Gender* × *Admission* occurs only in some departments. This can be tested expanding the three-way interaction term and selecting the departments of interest. For example, the following model would address whether department A presents some bias:

$$\log m_{ijk} = \mu + \lambda_i^A + \lambda_j^G + \lambda_k^D + \lambda_{ik}^{AD} + \lambda_{jk}^{GD} + \delta_{K=A}\,\lambda_{ijk}^{ADG} \qquad (5.4)$$

where $\delta = 1$ if $k = A$ or 0 otherwise. This model asserts that *Gender* and *Admission* are independent except in department A.

Model builder window. Figure 5.8 shows the operations that are possible with the model builder window. This figure represent each operation by using two pictures; the first one displays the menu before the user acts, and the second, after. We use the pictures to illustrate the functioning of the model builder window in LoginViSta.

The model builder window lists all the main effects and interactions of the variables in the model. The first illustration shows how the window works *hierarchically* (i.e., clicking on an item selects all the items in the hierarchy included in that item). Thus, clicking on the last item, the highest in the hierarchy, selects all the items. The second illustration provides another example; clicking on the *Gender* × *Department* term also selects the main effects *Gender* and *Department*. The window also has a *non-hierarchical* mode that allows selecting individual items in the list (and consequently, specifying nonhierarchical models). The third illustration illustrates *deselecting* of model terms. Again this function works hierarchically, which hinders deselecting an item if other items higher in the hierarchy are still selected.

Notice that as the process of building log-linear usually starts with the saturated model, deselecting terms is often the most used action. It is usual that the analysis sets off in the saturated model and proceeds by removing terms from the model. The process stops when a parsimonious model with satisfactory fit is reached.

The fourth illustration shows how to add a specific vector to the design matrix. Clicking with the right button on a nonselected item pops up a menu with a list of the parameters for the term. The specific example in Figure 5.8 displays the parameters for the three-way interaction *Admission* × *Gender* × *Department*. Selecting the parameter for department A adds it to the list of terms and interactions. This makes it possible to fit the nonstandard model shown in equation (5.4).

5.3.3 Evaluating the Global Fit of Models and Their History

For each log-linear model that we may fit a dataset, there are overall or global measures of goodness of fit, described below. In practice, we usually want to explore several alternative models and choose the simplest model that achieves a reasonable goodness of fit. In an interactive setting, we introduce the idea of a history plot, show-

Figure 5.8 Model builder window of LoginViSta.

171

ing a comparative measure of fit for all models contemplated. The interactive features of this history plot also allow you to return to a previous model (by selecting its point) or to make model comparisons.

Tests of global fit. The most commonly used measures of how well the model reproduces the observed frequencies are the familiar χ^2 Pearson statistic,

$$\chi^2 = \sum_i \frac{(n_i - \hat{m}_i)^2}{\hat{m}_i}$$

and the *likelihood ratio* or *deviance statistic*,

$$G^2 = 2\sum n_i \log(n_i / \hat{m})$$

where the n_i are the frequencies observed and the m_i are the expected frequencies given the model considered. Both of these statistics have a χ^2 distribution when all expected frequencies are large. The (residual) degrees of freedom is the number of cells minus the number of parameters estimated. In the saturated model the deviance (or Pearson χ^2) is zero and the degrees of freedom are also zero. More parsimonious models will have positive values of deviance but also more degrees of freedom. As a rule of thumb, nonsaturated models fit the data if their deviance is approximately equal to their degrees of freedom (or the ratio χ^2/df is not too large).

The deviance is unique in that it can be used to compare nested models. Two models are nested if one of them is a special case of the other. Comparison of models pro vides a way of focusing on the additional effects of the terms included in the larger model.

Plot of the history of models. Figure 5.9 is a plot of the history of the values of fit obtained along a session of modeling. The points in the plot represents the values of χ^2/df for five models for the Berkeley data (we could have used the deviance in this plot, too). Values close to 1 mean that the model fits well. Labels of the points identify the models considered. This plot allows us to see the process of model fitting at a glance.

The example in Figure 5.9 portrays the following sequence of analysis:

1.   The plot starts with the saturated model $[AGD]$, which fits perfectly.

2.   It continues with the model with all the two-way interactions, which does not fit well.

3.   The model that evaluates the hypothesis of no gender-bias effect is tested (no [AG] term), but its fit is still not satisfactory.

4.   As preparation for testing model 5, the category of reference for the variable *Department*, set at this moment to A, was changed to F. This produces model 4, which has the same fit and terms as model 3.

5.   Finally, the model testing the interaction of *Gender* and *Admission* was tested only for department A . This model has a good fit as shown in Figure 5.9.

The history plot in LoginViSta is not only a way of representing the successive values of fit of the models considered but a control panel for managing the process. The actions that can be performed from this panel are the following:

- Selecting a point in the plot changes all the plots in the spreadplot of Figure 5.7 to display the values of the model symbolized by such a point. This action allows you to review models fitted in the past.
- Selecting two points (two models) produces a model comparison test. Login-ViSta checks if the models are nested before the actual test is accomplished. If the comparison is appropriate, a test of significance of the difference in deviance is displayed in a separate window.

Other actions that could be added to this plot are the capability of removing points, rearranging them, and supporting different threads (using lines of different colors) and of other goodness-of-fit indexes. In summary, to find an adequate parsimonious model, the plot in Figure 5.9 could be extended to manage completely the process usually followed.

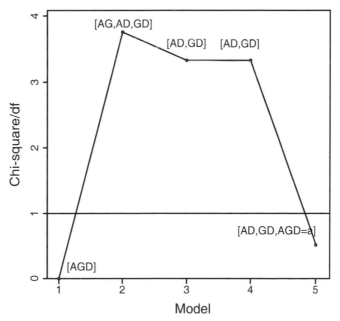

Figure 5.9 History of models applied to Berkeley data.

5.3.4 Visualizing Fitted and Residual Values with Mosaic Displays

For a log-linear model, residuals provide important information regarding the adequacy of the model. Outlier cells (with large absolute residuals) may indicate particular levels of some factor that are not well explained by the current model, as we saw in the Berkeley data. More generally, the pattern of signs and magnitudes of residuals may suggest a better model. Mosaic plots are an excellent way to display residuals from a fitted model in the context of a picture of the data.

Another use of mosaic plots is related to understanding the constraints imposed by models. Mosaic plots of fitted values (instead of observed) show the assumed relationship between variables in a graphical way, illustrating the consequences of introducing or removing terms in the model. Also, comparisons of fitted and observed mosaic plots provide a visual check of fit of the model.

In this section, we review the formulas for residuals, including an example of the problems that arise using point plots. Then we show how using predicted and observed mosaic displays can be a good alternative for visualizing these residuals.

Residuals for log-linear models. For a log-linear model, raw residuals are the difference between the values predicted and observed in the cells. However, raw residuals are of little use, as cells with larger expected values will necessarily have larger raw residuals. The *Pearson residuals*

$$e_i = \frac{n_i - \hat{m}_i}{\sqrt{\hat{m}_i}}$$

are properly scaled and approximately normally distributed, but they depend on the leverages of the matrix of the predictors. Therefore, one may define *adjusted residuals* (Agresti, 1990; Friendly, 2000b; Haberman, 1973):

$$r_i = \frac{n_i - \hat{m}}{\sqrt{\hat{m}_i(1 - h_{ii})}}$$

(where h_{ii} is the leverage or hat value computed as in any generalized linear model), which are standardized to have unit asymptotic variance.

To check for cells that deviate from normality, the adjusted residuals can be plotted using a normal probability plot. However, the constraints imposed by the model have consequences that make this type of plot less useful than it might be. We will provide an example using the Berkeley data introduced in Section 5.1.2. Table 5.5 contains the residuals of the model $[AD][GD]$ [see equation (5.3)], which asserted that there is an interaction for *Admission* and *Department*, and for *Department* and *Gender*, but not for *Gender* and *Admission* (i.e., there is no gender bias). As signaled by Agresti (1990), this model imposes the constraint that the sum of interactions predicted between the categories of *Department* and those of *Gender* be the same. A bit of manipulation shows that the residuals for each department are the same in absolute

Table 5.5 Residuals for Model [AD][GD]

			Admission	
			Yes	No
A		Female	4.15	4.15
		Male	-4.15	4.15
B		Female	0.50	-0.50
		Male	-0.50	0.50
Department	C	*Gender* Female	-0.87	0.87
		Male	0.87	-0.87
	D	Female	0.55	-0.55
		Male	-0.55	0.55
	E	Female	-1.00	1.00
		Male	1.00	-1.00
	F	Female	0.62	-0.62
		Male	-0.62	0.62

values, but with changed signs. Now, notice that a normal probability plot would be less than ideal in this situation because it would not reflect such a structure in the data. Mosaic displays, discussed in Section 5.2, are a good alternative to Table 5.5 because they can be laid out to reflect such structure.

The Predicted and observed mosaic plots. Figure 5.10 shows two mosaic displays for the model [AD][GD] for Berkeley data. These plots use transformation [see or discussion on relabeling in Section 5.2.2; in fact, Figure 5.10 right is the same as Figure 5.5 right except for the coloring].

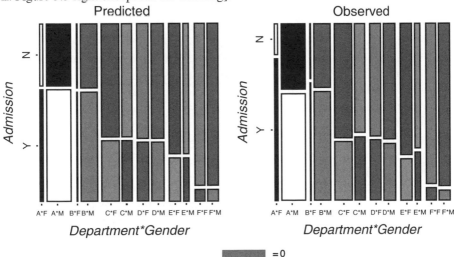

Figure 5.10 Mosaic plots for model [AD][GD] for Berkeley data.

175

The mosaic display on the left of Figure 5.10 has tiles proportional to the values predicted for the model, while the mosaic display on the right has tiles proportional to the values observed. Notice that as the model specified excludes interaction between *Gender* and *Admission,* the horizontal lines within each department are parallel in the mosaic plot for values predicted . Comparison of these lines with the lines in the display for the values observed reveals that the predictions are generally correct except for Department A (and perhaps B). This points to the aforementioned lack of gender bias except for department A (and perhaps B).

The previous impression can be reinforced by looking at the residuals of the model. Residuals are typically represented in mosaic displays using two colors (usually red for negative residuals and blue for positive residuals), the greatest intensity of the color representing the extreme of the values. Thus, pure blue would stand for very extreme positive residuals and pure red for very extreme negative residuals. Finally, cells with null residuals are displayed in white.

Nevertheless, as the previous scheme does not transfer well to black-and-white publications, we have chosen to employ one that uses pure black and pure white to represent the maximum and minimum residual values (those for department A), and a middle gray residual equal to zero. This scheme is not as effective as one based on colors but can work well if the patterns of residuals, as is our case, are simple.

Examination of Figure 5.10 portrays the residuals of the model $[AD][GD]$. Department A has the most extreme residuals under the model specified, being positive for admission of females and rejection of males, and negative, vice versa. The remaining departments are colored with similar, close to intermediate, levels of gray. In summary, the model $[AD][GD]$ copes well with departments B to F, but not with department A.

5.3.5 Interpreting the Parameters of the Model

Parameters of log-linear models are an important tool for interpretation of results based on log-linear analysis. We mentioned in Section 2.3 that log-linear models could be understood as linear models of the logarithm of the frequencies. Therefore, in much the same way that parameters provide an essential tool for understanding regression models, they are also fundamental for log-linear analysis.

However, parameters for log-linear have a reputation for being difficult to interpret, as "attested by the frequency of articles on the topic" and that "they devote considerable space to correcting the errors of their predecessors" (Alba, 1987). Therefore, it is not unusual that researchers do not take full advantage of their potentiality and that parameters remain underused.

Software for log-linear analysis has not helped to ease this situation. Many of the programs do not print the coefficients of the parameters for the models very clearly, so that to understand this part of the output, researchers need considerable diligence. The common flaws are excessively long outputs, without proper labels and with no indication of aspects of computation necessary for correct interpretation (e.g., the coding used).

The display included in LoginViSta is, on one hand, an attempt to organize this part of the output of log-linear analysis, but on the other hand, is also a way of supporting the task of interpreting the parameters of log-linear analysis. This last goal is reached by providing a set of interactive tools that interrogate the plot to obtain automatic interpretations of the meanings of the parameters. These explanations decrease the cognitive burden of interpreting the parameters.

Our plan in this section is to review the theory necessary for the interpretation of parameters of log-linear models and then proceed to a description of the plot of parameters.

Interpretation of parameters. The meaning of the parameters in log-linear models depends critically on the type of coding used. Two types of coding, *effect coding* and *dummy coding* are the most commonly used (see, e.g., Rindskopf, 1990). The fitted values are the same under both methods of coding, but the interpretation of parameters depends on the coding used. It is often claimed that dummy coding makes interpretation of parameters easier than effects coding because the parameters are interpreted as comparisons between the categories they represent and the reference category. For such this, we will restrict this exposition to dummy coding.

In dummy coding, one generates a number of vectors such that in any given vector membership in a given category is assigned 1, and nonmembership in the category is assigned 0. To avoid collinearity of the categories, the vector for one of the categories is removed from the design matrix. The category removed is called the *reference category* or *baseline category*. Therefore, the number of vectors for each of the main effects in the model is $k - 1$, where k is the number of categories of the variable. Vectors for interaction effects are built by multiplying vectors for main effects.

The parameters created as described above have a simple interpretation as a function of the values predicted for the cells. Using the Berkeley dataset introduced in Section 5.1.2 and setting *Admission* = No (n), *Department* = F (f), and *Gender* = Male (m) as reference categories for the saturated log-linear model in equation (5.2), the value of the intercept in this model is

$$\mu = \log m_{nfm}^{ADG}$$

(i.e., the logarithm of the value expected for the cell, corresponding to the reference categories for the variables in the model).

The coefficient for the main effect of *Admission* is

$$\lambda_y^A = \log \frac{m_{yfm}^{ADG}}{m_{nfm}^{ADG}}$$

which is the *odds* of *Admission* (holding the other variables constant at their reference category).

The coefficient for the interaction between *Admission* and *Gender* for *Department* = F is

$$\lambda_{yf}^{AG} = \log \frac{m_{yff}^{ADG} / m_{nff}^{ADG}}{m_{yfm}^{ADG} / m_{nfm}^{ADG}}$$

which is the logarithm of the ratio of odds ratio of admission of males relative to females in department F.

The interaction of *Admission* with *Department* produces five coefficients, which are the product of the $(k-1)(j-1)$ categories of these two variables. As an example, the coefficient for department B turns out to be

$$\lambda_{yb}^{AD} = \log \frac{m_{ybm}^{ADG} / m_{nbm}^{ADG}}{m_{yfm}^{ADG} / m_{nfm}^{ADG}}$$

which is again an odds ratio. Finally, there are five coefficients for the three-way interaction term. As a sample, the coefficient for department B is

$$\lambda_{ybm}^{ADG} = \log \frac{m_{ybf}^{ADG} / m_{nbf}^{ADG}}{m_{ybm}^{ADG} / m_{nbm}^{ADG}} / \frac{m_{yff}^{ADG} / m_{nff}^{ADG}}{m_{yfm}^{ADG} / m_{nfm}^{ADG}}$$

which is an odds ratio of odds ratios. The same structure can be applied to define higher-order parameters.

As can be seen from the description above, the difficulties of interpreting parameters of log-linear models are not conceptual, as the rules for composing the parameters for hierarchical models are rather simple, but practical, as it is necessary to keep in mind a number of details, such as the reference categories used and the general structure of the model. Also, the number of parameters in a log-linear model is often large, so they can be difficult to visualize readily. As we will see in the next section, dynamic interactive graphics techniques can diminish enormously the burden of interpreting the parameters.

Parameters plot. The parameters plot in LoginViSta is an interactive extension of a plot proposed by Tukey (1977) in the context of ANOVA. Figure 5.11 is a plot of parameters applied to the Berkeley data. In particular, the model displayed is the saturated or complete model of equation (5.2). This model includes all the main effects and interactions of the variables and can be a useful starting point in a session of modeling. The interactive tools described below enhance the plot to make it still more helpful for working with log-linear models.

Figure 5.11 displays the parameters grouped by terms. The value of each parameter can be read on the left side of the plot in the log scale, and on the right side in the original scale (odds, or odds ratios). Notice that the parameters excluded to avoid redundancy are set to 0 (1 in the original scale), so they can be visualized as standing on the dashed line.

Selecting a point in the plot reveals a 95% interval of confidence for the parameter. Two points are selected in this manner in Figure 5.11. First, we have selected the only parameter for interaction between *Gender* and *Admission* that happens not to be different from zero. This suggests removing this parameter from the model (but that would make the model nonstandard). The second parameter selected is the one for

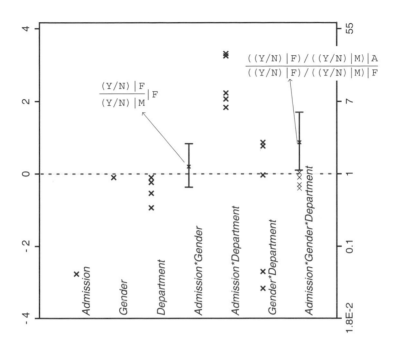

Figure 5.11 Plot of parameters for Berkeley's data-saturated model.

three-way interaction involving department A. The confidence interval for this parameter does not cross the zero line (i.e., it is different from zero).

The plot also shows the interpretation in terms of the odds ratios of the parameters selected. Focusing on the parameter selected in the three-way interaction term, we can see that this parameter basically tests the admission of females over males in department A versus department F. As mentioned above, this parameter is different (larger) from zero. Therefore, as the rest of parameters for this term are not different from zero (tests not shown), we can infer that the only department with bias gender is department A.

5.4   Conclusions

In this chapter we have presented techniques for interacting with visualizations of frequency data and with models of the same type of data. We have reviewed three topics: the manipulation of tables, dynamic mosaic plots, and the visualization of log-linear models. The methods considered are not an exhaustive inventory of the topics regarding the interactive dynamic exploration of frequency data, but we believe that they can be extended to many situations involving frequency data. For example, although the specifics of the history of models plot presented in Section 5.3.3 concern log-linear models, the general concept is of relevance to any other generalized linear model. Interactive manipulation of mosaic plots would be another example of concepts that could be applied to other plots, such as bar plots or pie plots. In summary, the ideas discussed in this chapter can be generalized to other types of data and other types of models..

6 Seeing Univariate Data

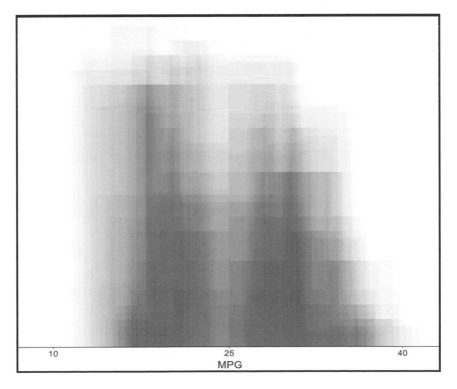

6   Seeing Univariate Data

Magnitude data are far and away the most common type of statistical data. This should not be surprising, since magnitude information provides a basis for scientific inference which is superior to that provided by other data. Thus, it should also not be surprising that there are more data analysis tools, including more visual tools, for understanding magnitude data than for any other type. Correspondingly, the largest topic in this book is the topic of visualization methods for magnitude data.

Visualization methods for magnitude data topics are covered in the next four chapters. The first three of these chapters cover, successively, graphical methods that visualize one numerical variable, thereby yielding unidimensional views of the data (this chapter), graphical methods that visualize two numerical variables, thereby yielding bivariate views of the data (Chapter 7), and graphical methods that visualize three or more variables, thereby yielding multivariate views of the data (Chapter 8). Note that we may, and usually should, use unidimensional or bidimensional methods even when the data have many more than one or two variables, since these methods give us detailed information about the individual variables or about pairs of variables that are not provided by the multivariate views. The fourth chapter (Chapter 9) on visualization methods for magnitude data deals with what to do when the data have missing values, a very common situation.

6.1   Introduction

Unidimensional methods for visualizing magnitude data are graphical methods that allow us to look at one numerical variable in detail. These methods can be used with data that have more than one variable, but only one variable at a time. We discuss these methods:

Dotplot. The simplest and most common way to represent the empirical distribution of a numerical variable is by showing the individual values as dots arranged along a line. The main difficulty with this plot concerns how to treat tied values. We usually don't want to represent them by the same point, since that means that the two values look like one. What we can do is "jitter" the points a bit (i.e., move them back and forth at right angles to the plot axis) so that all points are visible

Boxplot and diamond plot. A boxplot is a dotplot enhanced with a schematic that provides information about the center and spread of the data, including the median, quartiles, and so on. This is a very useful way of summarizing a variable's distribution. The dotplot can also be enhanced with a diamond-shaped schematic portraying the mean and standard deviation (or the standard error of the mean).

Histogram and frequency polygon. Histograms and frequency polygons display a schematic of a numeric variable's frequency distribution. These plots can show us the center and spread of a distribution, can be used to judge the skewness, kurtosis, and modicity of a distribution, can be used to search for outliers, and can help us make decisions about the symmetry and normality of a distribution. Although very popular, these techniques have severe problems. In this chapter we discuss dynamic interactive methods that can contribute to alleviating such problems.

Cumulative distribution plot. The cumulative distribution plot is a family of plots that show us whether the data were generated by a generating function commensurate with a specified generating function. The family includes normal distribution plots, for example, which test the assumption of normality.

Lag plot: A lag plot displays the values of the variable versus themselves but with a "lag" of a specified number of observations. A lag of 1 means that each value is plotted against the value directly following it in the observation order. A lag of 2 means that each value is plotted again the value two places behind it in the observation order. This kind of plot helps us check on whether a variable's values represent a random sample from a population. If it is such a sample, the plot should reveal no patterning (NIST/SEMATECH, 2004).

Sequence plot. Sequence plots are simply plots of a variable's values versus an index specifying the sequence order in the values were observed. This plot is proposed as a simple graphical summary of a variable's values (NIST/SEMATECH, 2004). It is commonly assumed that these values represent a random sample from a population. This implies that there should be no shift in location or scale from one sample to the next, or, in other words, that the values should all have the same location and scale. Sequence plots show such shifts, making them quite evident. They are also good for detecting outliers.

Note that these unidimensional methods are commonly used for data that have more than one variable, since they provide detailed information about the nature of each variable in the data. They do not, of course, provide any information about how any of the variables are related to any of the other variables, but when used with such methods (which are discussed in Chapters 6 and 7) can provide invaluable information on many aspects of a dataset.

6.2   Data: Automobile Efficiency

We illustrate our discussion with data about the fuel efficiency of 38 automobiles manufactured in 1978–1979 (Henderson and Velleman, 1981). Fuel efficiency is measured in miles per gallon of gasoline (*MPG*), a measure of fuel consumption indicating how many miles a car can move on a single gallon of gasoline. The metric equivalent is the number of kilometers each car can travel on 1 liter of gasoline (Km/L). The conversion is 1 MPG equals 0.425 km/L. Note that km/L is the reciprocal of the L/km (liters per kilometer) measure used in most of the world (1 MPG = 2.35 L/km). The complete dataset, which has measurements for each automobile on six variables, is presented in Chapter 7. For this chapter we consider only the fuel efficiency variable (Table 6.1.) Automobile names are used as labels.

Table 6.1 Miles per Gallon for 38 Automobiles Manufactured in 1978-79

Car	MPG
Buick Estate Wagon	16.9
Ford Country Squire	15.5
Chevy Malibu Wagon	19.2
Chrysler LeBaron	18.5
Chevette	30.0
Toyota Corona	27.5
Datsun 510	27.2
Dodge Omni	30.9
Audi 6000	20.3
Volvo 240 GL	17.0
Saab 99 GLE	21.6
Peugeot 694 SL	16.2
Buick Century	20.6
Mercury Zephyr	20.8
Dodge Aspen	18.6
AMC Concord D/L	18.1
Chevy Caprice Classic	17.0
Ford LTD	17.6
Mercury Grand Marquis	16.5
Dodge St. Regis	18.2
Ford Mustang 4	26.5
Ford Mustang Ghia	21.9
Mazda GLC	34.1
Dodge Colt	35.1
AMC Spirit	27.4
VW Scirocco	31.5
Honda Accord LX	29.5
Buick Skylark	28.4
Chevy Citation	28.8
Olds Omega	26.8
Pontiac Phoenix	33.5
Plymouth Horizon	34.2
Datsun 210	31.8
Fiat Strada	37.3
VW Dasher	30.5
Datsun 810	22.0
BMW 320i	21.5
VW Rabbit	31.9

6.2.1 Looking at the Numbers

Spend a moment thinking about what would be involved to substitute computers successfully for humans when the task is analyzing data. It has been claimed that subject-matter knowledge, or contextual knowledge, is the most difficult problem faced by attempts to develop expert systems for data analysis (St. Amant, 1997). Computers cannot reproduce the subtlities of the human brain when dealing with this type of knowledge. However, without appropriate ways of promoting that knowledge, humans can also fail to use and apply it correctly.

A way to increase that knowledge is to look at the numbers themselves. When you do, it is possible to see all sorts of small details that can be of great help to guide further analysis and interpretation. This type of exploration will hardly produce insights or summaries of much generality, but it can generate informal knowledge that will enrich the data analysis process. And it can reveal problems in the data as well.

Datasheets or spreadsheets are the basic tools for representing and manipulating tables of data. Columns of the datasheets represent variables and rows represent observations. Sometimes, the first column of the table takes the role of showing the label of each observation. Also, the first row will usually be variable names, and the next row or several rows may contain variable information, such as measurement type, role in the study, and so on.

We described some interactive techniques for spreadsheets in Section 4.2. Those techniques, especially when used with a dynamic, interactive datasheet, constitute a preliminary, albeit simple data analysis that can provide useful insights in their own right. ViSta has a datasheet that supports many of the important features just described (Figure 4.4). The File menu's Open Data item displays a dialog box for loading a data object. Opening the data object shows a datasheet that supports editing individual values and changing the type of variable. The Transform menu has a list of transformations that are useful for managing data. Sort-Permute, the first item in the menu, can sort the datasheet according to the values of one or more of the variables.

The datasheet in ViSta is not as functional as those in commercial packages. However, the Data menu's Create Excel Spreadsheet item gives you Excel's spreadsheet, and the File menu's Export Data item exports a text file that can be read with Excel or any other spreadsheet. The result can be imported using the File menu's Import Data item.

6.2.2 What Can Unidimensional Methods Reveal?

The fundamental fact about several observations of a single variable is that they vary. Even if we observe the response of the same individual repeatedly on the same variable in the same situation, the values will vary, perhaps because of measurement error, perhaps due to changes in the variable we are measuring, or commonly, due to both. Thus, what we should be looking for in a variable, numerical or not, is information about the way the values vary.

Since the values observed vary, the collection of values observed will have a distribution. The "shape" of the distribution of a variable's values has a large impact on what statistical methods can legitimately be applied to the variable. We need to determine particularly whether the shape is symmetric or asymmetric, and whether it is normal or nonnormal. In addition, we need to determine whether there are *outliers* — whether some of the values just don't seem to fit into the overall shape of the distribution of the rest of the values.

Visualization procedures are very useful at this point, because they can tell us a lot about the shape of the distribution and about the presence of outliers. In fact, before we do anything else, we must look for and deal with outliers. Then, once outliers have been dealt with, we can turn to shape of the distribution, first asking whether there is more than one mode, then asking whether it is symmetric, and when the data seem to be symmetric and unimodal, whether they are normally distributed. So we begin with a discussion of outliers and number of modes, then we to specific characteristics of shape, and finally, cover the overall general nature of shape.

Outliers. An outlier is an observation that does not appear to belong to the distribution of other observations falling outside their distribution. Outliers can have drastic effects on any analysis performed on the data. For this reason it is necessary to check data for outliers. See Barnett and Lewis (1995) for a complete reference on outliers.

Sometimes, it is completely obvious that an outlier is an erroneous value that can be corrected by finding and introducing the correct value. On other occasions, however, an outlier is an exceptional value that has been produced by a legitimate cause. In this situation the outlier deserves some type of special treatment whose nature you must determine from the situation. Here, rather than being a nuisance, the outlier should be considered to be very valuable, as it is, in some way, carrying more information than the other ordinary observations are carrying.

Basically, there are no guidelines here other than that you should always, as the first step in any data analysis, look for outliers, and when found, deal with them as seems best to you, before you proceed to analyze the data.

Modes. The modicity of a variable refers to the number of modes of the variable's empirical distribution. The mode of a variable's empirical distribution is the portion of the distribution which is, at least locally, the most frequent or most densely concentrated part of the distribution. Modicity does not have a strict mathematical definition, which makes it a troublesome and difficult characteristic to deal with. Regardless, it is quite important that a variable's empirical distribution be unimodal: have just one mode, as few multimodal variables are subject to appropriate analysis by statistical methods.

A distribution with more than one mode can be an indication that data come from a *mixture of distributions.* If there are two modes, perhaps the data come from two populations with different means and/or standard deviations. There are methods for estimating the means and standard deviations of the several distributions in this case, although we do not pursue them here.

Shape. When trying to determine shape, we obtain the most accurate understanding by making two separate types of considerations. First, our understanding is improved by considering several specific characteristics, such as the location, spread, and skewness of the distribution. But also, our understanding is helped by asking whether the overall general shape appears to be unimodal or multimodal, symmetric or asymmetric, normal or nonnormal. Thus, we turn to these two aspects of understanding shape: investigating the specific characteristics of shape, and looking at the overall nature of the shape.

When trying to understand the nature of the shape of a variable's distribution of values observed—of its empirical distribution—there are several specific characteristics we can consider. These include:

Location and spread. A variable's values tend to vary around a particular value on the real number continuum. The *location* is the value around which they tend to vary, and the *spread* is the amount of variation around that location. Both concepts can be quantified by numerical measures (mean and variance, for example).

Skewness. A variable can have values that are skewed: The values tend to pile up toward one end of the scale and taper offer gradually at the other end.

Kurtosis. This concept refers to whether a variable's values have a distribution which is very flat or very peaked (low or high kurtosis), or is somewhere in between.

These specific aspects of an empirical distribution have certain implications about the variable's shape. These implications are discussed in the next section.

Shape: is it symmetric? A variable's values can be distributed symmetrically or asymmetrically. A symmetric distribution is one such that the mirror image of one side of the distribution is identical to the other side, or at least nearly so. In Figure 6.1, the distribution on the left is symmetric, whereas the distribution on the right is not. A skewed distribution with the tail on the right-hand side (as in the right-hand part of Figure 6.1) is called positively skewed because the tail points toward the positive end of the X-axis. If the tail points toward the left, the distribution is negatively skewed. All symmetric distributions are unskewed. Not all symmetric distributions are unimodal, but the most useful ones are.

Symmetric distributions cannot be skewed, but they can have any type of kurtosis. But skew and high or low kurtosis are usually thought of as being not good, since many data analysis procedures assume that the data are not skewed nor very kurtotic. However, data visualization procedures do not require any particular skewness or kurtosis and can be used to check on the degree of each for any particular variable.

It is desirable to have a symmetric distribution, better to have a unimodal symmetric distribution, better yet if the unimodal symmetric distribution has no kurtosis, and best of all if the symmetric, unimodal, nonkurtotic distribution is also a normal distribution. So the next section is about normal distributions.

A symmetrically distributed variable that has only one mode can also be normally distributed (i.e., have a shape that well approximates the equation of the normal curve).

There are several reasons why symmetry is important for data analysis (Chambers et al., 1983):

- The center of a symmetric distribution is unambiguous since different ways of estimating it (mean, median) coincide.
- Symmetric distributions are easier to understand (the upper part is like the lower part)
- Whereas classical statistical procedures assume normality, a number of modern robust statistical methods only assume symmetry and can be used legitimally with nonnormal but symmetric data.

Shape: is it normal? Since statistical visualization procedures *do not* require normality, and since some statistical visualization procedures *can* be used to show normality (or lack thereof), and since many statistical analysis procedures *do* require normality, one important (perhaps the most important) usage of unidimensional statistical visualization to investigate whether it is reasonable to assume normality.

A normal distribution is shown on the left-hand side of Figure 6.1. The normal distribution is the most important distribution model. It has the following characteristics: A normal distribution is symmetric. A normally distributed variable has a sample mean and standard deviation, which are the best possible indicators of the population's center and spread. Normally distributed variables are not skewed and do not have extreme kurtosis. A normal distribution has one mode and no outliers.

Normal distributions are important because many aspects of nature are distributed normally. Examples include the height and weight of individuals, and the SAT scores of high-school students. Since the normal distribution is so pervasive, and since it has convenient mathematical properties, many of the statistical tests known as *significance tests* are based on an assumption of normality, with the significance level being incorrect if the distribution is nonnormal. For these reasons it is important to look at a distribution to assess its normality.

Normally distributed variables satisfy the assumptions of many inferential statistical procedures. However, real life does not usually offer data that behave exactly like the normal model. So it is an important data analysis task to investigate how adequately the normal distribution serves as a model for your data.

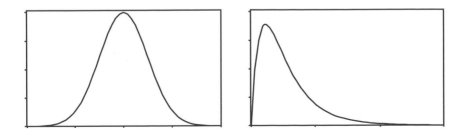

Figure 6.1 Left: symmetric, normal, right: asymmetric.

6.3   Univariate Plots

The most straightforward way to represent a variable's distribution is to show the individual values as dots arranged along a line according to the values of the variable. This simple plot, called a dotplot, allows us to investigate many of the issues mentioned Section 6.2. On the other hand, most of the unidimensional tools we have for visualizing magnitude data construct a schematic of the variable's distribution, the schematic designed with a purpose that varies from one tool to the next.

In our discussion we emphasize the dynamic features of the unidimensional tools for magnitude data, in addition to covering the traditional static versions of these plots whose descriptions are found in many places. In particular, we discuss dynamic features which are specifically designed to overcome limitations resulting from the nature of the static version of the plot. The plots we review are:

- Dotplots, boxplots, and diamond plots, the latter two being dotplots plus a schematic portraying features of the population distribution. Dotplots help us see modality, outliers, and symmetry and help us identify where specific observations or groups of observations fall in a distribution. Dynamic versions of these plots help us identify points more easily and provide more complete information about the characteristics of the data.
- Probability plots, which plot the values of the sample versus the quantiles of a probability distribution (such as the normal distribution) which could have generated the sample. These plots help us judge whether it is reasonable to assume that the data were generated by the specified theoretical distribution, such as the normal distribution model. Dynamic probability plots help identify the optimal values for parameters of the theoretical distribution.
- Histograms and the related plots, frequency polygons, employ schematics designed to visualize the density of the distribution at different points. Being able to visualize the density is another way to attempt to identify the population model and its parameters. Dynamic histograms help us better understand the data and its potential generating distribution.

6.3.1 Dotplots

The simplest and most common way to represent the empirical distribution of a numeric variable is by showing the individual values as dots arranged along a line. The line is oriented vertically as in Figure 6.2 to facilitate the display of point labels.

Dynamic versions of the dotplot are useful to identify individual points by brushing them. In the upper left plot in Figure 6.2 we see that the car with the maximum value for *MPG* variable (the most efficient car) is the Fiat Strada and the minimum *MPG* value (the least efficient) is the Ford Country Squire. This makes good sense, since if you know these automobiles, you will realize that the Fiat Strada is a very small, lightweight vehicle, whereas the Country Squire is very large and heavy.

A problem with dotplots is that when several observations have equal or close values, they obscure each other, as demonstrated in the upper-right plot in Figure 6.2.

This is a potentially very important problem for dynamic/interactive versions of the dotplot, where selecting observations is a critical task. The task becomes impossible when points are obscured. For example, when the user clicks on overlapping points and labels for the points are to be shown, they can become illegible, as can be seen in the figure, or some labels can obscured by others. This leaves the user with the difficult task of selecting points one at a time in order to read the labels.

A dynamic solution to this problem with static dotplots is called *jittering*, with examples of several different kinds being shown in the bottom row of plots in Figure 6.2 and in the plots in Figure 6.3. As we define it, jittering involves displacing a point in a dotplot by a random or systematic amount along the axis that is not the axis displaying the variable values. In our figures, the vertical axis displays the data and the horizontal axis is used for jittering. You will note that while the upper two figures in Figure 6.2 show the points falling along a straight line (they are unjittered), the points in the lower two plots do not fall along a straight line (they are jittered).

Dynamic jittering can further address the problem of hidden dots. The dynamism can take place in several different ways. One way enables the plot to generate a new round of jittering on the user's request, the user being expected to continue doing this

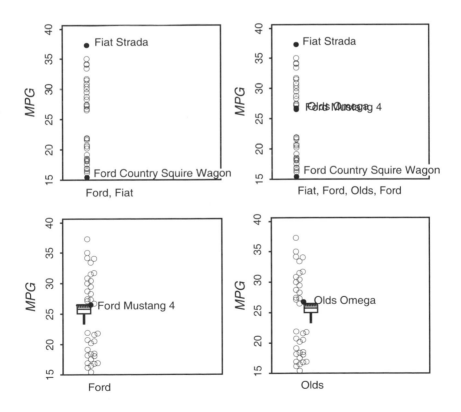

Figure 6.2 Four dotplots for *MPG*, the bottom two are jittered.

191

until getting a more satisfactory result. Another way is to rejitter automatically when the axis is stretched or shrunk, since this changes which points might be overlapping.

Jittering can be systematic as well as random. Either type of jittering can be applied selectively only to those that are involved in overlapping, rather than to all points. ViSta, for example, has an algorithm that displaces only those points that need to be displaced, and does the displacement in a systematic fashion that guarantees that every point is visible. The plots in Figure 6.3 use this systematic jittering method. Those in Figure 6.2 are randomly jittered.

In addition to permitting you to identify individual points, dotplots allow you to look into some of the distributional properties of a variable. For example, dotplots can be good for judging whether there are outliers. For the *MPG* data shown in Figure 6.2 there do not seem to be any: however for the *Age* variable[1] shown in the left plot of Figure 6.3, two students seem to be rather older than the rest.

Dotplots can also be good for looking for modality. Figure 6.2 shows that there is a concentration of values at the ends of the scales, with a gap in the middle. So we can see that the data are bimodal and thus that the data are not normal. Other types of non-normality, such as asymmetry, can be difficult to judge from the dotplot, although the middle plot in Figure 6.3, showing the distribution of self-reported *SATMath* scores, seems to be negatively skewed. The right-hand plot in Figure 6.3 shows the same data as the middle plot, but with jittering done differently.

Note that the left plot in Figure 6.3 is a systematically jittered plot of a variable that has many identical observations: This is the *Age* variable, and many of the students are 19 or 20 years old. With the appropriate jittering algorithm we can generate plots that show a dot for every observation, with the dots being lined up in a way that makes the overall plot somewhat like a barplot without the bars.

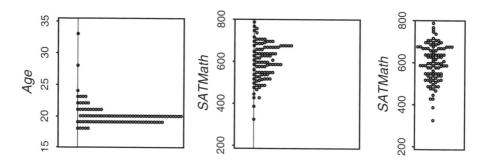

Figure 6.3 Three systematically jittered dotplots.

1.Data gathered by the first author from students in his course for introductory psychological statistics. The data were gathered on the first day of the course over several years. The variables concerned the attitudes of the students toward statistics, their experience with mathematics and computers, and their grade average and scores on the SAT (a nationally standardized test used widely in the United States as part of the university admissions procedure), along with their age, gender, and so on. Data available with ViSta.

6.3.2 Boxplots

Dotplots can be enhanced with the capability of adding or removing schematics, where the schematics provide a context in which we can evaluate the individual observations more completely. These schematics provide information about the center and spread of the data, such as the mean or median, the quartiles, the standard deviation, and so on.

We discuss two schematics that are commonly used to enhance dotplots: boxplots and diamond plots. Figure 6.4 shows two boxplots on the left and a diamond plot in the third position. The boxplot and diamond plot are shown superimposed on the right-hand side. We review boxplots in this section and diamond plots in the next.

The boxplot is the most useful plot that results from adding a schematic to a dotplot. The schematic of the boxplot is based on the median and other quantile measures, as is described below. Boxplots were first described by Tukey (1977), who added some important variations shortly thereafter (McGill et al., 1978). Figure 6.4 shows the dotplot in Figure 6.2 with a boxplot drawn on top of it. The elements of the boxplot are the following:

- *Box*: The horizontal line in the center is located at the median: thus, half the data are above this line, half below. The bottom and top edges of the box are located at the first and third quartiles, which, with the median, divides the data into quarters: thus, half the data are inside the box, half outside; one-quarter below the box, one-quarter above; and each half of the box contains one-quarter of the data. Notched boxplots add intervals of confidence around the median as shown in the first plot of Figure 6.4.

- The *whiskers* of the boxplot, which are the bottom and top horizontal lines, are located in the following way. The upper whisker is drawn at the largest observation, which is less than $q_3 + 1.5 \times (q_3 - q_1)$, where q_1 and q_3 are the first and third quartiles, respectively. The lower whisker is located at the smallest observation that is larger than $q_1 - 1.5 \times (q_3 - q_1)$. Points larger or smaller than whiskers are considered *outside values*. If the data were from a normal

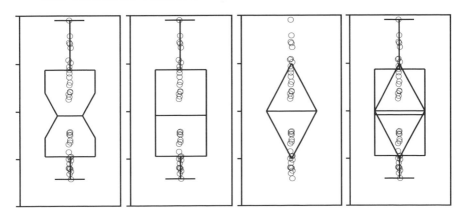

Figure 6.4 Notched boxplot, boxplot, diamond plot, and box + diamond plot.

193

distribution, such values would occur less than once in 100 times. Tukey also defined points that would fall farther away than $3 \times (q_3 - q_1)$ as *far-out values* but no indication of them is given in Figure 6.4.

As defined by Tukey (1977) boxplots only plot individually the outside or far-out values. We prefer, however, to show all the points because the dynamic-interactive capabilities of the plot would suffer if some of the points could not be selected or identified. Also, the information portrayed by the schematic can be better judged with the help of the individual points, as we will soon see.

The boxplot in Figure 6.4 shows that the *MPG* data are approximately symmetric, even though it is apparent that the upper whisker is somewhat longer than the lower whisker. We do not consider this difference large enough to declare that *MPG* is asymmetric. It is common, of course, to see variations in the boxplots, and in our experience it is normal to see small deviations like this for data that are normal. This point is demonstrated in Figure 6.5, which shows boxplots for 10 random normal variables set side by side with the boxplot for *MPG* (on the right).

Another aspect that can be addressed using the boxplot in Figure 6.4 is the presence of outliers. For *MPG* no outliers are apparent since no point extends beyond the whiskers. Therefore, if we limit our interpretation to the shape of the boxplot, we can conclude that we have a symmetric variable with no outliers. However, there are still other elements that can be explored in Figure 6.4.

We have pointed out before that the dotplot of *MPG* in Figure 6.2 shows a gap around the center of the distribution. The boxplot schematic shows that the median of the distribution is located exactly at this gap. This is a rather strange conclusion since it means that there do not seem to be any cars that have an average (as measured using the median) fuel use. Again, this suggests two modes in the data and, probably, two subgroups of cars that should be analyzed separately.

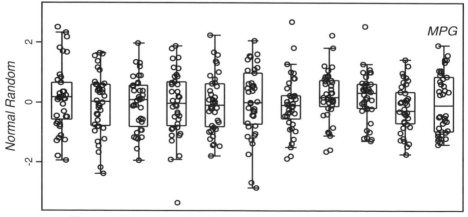

Figure 6.5 Boxplots for 10 normal distributions and for *MPG*.

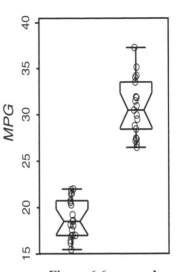

Grouped boxplots: Grouped boxplots are another aspect of boxplots that is discussed later. We will give you a teaser at this point, however, and show you a grouped boxplot for *MPG*. The plots that we have seen so far suggest that there are two subgroups of observations in our sample. We can separate the data into two groups, one with only the observations in the upper group, and the other with only the lower group's observations. We can then plot these two groups in a single window, showing two "grouped" boxplots, each including its own dotplot. The result is shown in Figure 6.6. Both boxplots still look rather symmetric when displayed by themselves and neither seems to have outliers. They certainly have different locations, and perhaps the upper group has a bit higher spread.

Figure 6.6 grouped boxplots of *MPG* data.

DIG: dynamic interactive boxplots. Dynamic interactive boxplots are most useful when we have more than one variable, as we will see in a later chapter. When we have but one variable, we can interactively identify the points with techniques discussed in Chapter 2 and compare the points to the information in the boxplot schematic. For example, using the same techniques as we used with dotplots, whose results are shown in Figure 6.2, we can see that the Fiat Strada is the car with the best *MPG* value. What we couldn't decide from the dotplot, but can from the boxplot, is that neither of these two observations can be considered to be outliers since neither one is shown by a point that is beyond the upper whisker.

Diamond plot. A diamond plot represents a variable's distribution by a schematic showing the data's mean and standard deviation, whereas the boxplot, as we have seen, uses the data's median and quartiles. The diamond plot for *MPG* is shown in the middle of Figure 6.4.

The diamond and box plots are quite similar. Both plots use measures of location and spread of a numeric variable to construct a schematic representation of the variable. Further, both plots use the location and spread in the same way to construct the schematic: In each plot, the measure of location is shown by a centrally located horizontal line, and the measure of spread is represented by the height of the schematic.

Diamond plots are necessarily symmetric by construction, but they can be used to judge symmetry when combined with a boxplot, as in the right-hand plot in Figure 6.4. For a normal symmetric distribution, the mean and median coincide. If data are positively skewed, the line for the mean is displaced to the upper part of the plot, with the reverse happening for negatively skewed data. Adding a diamond plot to a boxplot then becomes an interesting tool for exploring asymmetry in data. In Figure 6.4

we see that the line for the mean is somewhat above the line for the median, indicating positively skewed data. However, the difference is very small and we do consider it to be of much importance.

Notice that a dotplot with a boxplot and diamond plot simultaneously looks somewhat crowded. To be able to remove the schematics is a very welcome feature in these plots. In fact, clicking each one on and off is an effective way to use the interactive features of dynamic graphics to compare the locations of the mean and median.

6.3.3   Cumulative Distribution Plots

Cumulative distribution plots are a family of graphical techniques designed to help us determine the nature of the probability function that generated the data. These plots are commonly called *quantile plots* (Q-plots) or *probability plots* (P-plots). We can use a cumulative distribution plot to see if the data were generated by a normal process, by a chi Square process, and so on. The cumulative distribution plot shows the data plotted against a theoretical distribution in a way that will reveal a linear swarm of points if the data were generated by the theoretical distribution specified.

Normal probability plot. A normal probability plot (also called an NPplot) is the best known cumulative distribution plot, since it helps us assess whether a numeric variable is normally distributed. It does this by plotting the values of the numeric variable versus the score that would be obtained if the datum were generated by a normal process. If the resulting plot forms a relatively straight line, the values are normally distributed. If the line is not straight, the values are not normal.

The basic normal probability plot for *MPG* is shown in the left-hand graphic in Figure 6.7, which plots *MPG* on the vertical axis versus the values expected from a normally distributed process on the horizontal axis. The jagged line connecting the dots represents the variable's empirical distribution, while the straight line represents a

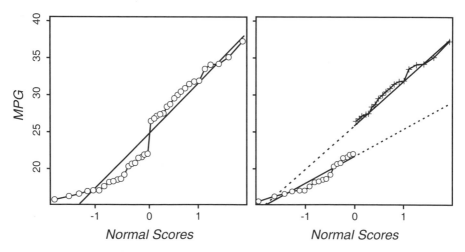

Figure 6.7 Normal probability plots for *MPG*.

normal distribution. Since it looks like the jagged line shows systematic departures from the straight line, we conclude that *MPG* is not generated by a normally distributed process. (We describe the right-hand figure momentarily.)

Of course, there can be various kinds of systematic departures from a straight line. Certain departures indicate certain kinds of nonnormality: For example, if the departure is fairly smooth with an observed line that is smoothly curved and is below the normal line, then the data are positively skewed (if it is above, the data are negatively skewed). If the data observed fit the line well except for a few observations, the data contain outliers. If the jagged line has horizontal segments, plateaus or gaps, the data may be discrete rather than continuous. If the jagged line seems to have several pieces that are linear, perhaps the data in the several pieces were generated by several different normal processes.

Quantile probability plot. The simplest probability plot is the quantile probability plot (usually called the quantile plot or Q-plot). A quantile plot for the variable *MPG* is shown in Figure 6.8. This plot represents a variable's distribution by plotting the value of a specific datum versus the fraction of the entire set of data values that is smaller than the specific datum. The resulting jagged line represents the variable's distribution.

The quantile probability plot is used to investigate the symmetry of a variable's distribution. For a symmetric distribution the points in the upper half of the Q-plot will stretch out toward the upper right in the same way that points in the bottom half stretch out toward the lower left. Again, the big gap in the center of the distribution is clearly visible. However, the two halves of the distribution seem to suggest that the distribution is symmetric.

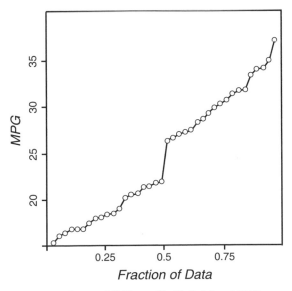

Figure 6.8 Quantile Pplot for *MPG*.

197

Generalized probability plot. Probability plots can be created for any given generating function, such as normal, chi Square, Weibull, and so on. In all cases, the plot shows the generating function's fitted values on the horizontal axis and the data on the vertical axis. Of course, for any probability plot the correlation can be computed between the generating function and the data observed to give us an indication of the fit of the generating function to the data. Probability plots for several alternative generating functions can be generated and a plot made of the resulting correlation coefficients. This helps us decide which generating function is most appropriate.

We can also make plots for the correlation coefficients of probability plots created for generating functions with shape parameters, showing the fit of the generating function model to the data as a function of the shape parameter's value. For example, with the chi-square generating function, we could make a plot of fit to the data as a function of the degrees of freedom of the chi-square distribution. This would help us decide on the optimal value of the shape parameter, in this case the degrees of freedom. Furthermore, estimates of the location and scale parameters of the generating distribution are provided by the intercept and slope of the fitted linear function.

DIG: subsetting observations. We can pursue several possible explanations using interactive dynamic graphics with the *MPG* variable. One possible explanation that is suggested by the swarm of points in the NPplot is that the data may consist of several linear pieces. With DIG, when we use our mouse to select a subset of data, information is added instantaneously to the graphic portraying the linearity of the subset selected: Depending on the specific software, we see a regression line, perhaps with residuals shown, and certainly with the measure of fit, in this case the correlation index. We can demarcate a specific subset of points, perhaps by coloring it or specifying a certain symbol for the points in the subset, all the while keeping the regression information for the subset. We can do this for multiple subsets of points, enabling us to see the linearity of the several subsets simultaneously, with all of the appropriate information being updated continuously as the cursor is moved over changing subsets of points. Thus, we can watch the information change dynamically, in real time, as we interact with the data.

The end result of such a process is shown on the right-hand side of Figure 6.7, where we see that two subsets have been identified, each with its own regression line. The result suggests that the data are fit remarkably well within each subset, suggesting that the data might have been generated by two separate normal processes. Perhaps, if we knew more about these data, we would discover something which would clarify what these processes might be. For the moment, we simply take this as a suggestion.

DIG: probability model selection. As mentioned above, probability plots can be created for any given generating function, of which the most appropriate are the distributions used for modeling, such as the normal, Cauchy, exponential, lognormal, Weibull, gamma, and Poisson functions. For any probability plot the correlation can be computed between the generating function and the data observed , giving us an indication of the fit of the generating function to the data. Probability plots for several alternative generating functions can be generated and the correlation coefficient can be calculated for each plot. Some of these functions are actually function families, with one or more parameters that control the specific details of the generating function. Since these details affect the shape of the generating function, the parameters are often called *shape parameters*.

To decide which generating function is most appropriate, we can make plots of the correlations obtained from each function versus the parameter values, comparing the fits for the various generating functions to see how the fit varies as the shape parameters vary. Some generating functions have two shape parameters, so for these we would have a surface plot showing the fit surface over the two-dimensional parameter space.

We can obtain a suggestion of how well other generating functions perform in comparison to the normal function as models of the generating process underlying our data. DIG methods can be used to search for the distribution that is the best generating function for any set of data, and can either show the data analyst the process of searching for the best function, as that process unfolds, or can provide the analyst with tools for doing the search himself or herself.

ViSta supports such interactive dynamic graphics, although we do not show them here. Two approaches are provided. With one of them, the data analyst can click a button to have ViSta generate a series of plots, watching a dynamic but not interactive "movie." With the other approach, the data analyst can "grab" the slider and interactive with the dynamic graphics directly, playing the "video game" version of the probability plot.

It is very important to mention, however, that the power of dynamic interactive graphics can easily be misused at this point. We should not mindlessly search for the best-fitting generating distribution. Rather, we need to have a theory that leads us to predict that a specific generating distribution is the one to consider. If that distribution is actually a family of distributions, we are justified in searching for the best possible member of the family. Thus, the probability plot can be used to determine if a given distribution provides a good fit to the data, to determine which distribution best fits the data, and to determine estimates of the location and scale parameters of the generating distribution chosen.

6.3.4 Histograms and Frequency Polygons

The histogram and frequency polygons are two closely related schematic representations of the distribution of a numeric variable. They can show us the center and spread of a distribution, can be used to judge the skewness, kurtosis, and modicity of a distribution, can be used to search for outliers, and can help us make decisions about the symmetry and normality of a distribution. Having all of these capabilities, it is no wonder that histograms and frequency polygons have a long history of use and development in statistics, always seeming to be the focus of much interest, among graphics developers and users alike.

Both histograms and frequency polygons represent the relative density of the sample distribution by counting the numbers of observations that fall into intervals of equal width. The intervals, which are called *bins,* are constructed by partitioning the range of the data. The number of observations in each bin is represented by histograms as bars with height proportional to frequency, whereas frequency polygons represent the same information by a polyline of connected dots. ViSta's default initial histogram and frequency polygon are shown in Figure 6.9.

Unfortunately, neither histograms nor frequency polygons provide a very accurate schematic of the data. In fact, they can present a seriously misleading picture of the distribution of a variable's values. However, when compared to other techniques, the histogram and frequency polygon are easy to understand and are easy to explain and communicate. They are particularly useful for understanding the concepts of a variable and a distribution. As a consequence, experts with advanced and thorough understanding of statistics use histograms when they address audiences without experience in statistics, even though the advanced statistician undoubtedly knows that histograms are not the best way of portraying the distribution of a variable's values. The problems with histograms derive from two of the decisions required for its construction: the bin width and the bin origin.

Bin width problem. The bin width problem is illustrated in Figure 6.10 for the data about *MPG*. In this figure we show six histograms, each with a different bin width. We can see that the main feature of the *MPG* variable, the gap in the center of the distribution, is visible in some of the histograms but not in all.

Bin width has an enormous effect on the shape of a histogram (Wand, 1996). Bins that are too wide produce an appearance of solid blocks. There will be no evidence of the gap that we have seen so clearly in the *MPG* data. Bins that are too narrow produce a jagged histogram with too many blocks. There will be almost no information about points that have a higher density of observations.

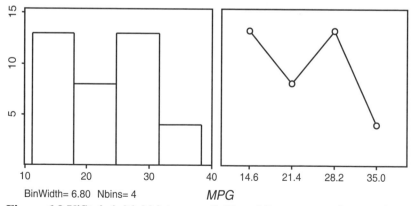
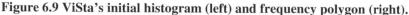

Figure 6.9 ViSta's initial histogram (left) and frequency polygon (right).

It is also the case that to produce nice tick marks, changing the width also changes the origin of the bins, producing an additional source of differences among the plots. Furthermore, scales on both axes must be modified to accommodate the bars in the space available.

In short, poor selection of bin width results in oversmoothing or undersmoothing of the histogram. This problem has been dealt with in the statistical literature, and a number of rules have been proposed that provide appropriate bin widths. We discuss some of these rules and we show that having the capability of interactively modifying the histogram can be useful to find values that are subjectively good.

Scott provides a historical account of the bin width selection problem (Scott, 1992). The earliest published rule for selecting bin width seems the be that of Sturges (1926) and amounts to choosing the bin width h as

$$\hat{h} = \frac{\text{range}}{1 + \log_2 n}$$

This rule is used as the default rule for many statistical packages even though it is known that it leads to an oversmoothed histogram, especially for large samples (Wand, 1996).

Scott (1992)) suggested the following rule

$$\hat{h} = 3.49\hat{\sigma}n^{-1/3}$$

This rule is known as the *normal rule* because it is tuned up to the normal distribution with a standard deviation of σ. This histogram is shown in Figure 6.10 in the fifth position, with four bins of width 6.80. Scott (1992) provided modifications of this

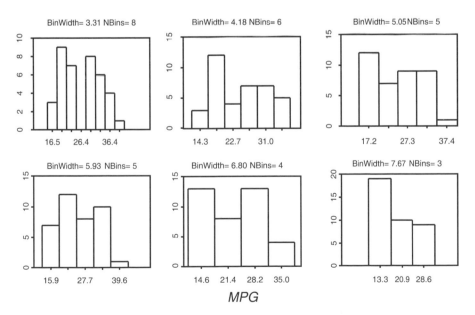

Figure 6.10 Series of histograms for *MPG*.

rule for distributions with varying skewness and kurtosis. Notice that this formula only gives the right answer to the bin width problem given an optimality criterion not discussed here. Other criteria could also be considered, and the data may not follow the normal distribution. Hence, exploring different versions of the histogram may be important to get a more thorough understanding of the data at hand.

Bin origin problem. In addition to the bin width problem, histograms suffer from what is known as the bin origin problem. The *bin origin* is the value at which the lowermost bin's lower boundary is located. Figure 6.11 displays four histograms with the same bin width but with slightly different origins. Notice that in this figure, just as in the preceding one, the main feature of the *MPG* variable, the gap in the center of the distribution, is visible in some of the histograms but not in all.

While the dependence of histograms on different bin widths is easy to comprehend, the dependence of the histogram on the origin of the bins is quite disturbing. Moreover, there does not seem to be any strategy available to cope with this problem. Our recommendation is to use kernel density estimation, an improvement over classical histograms that does not suffer from the bin origin problem. It is also a good idea to use dynamic interactive graphics to test different origins for the histogram as well as different bin widths. This gives one an understanding of the robustness of the features

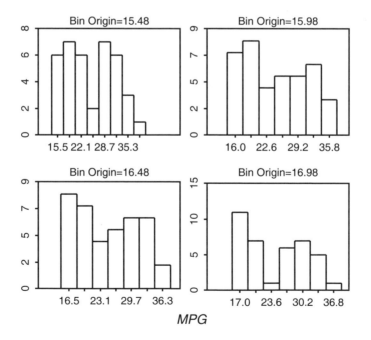

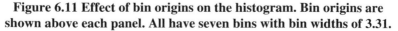

Figure 6.11 Effect of bin origins on the histogram. Bin origins are shown above each panel. All have seven bins with bin widths of 3.31.

of interest. We also recommend using the shaded histogram notion to see if it reveals anything of interest.

DIG: coping with histogram problems. ViSta provides a histogram that uses as a starting point the bin width given by the preceding formula. Using interactive tools, however, it is possible to explore other bin widths easily. Such an exploration of bin widths may give a more complete picture of the trade-offs required for the various choices. A slider is a particularly effective way to control the bin width of the histogram. Figure 6.10 was created through a selection of the histograms that an analyst could easily review using the slider (actually, only one of four plots is shown in Figure 6.10). Looking at these plots, it is easy to see why the histogram is not, nowadays, considered a very accurate tool for statistical analysis. The different histograms seem to tell different stories, not all of them coincident. Given that we know that there seems to be a gap in the middle of the data and that perhaps there are two separate distributions, we would choose the first plot left above. However, *not knowing what we know,* there does not seem any good reason to prefer one histogram over the others.

Fitting curves to the histogram. Curves for various functions can be fit to the histogram. These curves include the normal curve, shown on the left side of Figure 6.12, and the kernal density curve, shown on the right side.

Density traces. Density in data can be seen as the relative concentration of observations along the sections of the measurement scale (Chambers et al., 1983). Histograms can be considered a representation of density if they are modified slightly. The modification is to divide the frequencies by the number of cases so that the bars represent proportions or relative frequencies. Since the histograms would be redrawn on the new scale, this change does not alter their shape.

However, histograms are limited in that they inform about the density only in the middle of the interval and produce only one value for the entire interval. A more interesting approach would be to provide values that inform of the density at smaller

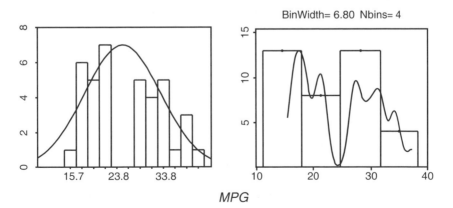

Figure 6.12 Histogram with normal (left) and kernal (right) curves.

intervals, while permitting the intervals to overlap. This is being done, conceptually, by the shaded histogram, but what we describe here is a mathematical way to attain the same goal. In this way, when we draw a line connecting the midpoint of each interval we will not see drastic changes like those we see with ordinary histograms.

The general formula for kernel density estimation is

$$\hat{f}_h(x) = \frac{1}{nh}\sum_i^n K\left(\frac{x - X_i}{h}\right)$$

where $\hat{f}_h(x)$ is the density estimated at point x given a size bin h, n is the number of cases in the data, $K(u)$ is a function denominated *kernel function,* and X_i are the sample data. An example of a kernal density curve is shown on the right side of Figure 6.12. The degree of smoothing in this formula depends on two factors:

- The bandwidth h, where shorter intervals will produce more local estimations and a rugged impression in the estimated curve, and longer intervals will produce smoother estimations.
- The kernel function $K(u)$, which is regarded as being a factor of less importance than the width h. These functions basically result in smaller values of $\hat{f}_h(x)$ for observations farther away from x and larger values for observations closer to x. The permissible functions vary in their weightings of the observations. As an example, a function commonly used is the *Gaussian:*

$$\frac{1}{\sqrt{2\pi}}e^{-(1/2)u^2}$$

Selecting an adequate bandwidth is generally seen as being more important than selecting an appropriate kernel. Selecting the best bandwidth can be performed using a slider connected to a display of the density. This permits the user to rely on intuition, interpretability, and aesthetics. In fact, the process of selecting bandwidth is probably as important as the final value chosen, since it serves to familiarize the user with the data.

Even though interaction techniques are well suited to the task of exploring for appropriate bandwidth values, we start with values that satisfy a given optimality criterion. An estimator that is commonly used is the *rule of thumb,* which in its robust version has the following formula (Silverman, 1986; Turlach, 1993):

$$\hat{h}_{\text{rot}} = 1.06 \, \min\left(\hat{\sigma}, \frac{\hat{R}}{1.34}n^{1/5}\right)$$

Determining a reasonable interval of values to explore is also a critical decision. Udina (1999) mentions the range $[h_0/9, \, 4/h_0]$ as being desirable.

Figure 6.12 shows a histogram that incorporates a curve indicating the density at the various points in the data. Notice that densities and frequencies of the histogram have normally very different scales, so displaying both in the same plot is usually meaningless. Scaling the densities to the frequency data is therefore necessary to visualize both simultaneously. This figure shows the optimal values for both histograms and kernel density traces computed according to the respective rules for optimal bin width. The kernel density trace uses the Gaussian as the kernel function.

The density trace in Figure 6.12 shows the two modes in the data with the big gap that we discussed previously. However, given the degree of smoothing chosen, these two modes are themselves composed by a succession of smaller bumps. Also, we see clearly that the histogram is oversmoothed compared with the density trace.

Figure 6.13 displays a selection of kernel density traces computed by varying the bandwidth. From left to right and up to down, the bandwidth is larger, with the consequence of producing smoother density traces. This results in smaller bumps going away and only two modes remaining visible. The final curve agrees with the previous interpretations carried out using other plots, but the smaller bumps are also of interest. For example, the first plot in the series shows a smaller gap set in the middle of two peaks for cars with a lower *MPG*. This gap distinguishes between the lower-low cars and the upper-low cars and can also be seen, for example, in the dotplot of Figure 6.2 and the boxplot of Figure 6.4.

Figure 6.14 shows a series of plots that keep the kernel density trace fixed but changes the histogram. As histograms are often the way used to present information of this type, it is of interest to find one that agrees with the impressions obtained using the kernel density. The final plot in the series represents quite well the two data modes that we have considered most appropriate for our data, and it could be used for nonexperts in statistics who know only about histograms. Notice that histograms do not have the property of kernel density traces that changes in interval width result in smooth variations of the display. Whereas kernel density traces always seem to produce the same answer but with different degrees of smoothing, histograms result in different answers at each step.

Shaded histogram. ViSta has the option to compute a shaded histogram, a computationally intensive method that may give a good picture of the sample's population distribution. The shaded histogram for *MPG* is shown in Figure 6.15 where we very clearly see the two separate distributions that we have inferred from previous investigations.

The idea for the shaded histogram came while watching the various histograms being displayed by the dynamic interactive graphical method described above. As you watch the series of histograms, you realize that some parts of the plot area are more likely than other parts to be "inside" the histogram. We took this idea and devised an algorithm to determine the probability that a pixel is "in" a histogram. These probabilities are rendered as gray values, darker grays for larger probabilities. The algorithm is as follows.

1.   Set the *Y*-axis to show proportions rather than frequencies.

2.   Determine the smallest range for the *X*-axis that contains all bins for all histograms.

3.   For a specifically sized plot, determine *n* and *m*, the number of row and col-

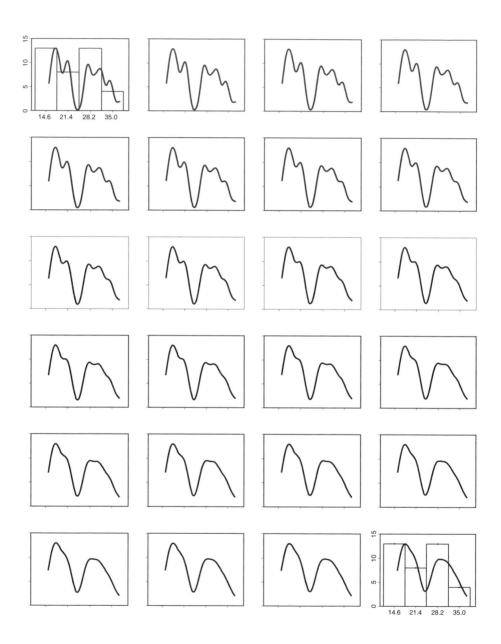

Figure 6.13 Modifying the bandwidth of the kernel density trace.

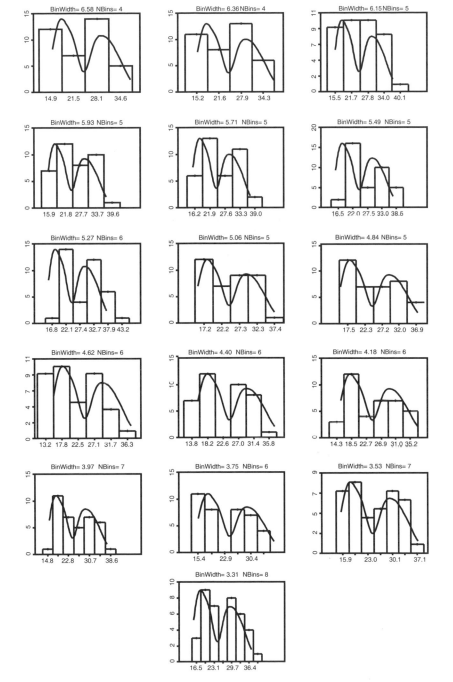

Figure 6.14 Histogram with kernel density trace.

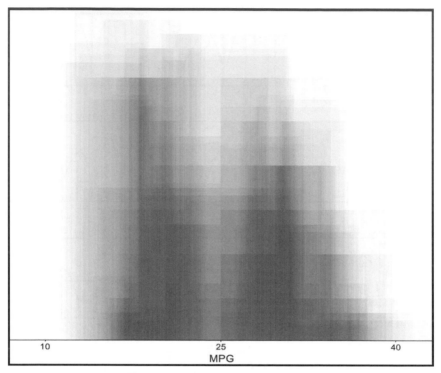

Figure 6.15 Shaded Histogram.

umn pixels of the rectangle formed by the two ranges.

4. Generate an $n \times m$ matrix of frequencies F where each element f_{ij} is the number of histograms covering the pixel.

5. Convert the frequencies to proportions p_{ij}. Render the shaded histogram by using the proportions for gray-scale values (e.g., the red, green, and blue values are $R_{ij} = G_{ij} = B_{ij} = (1\text{-}p_{ij})$).

6.3.5   Ordered Series Plots

Ordered series plots are plots of the values of the variable versus other information that is related to the order in which the data are arranged in the dataset. We briefly mention two plots, lag plots and sequence plots. As described here, these plots are simple versions of time-series plots, a type of plot not, unfortunately, otherwise covered in this book.

As pointed out in the NIST handbook (NIST/SEMATECH, 2004), these plots are useful for determining whether the values observed represent random samples from a population and whether the random sampling meets the standard conditions that are normally posited. For these types of data the plots should show no specific patterning.

If they do, it is the case that the assumption of random sampling, which underlies nearly all statistical testing, can be called into question.

Lag plots. A lag plot displays the values of the variable versus themselves, but with a *lag* of a specified number of observations. A lag of 1 means that each value is plotted against the value following it directly in the observation order. A lag of 2 means that each value is plotted against the value two places behind it in the observation order. This type of plot helps us check on whether a variable's values represent a random sample from a population. If it is such a sample, the plot should reveal no patterning.

Sequence plots. Sequence plots are simply a plot of a variable's values versus an index specifying the sequence order in which the values were observed. This plot has been proposed as a simple graphical summary of a variable's values. It is commonly assumed that these values represent a random sample from a population. This implies that there should be no shift in location or scale from one sample to the next, or, in other words, that the values should all have the same location and scale. Run sequence plots show such shifts, making them quite evident. They are also good for detecting outliers. We have shown examples of these plots in the visualization at the beginning of this chapter.

6.3.6 Namelists

A *namelist* is a "graphic" consisting of a list of names that can be linked with other graphics. For example, the namelist could be a list of observation labels. When it is linked with a plot with points corresponding to observations, such as a dotplot or boxplot, the namelist provides a way to identify individual points.

6.4   Visualization for Exploring Univariate Data

A univariate visualization of *MPG* is shown in Figure 6.16. The visualization consists of six graphs and a namelist. The graphs include examples of each of the univariate plots mentioned at the beginning of this chapter, including two ordered series plots (the sequence plot and lag plot in the upper row), a boxplot and the closely related dotplot, a cumulative distribution plot, and a histogram (which can be changed into a frequency polygon).

This visualization is constructed using the spreadplot technology developed by Young et al. (2003), which allows us to see several different graphs of the data simultaneously. The individual graphs each support dynamic interactive graphics and can be linked with each other via their observations. This means that when the user brushes or selects points or tiles or names in one plot, the corresponding points or tiles or names in the other plots are highlighted. Thus, for example, if we wish to see which observation is best in terms of *MPG* (i.e., has the highest *MPG* value), we can select it, say, in the dotplot, and see its label highlighted in the namelist and its label displayed in the other plots. This is what is happening in Figure 6.16, where we see that the brush cursor's rectangle is encompassing a point and that the point's label (Fiat Strada) is shown. Note that each of the other graphs also has a highlighted element labelled Fiat Strada. Of course, we can do the reverse: If we are interested in *MPG* for Fiat Strada, we can select it in the namelist and see where it falls in the other plots. Brushing the boxplot lets us see that the car that goes the farthest on a gallon of gasoline is the Fiat Strada, and the car that goes the shortest distance on that gallon is the Ford Country Squire Wagon.

The visualization shown in Figure 6.16 is a generalization of the four-plot concept presented in the NIST handbook (NIST/SEMATECH, 2004). The four-plot is a set of

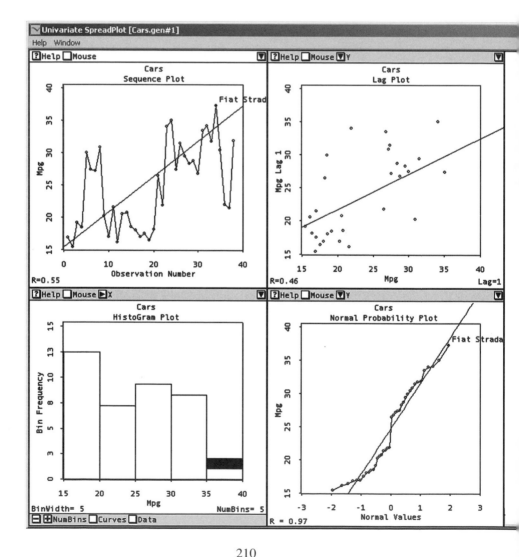

four plots that allow us to investigate the four fundamental assumptions underlying essentially all measurement processes. Although the four-plot is most appropriate for time series data, its usefulness extends to essentially any kind of numeric data. A four-plot consists of a run-sequence plot, a lag plot, a histogram, and a normal probability plot, plots that allow us to investigate the assumption that each value of a variable is generated by a process that behaves like (1) a random sample, (2) from a fixed distribution, (3) with a fixed location, and (4) a fixed variation.

If the four assumptions above hold, the variable will support probabilistic predictability. That is, the variable will permit legitimate probability statements about the

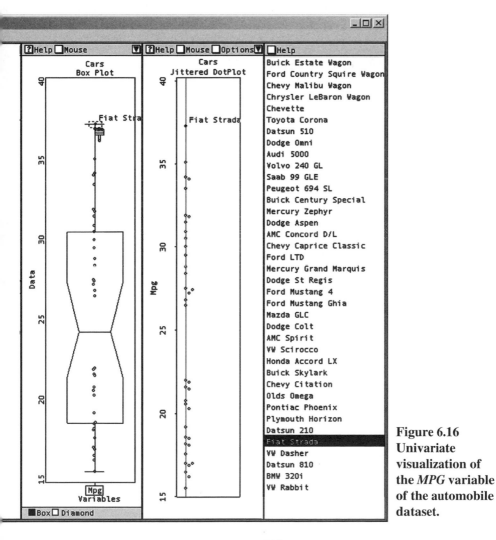

Figure 6.16 Univariate visualization of the *MPG* variable of the automobile dataset.

future as well as about the past. Since the validity of any conclusions we draw from a variable depends on the efficacy of the variable with regards to these four assumptions, it allows us to test these four assumptions. Since the graphics make no assumptions of their own, they provide an ideal test mechanism. And since the four-plot addresses the four underlying assumptions directly, it is the ideal way to do this.

The visualization shown in Figure 6.16 is an augmented four-plot. As originally proposed, the four-Plot is formed by the four square plots that constitute the left-hand two-thirds of Figure 6.16. The boxplot, dotplot and namelist taking up the right-hand one-third of Figure 6.16 are the augmentation. The additional boxplot and namelist enhance the basic four-plot by adding basic DIG capabilities of selecting, labeling, coloring, and so on. We interpret the four-plot as follows for our *MPG* data.

1. *Random sample*. If the assumption that *MPG* contains values that constitute random samplings, the lag plot should show no structure. This is decidedly not the case, as there is a gap in the middle of the distribution.

2. *Fixed distribution*. If the assumption that *MPG* contains values that constitute samples from the same fixed distribution were to be met, we would not expect to find a normal probabililty plot with a gap and with what looks like two linear sections.

3. *Fixed location*. If the assumption of fixed location holds, the run-sequence plot (upper left) will be flat, showing no drift, and the regression line it contains will be horizontal, with a correlation of zero. This is clearly not the case.

4. *Fixed variation*. If the assumption of fixed variation (scale) holds, the sequence plot will have the same vertical spread everywhere. This is clearly not the case.

Thus, our *MPG* variable does not seem to be a random sample from a fixed distribution with a fixed location and a fixed variation. Of course, given the nature of this variable, we should not worry about this. However, if we found that this variable is related in some way with a sequence, we should proceed to analyze this structure with more detail.

6.5 What Do We See in *MPG*?

Exploration of *MPG* reveals that the normal distribution is not a good model of this variable's distribution. Moreover, several of the plots, especially the histogram's kernal density function and the shaded histogram, show that there are two peaks in this variable's distribution. The first peak corresponds to a concentration of cars that have the poorest performance, whereas the second peak is for cars with better performance. Also, it looks like there may be a normal distribution around each peak.

Considering all of the evidence, it seems plausible that the cars come from different groups or populations. We could explore this possibility by selecting the groups in the boxplot and saving this information in a new datafile. The groups could be compared with the rest of the variables to see if the differences are present among all of them.

But before that, in Chapter 7 we examine one of the most obvious explanations of what we see here, which is the relationship of the variable *MPG* to the weight of the cars.

7 Seeing Bivariate Data

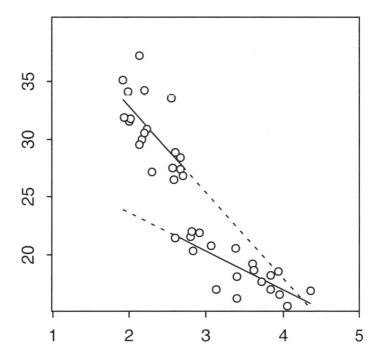

7 Seeing Bivariate Data

As we noted at the beginning of the Chapter 6 magnitude data are clearly the most common type of statistical data, the reason being that they provide the strongest basis for scientific investigation. It follows that there are more data analysis tools, including more visualization tools, for magnitude data than for other types of data. Due to the large number of such tools, we have divided our coverage of them into several chapters. In chapter 6 we covered univariate visualization methods—methods that produce plots of a single variable of magnitude data. This chapter covers bivariate methods — including methods that involve multiple bivariate plots—of bivariate magnitude data. The Chapter 8 covers multivariate methods.

7.1 Introduction

The fundamental fact about several observations of two variables is that the values of the two variables may have some sort of relationship. The material in this chapter helps us understand this relationship.

7.1.1 Plots About Relationships

Scatterplot. A scatterplot reveals the strength and shape of the relationship between a pair of variables. A scatterplot represents the two variables by axes drawn at right angles to each other, showing the observations as a cloud of points, each point located according to its values on the two variables. Various lines can be added to the plot to help guide our search for understanding.

The scatterplot is the most important of the three plots we consider in this chapter: It can reveal distribution shape, show strength of relationship, reveal trends, show clusters, uncover outliers or isolates, and reveal patterns. The scatterplot is well complemented by the comparison plot, since it can help us decide if the two variables represent samples from distributions with similar or different shapes. The parallel-coordinates plot can sometimes reveal additional information that cannot be gleaned from the other two plots, although it is seldom the source of insight, in our experience.

Consider panel A of Figure 7.1, which shows a small example dataset on the left, with a schematic of its scatterplot next to it. The scatterplot uses the standard Cartesian coordinate representation of a data matrix: The columns of the matrix are represented as dimensions of the scatterplot, and the rows as points in the scatterplot. Thus, the data elements are the coordinates of points in a plane. For example, the first row of data, which is $[-1.8, -3.1]$ is represented by point a in panel A. If you project the point vertically onto the X-axis, the projection arrives at the value -1.8, as is shown by the vertical arrow. Similarly, point a projects horizontally onto the Y-axis at -3.1, as shown by the horizontal arrow.

Distribution comparison plot. The distribution comparison plot shows whether two variables come from distributions which are similarly or differently shaped. Like the scatterplot, the distribution-comparison plot represents the two variables by axes drawn at right angles, but the values on each variable are sorted into order and plotted as a line of connected dots. A straight line is added to guide our interpretation.

Scrutiny of the comparison plot schematic in panel B of Figure 7.1 reveals that the location of its X values are in fact the X values of the scatterplot after having been sorted into order. Inspection of the figure reveals that the same is true for the Y values. Then, if one takes the two sorted values to be the coordinates of points in a plane, one obtains the configuration of dots that is shown in the "comparison" figure. You can also reach the same conclusion if you compare the X column of the data matrix and the X column of the sorted data matrix, and also to compare the Y columns of the two matrices. You will see that the comparison plot involves replotting the coordinates of the scatterplot after they have been sorted into order on each axis, giving the new configuration of dots shown in the comparison plot schematic.

Note that the new "dots" are not associated with a specific observation; rather, their horizontal position is associated with one particular observation, and their vertical position is associated with (what is usually) a different observation. So the lower-left point has two identifying labels; a and b, since its horizontal coordinate came from point a and its horizontal coordinate from point b. The a is written vertically to emphasize that it is the label of the observation contributing the vertical (Y) coordinate. Because of this loss of correspondence with the original data-points, the comparison plot is often just shown as a line, with no points. However, we choose to show the plot as connected points to take advantage of our dynamic graphics capability of linking the plot with other plots via the points. You may be familiar with the comparison plot but know it by a different name, since the name we use, although more descriptive, is not the standard name. The comparison plot described here is a special case of the slightly more general quantile–quantile (QQ) plot.

Parallel-coordinates plot. The parallel-coordinates plot (originated by Inselberg, 1985) shown in panel C of Figure 7.1 represents the two variables by axes that are parallel rather than by axes that are perpendicular, as is done by the other two plots. Each observation is represented by a line segment instead of a point. The value of an

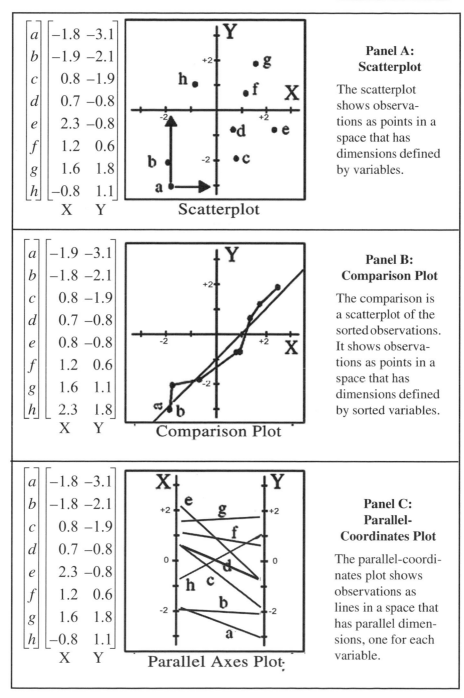

Figure 7.1 Schematics of three bivariate plots.

observation on each variable is used to locate the endpoints of the observation's line segment. Panel C of Figure 7.1 shows how our little dataset is represented by the parallel-coordinates plot. Careful consideration of the schematic reveals that the right end of each line segment is located at the same position on the *Y*-axis as is the point in the scatterplot schematic shown in panel A of Figure 7.1. Similarly, the left end of each line segment is located at a position that corresponds to the *X* position of the point in the scatterplot.

Let's consider the general similarities and differences between the three plots. All three plots represent the data by a two-dimensional arrangement of plotting elements, with dimensions corresponding to variables and the plotting elements corresponding to observations. For both the scatterplot and the distribution comparison plot, the dimensions are drawn at right angles, whereas for the parallel dimensions plot the dimensions are drawn in parallel, as the name suggests. Both the scatterplot and the parallel-coordinates plot portray the observed values of the *X* and *Y* variables by locating the plot symbols on the dimensions according to the observation values: For the scatterplot the symbol is a point, whereas for the parallel-coordinates plot the symbol is a line. In contrast, the comparison plot portrays the values of the *X* and *Y* variables by positioning the ends of a line segment according to the order of their sorted values.

We do not spend much time here with the parallel-coordinates plot, because it is seldom useful with bivariate data and contributes little to this chapter. You can imagine, however, that it is very easy to extend the definition to the multivariate situation, simply by adding additional axes, all drawn in parallel, to the figure. We will see in Chapter 8 that the high-dimensional parallel-coordinates plot can be very useful.

   Guidelines. While the information revealed by these plots can be clear, it is often the case that we are left in some doubt as to just what we can see. There are, therefore, various tools that have been developed to help us to see more clearly the more subtle effects. These tools, which we call guidelines, are functions that add reference lines to the scatterplot that show what a particular type of effect would look like if it were present. These functions often have parameters that can be manipulated by dynamic graphical techniques, permitting us to see how well the function describes the data. There are often alternative functions that may be tried, giving us several different possible explanations. It is also often possible to plot the residuals of the fit of the function to the data, which may provide additional insight.

7.1.2 Chapter Preview

We begin the chapter by introducing the data we use throughout. Then we review the bivariate plots that are useful for bivariate visualization. Our discussion emphasizes the scatterplot and the guideline tools available to help us interpret the scatterplot, since the scatterplot and its guidelines can tell us a lot about the relationship between two variables. We then turn to bivariate visualization spreadplot methods, including spreadplots for bivariate visualization and for simple regression analysis. Then we

discuss plot matrices and other closely related multibivariate visualization tools, concluding with a presentation of dynamic visualizations for Box–Cox transformations.

7.2 Data: Automobile Efficiency

We continue to use the automobile data described in Chapter 6. You will recall that these data are about 38 automobiles sold in the United States during 1978–79. In Sections 7.3 and 7.5 we look at two variables: *MPG*, the measure of fuel efficiency we focused on in Chapter 6, and *Weight*, a measure of the weight of the automobile. We choose these two variables because we anticipate a reasonably clear but nonlinear relationship between them: The more the automobile weighs, the less efficient (fewer miles per gallon) it should be. Since all variables are used in Section 7.4, the entire dataset is shown in Table 7.1. The variables are described as follows:

MPG. As we pointed out in Chapter 6 the fuel efficiency of automobiles is measured using miles per gallon (*MPG*), the standard measure used in North America. *MPG* indicates the distance, measured in miles, that the automobile can travel on a single gallon of gasoline. The metric equivalent would be the number of kilometers a car can travel on 1 liter of gasoline (km/L). The conversion is 1 *MPG* corresponds to 0.425 km/L. Notice that km/L is the reciprocal of the L/km (litres per kilometer) measure commonly used in most of the world. Thus, 1 *MPG* corresponds to 2.35 L/km.

Weight. The weight variable is defined in terms of the standard unit of weight used in North America, which is the pound (1 pound = 2.2 kg). The variable is actually defined in units corresponding to 1000 pounds (a kilo-pound or klb). Thus, the first automobile in the data (the Buick Estate Wagon) weighs 4360 lbs (4.36 klb).

The additional variables shown in Table 7.1 are:

Drive ratio. The drive ratio of an automobile is the number of times the wheels revolve per revolution of the engine in high gear (without overdrive). The larger the drive ratio, the less efficient the engine.

Horsepower. The horsepower of the engine, a measure of power of an engine, may be measured or estimated at several points in the transmission of the power from its generation to its application, giving somewhat different measurements, but in all cases, higher horsepower means lower fuel efficiency;

Displacement. The displacement of the engine is defined as the total volume of air/fuel mixture an engine can draw in during one complete engine cycle; it is normally stated in cubic inches, cubic centimeters, or liters. Our measure is in cubic inches, which if multiplied by 16.38 provides cubic centimeters, and if then divided by 1000 yields displacement in liters. All other factors being equal, a larger-displacement engine is more powerful and less efficient than a smaller engine.

Number of cylinders. The meaning of number of cylinders should be obvious. We

221

Table 7.1 Complete Datasheet for the Automobile Data

	Weight	MPG	DrRatio	HrsPwr	Dsplcmnt	Cylinders	Country
Buick Estate Wagon	4.36	16.9	2.73	155	350	8	U.S.
Ford Country Squire Wagon	4.05	15.5	2.26	142	351	8	U.S.
Chevy Malibu Wagon	3.61	19.2	2.56	125	267	8	U.S.
Chrysler LeBaron Wagon	3.94	18.5	2.45	150	360	8	U.S.
Chevette	2.16	30.0	3.70	68	98	4	U.S.
Toyota Corona	2.56	27.5	3.05	95	134	4	Japan
Datsun 510	2.30	27.2	3.54	97	119	4	Japan
Dodge Omni	2.23	30.9	3.37	75	105	4	U.S.
Audi 5000	2.83	20.3	3.90	103	131	5	Germany
Volvo 240 GL	3.14	17.0	3.50	125	163	6	Sweden
Saab 99 GLE	2.80	21.6	3.77	115	121	4	Sweden
Peugeot 694 SL	3.41	16.2	3.58	133	163	6	France
Buick Century Special	3.38	20.6	2.73	105	231	6	U.S.
Mercury Zephyr	3.07	20.8	3.08	85	200	6	U.S.
Dodge Aspen	3.62	18.6	2.71	110	225	6	U.S.
AMC Concord D/L	3.41	18.1	2.73	120	258	6	U.S.
Chevy Caprice Classic	3.84	17.0	2.41	130	305	8	U.S.
Ford LTD	3.73	17.6	2.26	129	302	8	U.S.
Mercury Grand Marquis	3.96	16.5	2.26	138	351	8	U.S.
Dodge St Regis	3.83	18.2	2.45	135	318	8	U.S.
Ford Mustang 4	2.59	26.5	3.08	88	140	4	U.S.
Ford Mustang Ghia	2.91	21.9	3.08	109	171	6	U.S.
Mazda GLC	1.98	34.1	3.73	65	86	4	Japan
Dodge Colt	1.92	35.1	2.97	80	98	4	Japan
AMC Spirit	2.67	27.4	3.08	80	121	4	U.S.
VW Scirocco	1.99	31.5	3.78	71	89	4	Germany
Honda Accord LX	2.14	29.5	3.05	68	98	4	Japan
Buick Skylark	2.67	28.4	2.53	90	151	4	U.S.
Chevy Citation	2.60	28.8	2.69	115	173	6	U.S.
Olds Omega	2.70	26.8	2.84	115	173	6	U.S.
Pontiac Phoenix	2.56	33.5	2.69	90	151	6	U.S.
Plymouth Horizon	2.20	34.2	3.37	70	105	4	U.S.
Datsun 210	2.02	31.8	3.70	65	85	4	Japan
Fiat Strada	2.13	37.3	3.10	69	91	4	Italy
VW Dasher	2.19	30.5	3.70	78	97	4	Germany
Datsun 810	2.82	22.0	3.70	97	146	6	Japan
BMW 320i	2.60	21.5	3.64	110	121	4	Germany
VW Rabbit	1.93	31.9	3.78	71	89	4	Germany

do not include this variable in our analyses here. Note that it can be treated as numerical or categorical.

7.2.1 What the Data Seem to Say

So what will we see in these data? Let's take a quick look, even though we are getting ahead of ourselves. Of course, we haven't yet explained what these graphics are about. But you can rely on your visual intuition to see what there is to be seen. And rest assured that in the rest of this chapter we will be showing you the details of how to use these graphics.

One of the most important graphics in this chapter is shown in Figure 7.8. We reproduce it here as Figure 7.2, so that we can show you what we conclude from it. The

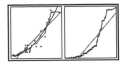

graphic is a scatterplot of *Weight* versus *MPG*, with several guidelines added to aid our understanding. We explain the details in the next section.

Our first impression is that there is a strong negative relationship between weight and efficiency: The heavier cars are less efficient, just as we would expect. But let's look at the relationship in more detail. When we do, we notice that there is a gap in the *MPG* variable: There just are no middle-efficiency automobiles. There are autos above 26 *MPG* and autos below 22, but none in between. So, let's say that we have heavy cars and light cars. Of course, the heavy ones are less efficient than the light ones.

When we study the relationship between weight and efficiency for the heavy cars, we see that it is different from the relationship for light cars. In the plot this is shown by two different regression lines (the straight lines that are partially solid, partially dashed), the one for the lightweight cars being steeper than the one for the heavy cars. So for lightweight cars, changing their weight has greater impact on efficiency that

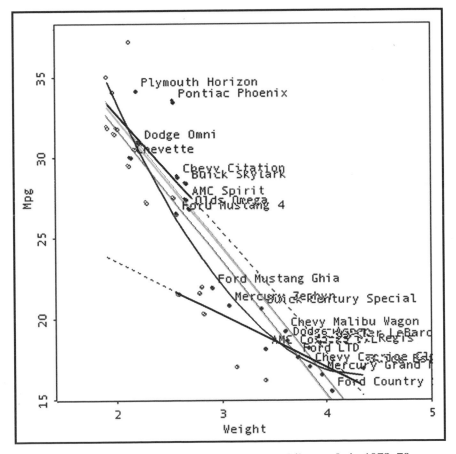

Figure 7.2 *MPG* and *Weight* of 38 automobiles made in 1978–79.

for heavy cars. This may seem counterintuitive, but it's telling us that if we build a lightweight car and can shave off another 100 kg or lb of weight, we will improve the efficiency of the automobile by a greater amount than if we did the same thing for a heavy car. We note that rather than modeling this effect with two separate linear regressions, we can model it with one quadratic regression, represented by the curved line shown in the figure. We can't be sure which is right, and would need to look further, or develop theory to draw a conclusion.

Finally, we have selected and labeled the cars that were manufactured in the United States, and we notice that they are consistently more efficient than we would expect. They are above the quadratic regression line, meaning that the quadratic regression underpredicts their efficiency as a function of weight. They are more efficient than we expect. Another way of seeing this is shown by the two essentially parallel gray lines. One of these lines (the lower one) is the overall regression between *Weight* and *MPG*, whereas the upper one is the regression for *Weight* and *MPG* for the American cars only. We see that they are a bit more efficient, for their weight, than cars in general: An American car of a specific weight, which was built in 1976, could go about 1 mile further on a gallon of gasoline than cars in general of the same weight. This also is an unexpected result.

7.3 Bivariate Plots

As mentioned at the beginning of this chapter, the two main bivariate visualization methods are the scatterplot and the distribution comparison plot. The parallel-coordinates plot and parallel boxplot can also be used. We discuss all four plots next.

7.3.1 Scatterplots

The scatterplot shows the strength and shape of the relationship between two variables. Two scatterplots are shown in Figure 7.3. Each scatterplot shows the *MPG* and *Weight* variables. In each plot the variables are represented by the *X*-axis (drawn horizontally) and the *Y*-axis (drawn vertically), and the observations are represented by the points in the scatterplot, each point being located according to the observation's values on each of the two variables. These values can be approximately determined from the plot by seeing what value of the *X*-axis is below the point and what value of the *Y*-axis is to the left of the point.

Sometimes, we think of two variables as though one is caused (at least partially) by the other, sometimes we do not, depending on the details of the empirical situation. When we do think there is a causal relationship, we commonly call the *caused* variable "dependent," "response," or "effect," and we call the *causing* variable "independent," "predictor," or "factor." Of course, when we believe causality, our goal is to study how the response depends on the predictor. In this case, it is conventional to put the response variable on the *Y*-axis and the predictor variable on the *X*-axis. In those cases when neither variable is thought of as being effect or response, the goal is to understand how they are related, and either variable can be on either axis.

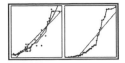

For our example of *Weight* and *MPG*, we would consider the *Weight* variable to be the predictor and the *MPG* (efficiency) variable to be the response, since it is well known that the distance that a car can travel on a given quantity of fuel is at least partially determined by the weight of the car, with heavier cars needing more fuel than lighter cars to travel the same distance. That is, it is reasonable to think of increased weight causing decreased efficiency, and not the other way around. Thus, *Weight* should appear on the *X*-axis and *MPG* on the *Y*-axis, which is the orientation shown in the right image of Figure 7.3.

7.3.1.1 What we can see with scatterplots. It is important to pay attention to the first impression one gets from a scatterplot. For the plot showing the relationship between *MPG* and *Weight*, shown as the right image of Figure 7.3, the first impression is that the weight of the cars is inversely related to their efficiency and that the relationship is very strong. It appears that the relationship is nearly, but not quite linear, there being a slight curve in the trend.

The scatterplot can show us information about the shape, strength, direction, and density of the relationship between the two variables. It can also tell us about skedasticity and about outliers and isolates. Many of these are illustrated in Figure 7.4, where we have plotted artificially generated data to illustrate various features.

Shape. The shape of the relationship between two variables refers to the overall pattern of that relationship. Some of the more common shapes of the many that we may encounter are shown in Figure 7.4. All but the last of these scatterplots are based on simulated data where the *y*-values are related to the *x*-values by an equation. The actual *y*-values represent error-containing samples from a distribution whose values are calculated the *x*-values by an equation that is (a) linear, (b) quadratic, (c) cubic, (d) exponential, or (e) sinusoidal. Each graph has a title indicating which equation defines the relationship between *X* and *Y*. The equation itself is shown as the label of the *Y*-axis. The line in each figure is the actual equation; the points represent the error-containing samples from the indicated relationship. Plot (f) is an empirical scat-

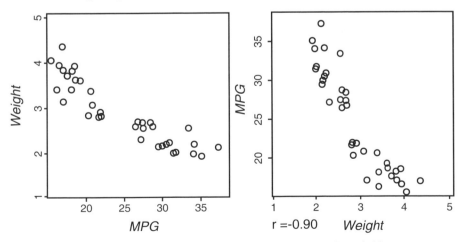

Figure 7.3 Scatterplots for the *Weight* and *MPG* variables.

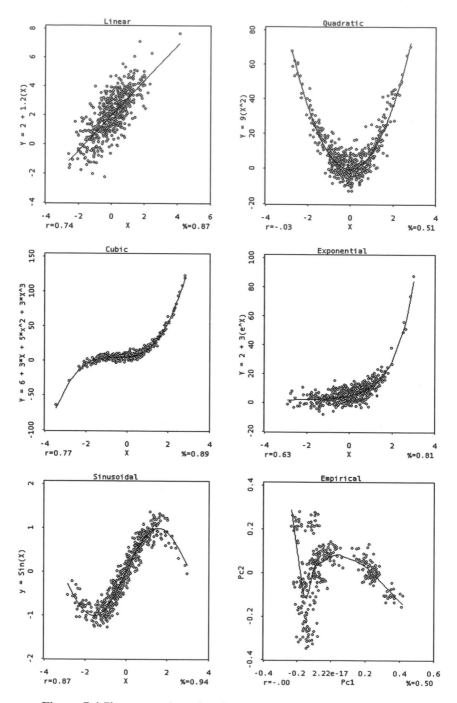

Figure 7.4 Six scatterplots showing various types of relationships.

terplot, based on real data. Here, the line that is drawn is a Lowess smoother, a type of running local average. We would describe its shape as irregular and clump. The shape of the relationship between *Weight* and *MPG* (Figure 7.3) is linear, with a hint of curvilinearity. Note that many statistical tests require linearity.

Strength. The strength of the relationship concerns how close the points in the figure are to their presumed underlying error-free functional relationship. The strength may be described by words such as *none, weak, moderate, strong*, or *very strong*, or it can be summarized quantitatively by computing the correlation (or a nonlinear measure of association such as η^2) between the two variables. The top two plots show a fairly strong relationship, the cubic relationship is very strong, whereas the next two are reasonably strong. The strength of the relationship between *MPG* and *Weight* is very strong.

Of course, we can use the coefficient of correlation to summarize the strength of relationship between our two variables. This value, which is shown in the lower-left corner of each plot, tells us the strength of *linear* relationship between the values observed X and Y. The most straightforward interpretation of r is to interpret R^2, which tells us the proportion of variation in the response that is predicted by the predictor, assuming a *linear* relationship between the two measures.

We can see in Figure 7.3 that vehicle weight predicts 81% (= 0.90^2) of the variation in fuel efficiency, assuming that their relationship is linear. But when we look at Figure 7.4 we can see that the assumption of linearity underlying the calculation of the correlation coefficient can make its value very misleading. The quadratic relationship, which we know to be strong, has a linear correlation coefficient of zero.

Direction. This concept applies unambiguously only to monotonic (strictly increasing or decreasing) relationships, which, of course, include linear relationships. A relationship is positive (or increasing) if as X increases, so does Y. It is negative (or decreasing) if as X increases, Y decreases. The top-left plot displays a positive direction. We can also apply the concept without much difficulty to the cubic and exponential (center) figures, both of which have a positive direction. We cannot, however, apply it to the bottom two figures. The relationship between *Weight* and *MPG* in Figure 7.3 is negative or decreasing.

Density. The mathematical concept of density (which we refer to here, and which is not the same as the statistical concept of density) concerns the relative proximity of the points in the point cloud. If it is the case that the relative proximity remains the same throughout the point cloud, the density is said to be "everywhere equally dense." Alternatively, it is often the case that the density gradually tapers off as we proceed from the center to the edge of the cloud, creating what is known as a *tapered density*. Finally, it is not uncommon to see scatterplots where the density is very clumpy. In Figure 7.3 the five simulated distributions are equally dense; the empirical one is clumpy.

Many commonly assumed distributions (normal, for example) are smoothly tapering or equally dense, not clumpy, so clumpiness is a warning that standard assumptions may not apply and that multiple generating functions may be at work. We may wish to form groups of points that seem to form a clump and keep our eye on it as we progress through the analysis.

Outliers/isolates. We should always look for outliers, as they may represent interesting or erroneous data, as discussed in Chapter 5. Note that it is possible to have points that do not appear to be outlying on either of the two variables when they are looked at individually, but will be clearly seen as being outliers when the two variable are looked at together. The points at the top of the cubic and exponential distributions may be outliers. Outlying points can drastically affect the results of analyses, and should be carefully attended to before further analyses are performed.

Skedasticity. *Skedasticity* refers to whether the distributional relationship between two variables has the same variability for all values of X. A distribution which does have this characteristic is called *homoskedastic*, one that does not is called *heteroskedastic*. Some statistical tests require homoskedasticity. The linearly based relationship looks homoskedastic, whereas the others do not.

7.3.1.2 Guidelines. The task of detecting trends in scatterplots can be enhanced by adding lines of various kinds, These lines, which we call guidelines, since they help guide our interpretation of the plots, include the principal axis line, two different kinds of regression lines (one linear and the other monotonic), and two families of smooth lines (Lowess smoothers and kernal density smoothers).

Principal axis. The principal axis line (left panel of Figure 7.5) shows us the linear trend between the X and Y variables. It is the "longest" direction in the plane: It is the (straight) line such that when the points are projected onto it orthogonally, their projected values have the maximum possible variance. There is no other direction in the space which when the points are projected orthogonally onto the line for that direction, would have a greater variance. The principal axis line is also the line that has the shortest possible set of residual lines (these are the short lines drawn at right angles — orthogonal to—the principal axis). There is no other direction through the space that

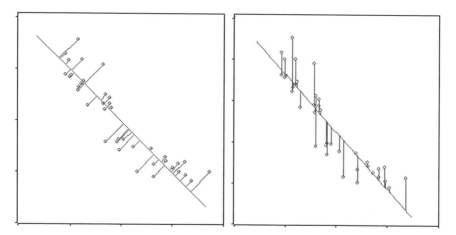

**Figure 7.5 Left: principal axis line. Right: regression line.
Each line is shown with its residual lines (the short lines).**

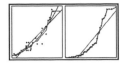

has a smaller sum of squared lengths of the residual lines (i.e., it is the best fitting line). It is the line that is the best linear summary of the *XY* plane.

Regression Lines: Regression lines provide information about how the *Y* variable responds to the *X* variable. These lines only make sense when we believe that the values on the *Y*-axis variable are a response that depends, at least in part, on the values of the *X*-axis variable.

Linear regression. The linear regression line (right panel of Figure 7.5) shows us the least squares linear prediction of the *Y* variable from the *X* variable. This is not the same line as the principal axis, except in unusual circumstances. The regression line maximizes fit to the dependent variable only, whereas the principal axis maximizes fit to both variables simultaneously. Whereas the principal axis measures fit orthogonally to the line (and thus uses information about lack of fit to both *X* and *Y*), the regression line measures fit vertically, only using information about lack of fit to *Y*. This difference is shown in the figure by the differently oriented residual lines: orthogonal to the principal axis in the one case, and vertical in the other case.

Quadratic regression. The quadratic regression line (lower-left panel of Figure 7.6) shows us the least squares quadratic prediction of the *Y* variable from the *X* variable. This regression is done by constructing a variable which is the squares of the *X* values and then fitting *Y* with a linear combination of X and X^2.

Cubic regression. The cubic regression line (lower-right panel of Figure 7.6) shows us the least squares cubic prediction of the *Y* variable from the *X* variable. This regression is done by constructing two variables, one of which is the squares of *X* and the other the cubes, and then fitting *Y* with a linear combination of X, X^2 and X^3.

Monotone regression. The monotonic regression line (upper-left panel of Figure 7.7). is the line that shows the order-preserving transformation of the *X* variable that has the maximum least squares fit to the *Y* variable. Monotone regression may degenerate into a *step function*, resembling a staircase. In such a case the function may be artifactual or may represent overfitting of the *X*-variable to the *Y*-variable.

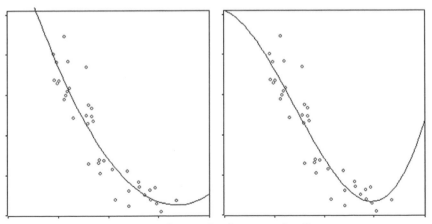

Figure 7.6 Left: quadratic regression line.
Right: cubic regression line.

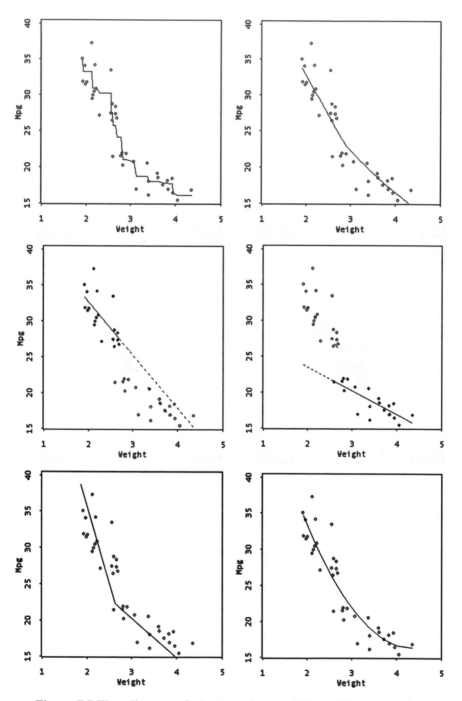

Figure 7.7 Five glimpses of what may be two different linear trends, plus one glimpse suggesting a single nonlinear trend.

Subset regression: Subset regression produces separate regression lines for subsets of the data. The subsets can be based on category variables or on any other groupings the user wishes to make. These groupings can be indicated by point color, point label, or by the name or shape of the object used to represent the point.

Smoothers. Smoothers, including Lowess smoothers and kernel density smoothers, provide us with an approximate idea about where the vertical center of the points is for each part of the scatterplot as we move horizontally from left to right along the *X*-axis. They are, loosely stated, a kind of running weighted regression, where a value of *Y* is computed for each value of *X* by using a weighted regression where nearby observations are most heavily weighted.

There are two commonly used smoothers: Lowess smoothers and kernel density smoothers. Each has a parameter that must be manipulated to search for the specific member of the family that seems to be the best. Dynamic graphical techniques can be used for this search.

Lowess is an acronym for *LOcally Weighted regression Scatterplot Smoothing*, a method designed for adding smooth traces to scatterplots that works basically by selecting a strip of neighbors to each Y_i value and then using them to predict Y_i'. The process is iterative, so several passes are necessary to compute the values of the Lowess fit. There is a parameter for Lowess that you can control. It is the proportion of cases included in the strip. Narrow strips means that the smooth follow minor wiggles in the data. Wider strips provide a smooth trace that changes only gradually. The weight versus residuals plot used a strip of 0.75. In general, what we try with smoothers is to balance smoothness versus goodness of fit. A smoother that follows every spike in the data is not what is normally desired.

A *kernel density smoother* is also a locally weighted regression method which employs a kernel function to generate weights for a series of weighted regression analyses, one performed for each observed value of *X*, each generating an estimate of *Y* for that value of *X*. The kernel function is usually a symmetric probability function such as a normal density function. Although other functions can be used, they generally produce similar results. As with Lowess, kernel density smoothers have a parameter that controls the smoothness of the resulting function. In this case, the parameter controls the width of the kernel function.

Black threads: In their introductory statistics book, Mosteller et al. (1983) argued in favor of fitting lines by eye, their *black thread method*. As they noted, when several points fall on a straight line, the line clearly fits them. Nevertheless, when the points do not fall along a straight line, we may wish to summarize their positions by a straight line. The problem is how to choose the line. Of course, we can use the methods described above, but we should also be able to draw lines by eye. As they note "drawing a straight line 'by eye' can often be as helpful as the formal methods." Their name for the method comes from the fact that they recommended stretching black threads across the plot to locate the best line until it passed the IOI (Interoccular Impact) test. This was, after all, before the advent of computers.

We recommend the availability of a "piecewise" black thread method, one that allows the user to locate connected pieces of black thread, each piece fitting the local points as well as possible, according to the eye. Such a line is shown in the lower-left corner of Figure 7.7.

7.3.1.3 Using the Guidelines. The principal axis line, shown in the left panel of Figure 7.5, fits our automobile data very well. We should expect this, of course, from the general appearance of the plot, it being a compact distribution of points varying in essentially just one direction, a direction that is not oriented along just one of the dimensions of the scatterplot, but which involves both. The principal axis accounts for 95% of the variation of the two dimensions of the scatterplot. Although the principal axis accounts for nearly all of the variation, there are several hints that the principal axis does not tell the entire story: These are summarized graphically in Figure 7.8.

- One hint, first encountered in Chapter 6, is that there is a gap in the values of *MPG*—there are no automobiles with middle values on *MPG*.
- A second hint is that the relationship between *MPG* and *Weight* appears to be different above the gap than below the gap. We can see this in several ways:
  - The monotonic line, shown in the top-left panel of Figure 7.7, seems to

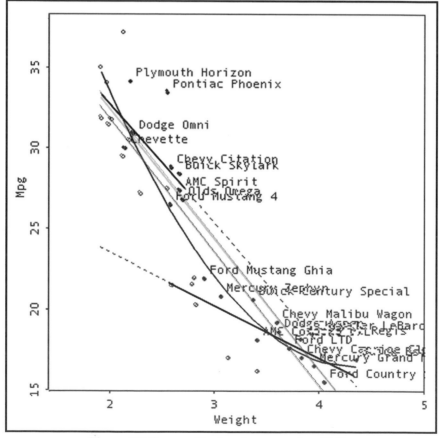

Figure 7.8 Composite showing several guidelines.

bend at the gap and be relatively straight on both sides.

- The Lowess smoother, shown in the top-right panel of Figure 7.7 for a parameter of 1.0, also seems to change slope at the gap.
- In the middle-left panel of Figure 7.7 we show a regression line for the subset of points that are above the gap (the points in the subset are shown as solid black points, those not in the subset as hollow circles). In the middle-right panel of Figure 7.7 we have also done this for the points below the gap. The regressions for these two subsets look quite different.
- The piecewise black thread easily passes the IOI test, as is seen in the lower-left panel of Figure 7.7.
- A third hint that the relationship between the two variables is not linear is shown by fitting a quadratic function to the relationship between the two variables, as is shown in the lower-right panel of Figure 7.7.
- A fourth hint that the principal axis does not tell the entire story of the relationship between *Weight* and *MPG* is seen by looking at the departures from fit to the axis. We have already discussed the two ways of looking at this information that are shown in Figure 7.5. You will recall that on the left is the principal axis along with the orthogonal projection of each observation onto the principal axis (the dashed lines), thus showing us the residuals from fit for each observation (each dashed line). Studying this plot shows that the residuals at the upper and lower portions of the range are on one side of the axis, whereas those in the middle are on the other side. Thus, we see that there are systematic departures from linear fit.

The several guidelines just discussed are combined to produce Figure 7.8. Included are two subset regression lines (with dashed portions), a monotonic regression line (its jagged), a black thread, and a least squares quadratic regression line.

The bottom line (so to say) is this: When we try several different ways of looking at our data, and with each way we see roughly the same thing, we have what is called *convergent validity*, and we can be more confident about what the data seem to say. But we must always remember that the convergence is on what the data "seem to say." not necessarily on what the data actually "do say," which we will never know.

7.3.2 Distribution Comparison Plots

The distribution comparison plot shows whether two variables come from distributions that are similarly or differently shaped. Like the scatterplot, the distribution-comparison plot represents the two variables by axes drawn at right angles to each other. However, unlike the scatterplot, the distribution-comparison plot sorts each variable into order and then plots the sorted values, whereas the scatterplot plots the unsorted values. Because each variable has been sorted into order, the plot is not a cloud of points but an ordered series of points. For emphasis, the points are connected together, creating a plot that shows a jagged monotonic line of connected points. This line represents the relationship between the two distributions.

In Figure 7.9 we present several distribution comparison plots. By way of preview, we conclude that the upper-left plot is of two variables which are similarly distributed: This means that they have distributions that are of the same shape. On the contrary, the other five plots all represent two variables which have distributions that are not similarly shaped and thus should be used cautiously for statistical significance testing. The conclusions are based on the apparent linearity of the jagged monotonic line of connected points: If the line is straight, the distributions are similar; otherwise they are not.

We interpret each of these plots as follows: In each of these plots there are three lines: The jagged monotonic line of connected points is the actual plot of the data sorted, while the two straight lines, one solid and the other dashed, are added to the plot as guidelines. The jagged line tells us whether the two variables have distributions that have the same shape. If the line is roughly straight, the two variables have roughly the same shape. Such variables are said to be *similar*, or to be *similarly distributed*. If the line is not so straight, the two variables are *not similarly distributed*.

It is important to know whether two variables are similarly distributed since many significance tests make such an assumption. Note that "similarly" distributed means that the two variables have the same shape but not necessarily the same means and variances. When two variables are both normally distributed, for example, they have the same shape and are said to be similarly distributed. If, in addition, they both have the same means and variances (as is the case, for example, when each is from a *standard* normal distribution), then they are said to be identically distributed.

The solid straight line is a guideline that represents two similarly shaped variables which have measures of center and spread that are the same as those of the observed variables. The dashed straight line is a guideline that represents two identically shaped variables, variables that have not only the same shape but also the same means and same variances. Some statistical significance tests require identically distributed variables, a stricter requirement than that of similarly distributed variables.

The six plots in Figure 7.9 correspond to the six scatterplots presented in Figure 7.4. The first five (i.e., all but the lower-right plot) are for simulated data, with the name of the function used to generate the data appearing as the title of the plot, and the equation that relates the Y-axis to the X-axis being used as the label of the Y-axis. The sixth plot, in the lower-right corner, is an empirical distribution.

Note that, because these data are simulated, we know which of the first five plots have variables that come from similarly shaped distributions: only the first one. And it is the only one with a straight-appearing jagged line of connected points.

The distribution comparison plot is, as just described, a special case of the well-known quantile-quantile (QQ) plot. In a QQplot, the quantiles of two variables are plotted against each other, forming the jagged line. Since, for these data, the two variables have the same number of observations, the jagged line is simply a plot of one sorted variable against the other sorted variable.

Distribution comparison plots have other uses apart from comparing two variables. For example, we can generate simulated data following a number of probability distributions and use the simulated variable, plus a real variable in a distribution compar-

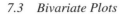

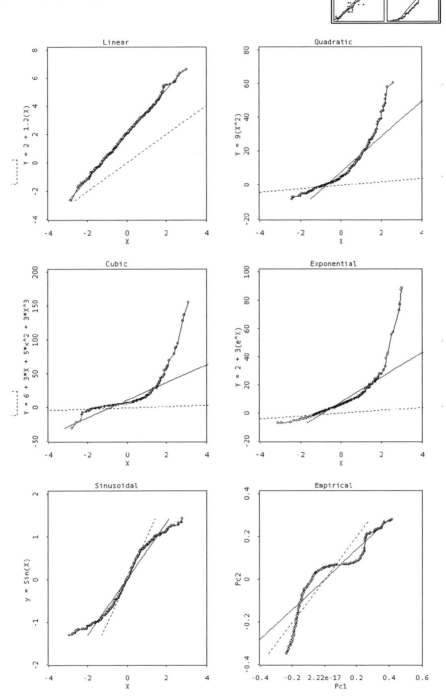

Figure 7.9 Six distribution comparison plots corresponding to the six scatterplots shown in Figure 7.4.

ison plot to decide if the variable observed has the same distribution as the variable simulated. Also, distribution comparison plots can be used for comparing the values of a variable across different groups.

7.3.3 Parallel-Coordinates Plots and Parallel Boxplots

We mentioned above that the parallel-coordinates plot, and its cousin the parallel boxplot, can be used as bivariate plots but that they are not often useful. They are quite useful, however, for high-dimensional applications, so we will skip them here and discuss them in Chapter 8.

7.4 Multiple Bivariate Plots

A plot matrix is a matrix of plot cells with a row and column for each variable being plotted. Each plot cell of the matrix contains a plot of the relationship of the cell's row and column variables. Thus, a plot matrix contains plots showing the relationship between all possible pairs of a set of variables. As such, it is a multiple bivariate plot, being neither truly bivariate nor truly multivariate. In addition, the plots can be multivariate as well. In this chaper we cover plot matrices with cells that are univariate or bivariate. In the next we cover those with multivariate plots.

The first plot matrix that was proposed was the scatterplot plot matrix, a plot matrix whose cells are all scatterplots. An early discussion of the scatterplot plot matrix can be found in Chambers et al. (1983), although they do not claim discovery, saying that they don't know who invented it. It seemed to be in the zeitgeist of the times.

The concept of a scatterplot plot matrix is (and was) easily generalized to that of a plot matrix where the individual plots are formed from pairs of variables, but the plots do not have to be scatterplots. Since the information being plotted is bivariate (except on the diagonal), the plots that can be used are generally one of the bivariate plots types discussed above. Univariate plots are often used on the diagonal of plot matrices, since the diagonal represents the relationship of a variable with itself.

If there are n variables, a plot matrix has n^2 plots arranged into a square $n \times n$ matrix. The i'th row contains plots P_{ij} of variable i versus each of the variables j, including itself. Usually, each plot is square. Often, a plot matrix has one or more additional tall but thin namelists appended to the left or right side of the matrix of plots. They help identify individual observations in the dataset.

The plot in cell ij does not have to be the same as the plot in cell ji, even though the information being plotted (variables i and j) is the same. The plot in cell ji could, for example, be the same type of plot as the plot in cell ij (e.g., both could be scatterplots), but with the axes switched. Or the plot in cell ji could be a totally different plot (e.g., a comparison plot) than the plot in cell ij (which might be a parallel-coordinates plot). Of course, the plots in the two cells can also be identical. The plots on the diagonal can be either bivariate or univariate. When the plot is bivariate, it is showing the relationship of a variable with itself, which may or may not be informative, depending on which specific plot is being used (the scatterplot of a variable versus itself can be

informative, but the comparison and parallel-coordinates plots for a variable paired with itself is seldom useful).

There is no mathematical or statistical reason why all of the plots from which the plot matrix is constructed must be the same. Although it can be very difficult to interpret a plot matrix that has an unsystematic selection of plots in the plot cells, we do discuss plot matrices for which all of the plots above the diagonal are the same type of plot, and all of those below the diagonal are also the same type of plot, but those above are a different type than those below.

A plot matrix can be viewed as a specialized type of trellis plot that visualizes the relationship between all possible pairs of a set of variables and has the very proscribed arrangement of square plot cells described above. For more information see Cleveland (1994a and b) and Cleveland and McGill (1988).

7.4.1 Scatterplot Plot Matrix

In Figure 7.10 we show a scatterplot plot matrix for the automobile data. Note that in the version we show, even the diagonal is formed from a scatterplot of a variable with

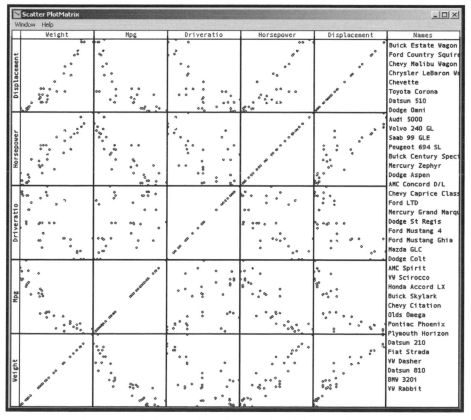

Figure 7.10 Scatterplot plot matrix with namelist.

itself, usually a plot that you would not be interested in seeing, but, in the context of dynamic interactive graphics, a plot that can be informative. Note also that the off-diagonal plots are not identical to their counterparts on the other side of the diagonal. That is, the plot in cell *ji* is not the same as that in cell *ij*. In fact, what we have is an example of how the plot in cell *ji* plots the same variables, but with their assignments to the axes switched.

7.4.2 Quantile Plot Matrix

In Figure 7.11 we show a plot matrix constructed from comparison plots (off-diagonal) and normal probability plots (on diagonal). Since these plots are also commonly called quantile-quantile plots, and quantile plots we call it the quantile plot matrix.

7.4.3 Numerical Plot-matrix

In Figure 7.12 we show a plot matrix constructed from scatterplots (below the diagonal), comparison plots (above the diagonal), and probability plots (on the diagonal). It can be argued that these are the three plots that form the most useful and informative

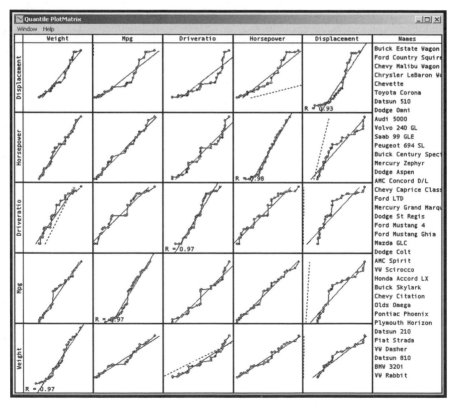

Figure 7.11 Quantile plot matrix, with namelist.

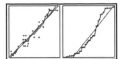

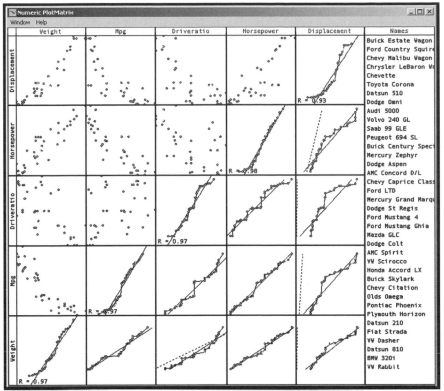

Figure 7.12 Numerical plot matrix.

set of bivariate and univariate numerical plots. Thus, we call it the numerical plot matrix.

7.4.4 BoxPlot Plot Matrix

The final plot matrix we present, shown in Figure 7.13, is a boxplot plot matrix. It is actually somewhat more than a plot matrix, since it is augmented with an extra row and two extra columns (as well as the namelist column that appears in all the other plot matrices). The extra row, at the bottom, is a parallel boxplot of all variables in the dataset. Each extra column contains grouped boxplots, where the grouping is according to the categorical variables in the dataset.

What we can see with plot matrices. A plot matrix can be very useful for investigating the linearity of the relationship between pairs of variables, an important point to pursue since bivariate linearity is implied by multivariate normality. That is, if we wish to perform some sort of statistical test, which assumes that the variables are each normally distributed (i.e., that the entire set of variables is multivariate normally distributed), an implication of such an assumption is that every pair of variables has a linear relationship.

239

The human eye is very good at seeing nonlinearity in the relationships shown in plot matrices. Indeed, a quick glance at the automobile data shown in these plot matrices is all it takes to see the nonlinearities. We already noticed that *Weight* and *MPG* are non-linear. We now see, particularly in the numerical, quantile, and scatterplot plot matrices, that displacement is related nonlinearly to the other variables

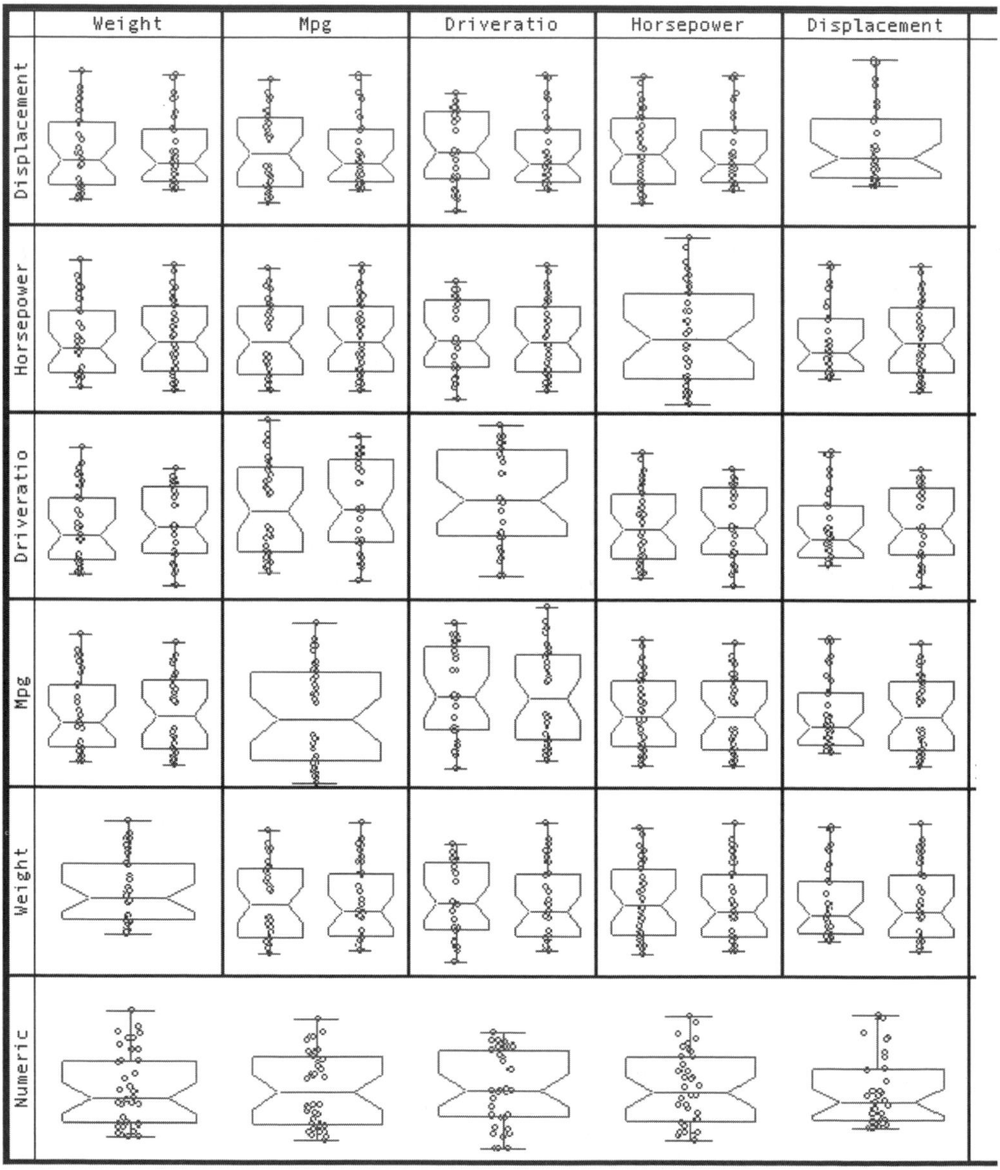

**Figure 7.13 Boxplot plot matrix with parallel and grouped boxplots
(parallel in bottom row, grouped in two right-hand columns).**

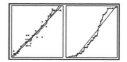

7.5 Bivariate Visualization Methods

Up to this point, we have discussed in the chapter bivariate and multiple bivariate plots. These plots, which include scatterplots, comparison plots, and a variety of plot matrices, all help us understand the relationship between two variables. (Boxplots and parallel-coordinates plots were mentioned but discussion was deferred until the chapter 8).

Cylinders	Country	Names
		Buick Estate Wag
		Ford Country Squ
		Chevy Malibu Wag
		Chrysler LeBaron
		Chevette
		Toyota Corona
		Datsun 510
		Dodge Omni
		Audi 5000
		Volvo 240 GL
		Saab 99 GLE
		Peugeot 694 SL
		Buick Century Sp
		Mercury Zephyr
		Dodge Aspen
		AMC Concord D/L
		Chevy Caprice Cl
		Ford LTD
		Mercury Grand Ma
		Dodge St Regis
		Ford Mustang 4
		Ford Mustang Ghi
		Mazda GLC
		Dodge Colt
		AMC Spirit
		VW Scirocco
		Honda Accord LX
		Buick Skylark
		Chevy Citation
		Olds Omega
		Pontiac Phoenix
		Plymouth Horizon
		Datsun 210
		Fiat Strada
		VW Dasher
		Datsun 810
		BMW 320i
		VW Rabbit

We used the *Weight* and *MPG* variables of the automobile data to demonstrate the use of scatterplots and comparison plots, concluding that weight is a strong predictor of fuel inefficiency, and that the relationship is not linear.

We used all five of the numerical variables in the automobile data to demonstrate the use of plot matrices, concluding that there are strong non-linear relationships among several of the variables, not just *Weight* and *MPG*.

Although the individual plots presented in this and Chapter 8 are useful in themselves, they are even more useful when used in bivariate visualizations.

In the next three sections we introduce bivariate visualizations for exploring, transforming, and modeling bivariate data. Since exploration of the automobile data continues to raise concern about nonlinearities, we transform the data visually to improve its linearity. We then use regression to model the transformed and untransformed data, concluding that the square root of the weight of an automobile predicts the number of gallons of gasoline consumed for each 100 miles traveled (reciprocal of *MPG*).

7.6   Visual Exploration

In this section we present two data visualizations. One is the default bivariate data visualization, consisting only of bivariate graphics and namelists. The other is an optional data visualization that also has several univariate graphics. This visualization, while potentially much more useful, is certainly more complete than the bivariate-only visualization, but it requires a larger screen for all of its plots to be large enough to be useful.

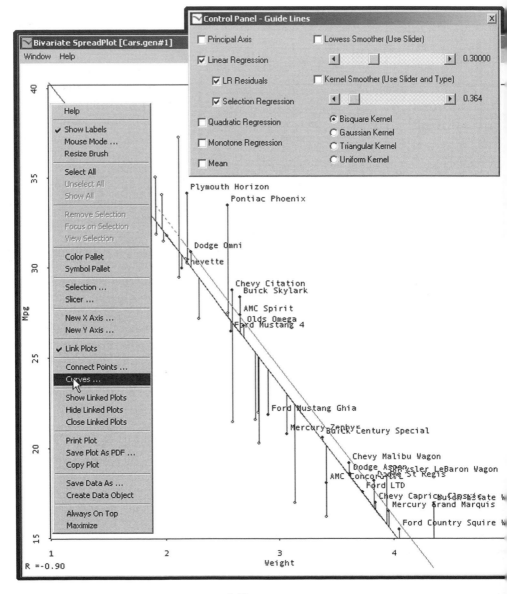

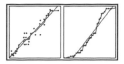

7.6.1 Two Bivariate Data Visualizations

A bivariate visualization for exploring magnitude data is shown in Figure 7.14 for the *MPG* and *Weight* variables. This visualization involves the bivariate graphs introduced above (scatterplot, comparison plot, and parallel boxplot). They are the essential tools for bivariate visualization of magnitude data, helping us understand the relationship between two of the variables. Of these three plots, the scatterplot is certainly the most useful, so it is shown larger than the other plots. The visualization also includes a namelist of the observation labels. If there are categorical variables, there

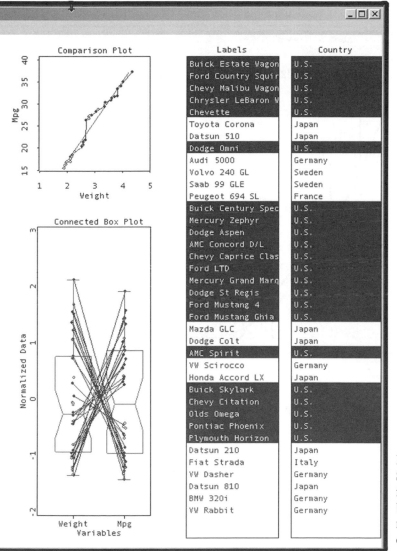

Figure 7.14 Spreadplot for seeing bivariate magnitude data.

can also be a namelist window for each of them. In Figure 7.14 there is a namelist for *Country* (the *Cylinders* variable is not included in the visualization).

The bare minimum for a bivariate data visualization consists of the three core bivariate plots and the namelist of observation labels. The example shown in Figure 7.14 is a bit more than this, but not much. Ideally, we would also like the visualization to include univariate plots, since then we could see the distributions of the two variables as well as their relationship. Accordingly, unless the screen on which the visualization is being presented is too small, there is another visualization which includes the plots used by the default spreadplot, plus a histogram and probability plot for each variable. This visualization is shown in Figure 7.15. Of course, a separate univariate numerical

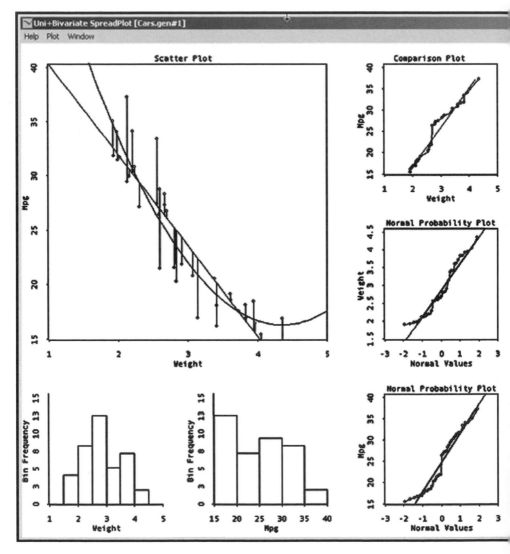

visualization can be used in coordination with the bivariate-only visualization, but awkwardly, since two spreadplots have to be managed.

7.6.2 Using These Visualizations

Using these or any other visualizations involves, in part, using menus and dialog boxes to se tup desired conditions for the visualization, and in part, using the cursor to activate or deactivate plotting elements such as points and labels. Thus, the first step is to use the pop up menus, one of which is shown in Figure 7.14 (normally, it would no longer be showing, but we present it in the figure for pedagogical reasons). Note that the Show Labels item is checked and we see that labels are showing in the scat-

Connected Box Plot	Labels	Country
	Buick Estate Wagon	U.S.
	Ford Country Squir	U.S.
	Chevy Malibu Wagon	U.S.
	Chrysler LeBaron W	U.S.
	Chevette	U.S.
	Toyota Corona	Japan
	Datsun 510	Japan
	Dodge Omni	U.S.
	Audi 5000	Germany
	Volvo 240 GL	Sweden
	Saab 99 GLE	Sweden
	Peugeot 694 SL	France
	Buick Century Spec	U.S.
	Mercury Zephyr	U.S.
	Dodge Aspen	U.S.
	AMC Concord D/L	U.S.
	Chevy Caprice Clas	U.S.
	Ford LTD	U.S.
	Mercury Grand Marq	U.S.
	Dodge St Regis	U.S.
	Ford Mustang 4	U.S.
	Ford Mustang Ghia	U.S.
	Mazda GLC	Japan
	Dodge Colt	Japan
	AMC Spirit	U.S.
	VW Scirocco	Germany
	Honda Accord LX	Japan
	Buick Skylark	U.S.
	Chevy Citation	U.S.
	Olds Omega	U.S.
	Pontiac Phoenix	U.S.
	Plymouth Horizon	U.S.
	Datsun 210	Japan
	Fiat Strada	Italy
	VW Dasher	Germany
	Datsun 810	Japan
	BMW 320i	Germany
	VW Rabbit	Germany

Weight Mpg
Variables

Normalized Data

Figure 7.15 Spreadplot for exploring bivariate magnitude data on a large screen.

terplot. The other two plots do not show labels because, apparently, their Show Labels menu item was not selected. We also note that the Link Plots menu item is checked, meaning that when the scatterplot's plotting elements (points) are selected, they are highlighted in the other plots (i.e., drawn as solid circles in the partially covered comparison plot, and as lines in the parallel-coordinates plot). Finally, the Curves menu item is being choosen, displaying the dialog box shown in the figure. This dialog gives access to the guidelines described above.

The next step is to see what the guidelines tell us about the cloud of points in the scatterplot. We have already gone through this process in Section 7.3.1.3 for the cluster of heavy–inefficient cars compared to the light–efficient ones, with the results summarized in Figure 7.8. The conclusion there was that there seems to be a different linear relationship for the two groups, or perhaps, there is one overall nonlinear relationship.

Using the data visualization spreadplot and its enhanced ability to identify points reveals a second interesting aspect of the data, which is shown in Figure 7.16. To see this new aspect, we need to point out that the dialog in the figure shows that linear regression, with residuals and selection regression options, is activated for the scatterplot. When you actually are interacting with the plot, it would be obvious to you that the lower of the two apparently parallel lines is the overall regression line for *Weight* predicting *MPG*, since that line goes on and off as you check and un-check the Linear Regression dialog option. On the other hand, the upper of the two apparently parallel lines is for the Selection Regression (i.e., it is for whatever subset of points is selected).

In the visualization we have selected the American-manufactured cars; thus, their points are highlighted in each plot. More important, since Selection Regression is activated in the scatterplot, there is a regression line for the subset. We see, then, that

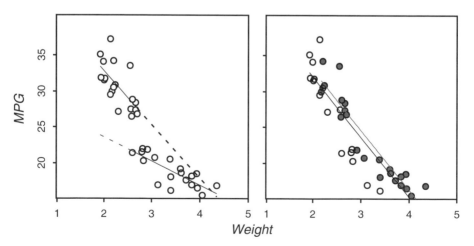

Figure 7.16 Two subset regressions, showing two patterns in the data.
Left: separate regression lines for light vs. heavy cars.
Right: separate regression lines for U.S. cars vs. all cars.

American automobiles, represented by the line with dashed ends (the upper line), are more efficient, for a given weight, than the entire group of automobiles. They can travel about 1 mile farther on a gallon of gasoline than the typical automobile of the same weight—a very interesting and counter-folkloric finding.

We found this result by using the spreadplot's data exploration capability, which is very straightforward to use: You just "rub" your cursor back and forth over the screen. As you do this you see that the cursor, which is a brush with a rectangle attached to it, highlights a feature of the graph whenever the brush's rectangle covers an example of that feature. For most of the graphs the feature is the set of points shown in the graph, and the highlighting consists of drawing the point with greater emphasis, and if the labels option is active, showing the point's label. However, for two of the graphs the highlightable feature is something other than a point:

1. For the histogram the highlightable feature is a narrow horizontal slice of a bar of the graph, and the highlighting consists of filling the slice with color.

2. For the namelist the feature is the name itself, and the highlighting is to display the name in reverse video [i.e., the name appears as white letters on a black (or colored) background].

When you rub your cursor brush back and forth across the visualization you also notice immediately that features of graphs other than the one under your brush are being highlighted. This is because the graphs are linked via their observations, so that a point, slice, or name of one graph corresponds to a point, slice, or name of another. As your brush highlights a feature of one graph, the corresponding feature of each of the other graphs that are linked to the one being brushed is also highlighted.

7.7 Visual Transformation: Box–Cox

Because of the wide variety of plots and methods we can choose from, we can design spreadplots to address a wide variety of specific statistical issues. Consider the normal distribution: When we have multiple variables it is very common to assume that they are *multivariate normal*, meaning that each of the variables is normally distributed. When a set of data are, in fact, multivariate normal, it is the case that every variable in the data is linearly related to every other variable. This property is called *bivariate linearity*. Thus, if we could assess whether all pairs of variables in a set of data were linearly related, we would be indirectly assessing whether the data were multivariate normal: Data that are not bivariate linear are not multivariate normal.

In this section we show a visualization method that we can use to transform variables to be more nearly bivariate linear, the motivation being that the data are also being transformed to be more nearly multivariate normal. This visualization method uses a scatterplot plot matrix to show us all bivariate relationships of the variables in a set of data, adding guidelines to each scatterplot to emphasize the plot's linearity or lack thereof. The visualization has tools that are used by the data analyst to transform the variables to be related to each other more linearly. These tools are highly interactive and very dynamic. With these tools the user focuses on one of the variables, transforming it to be more linear with the other variables. The user then focuses on another variable to improve the linearity of its bivariate relationships. Gradually, as the user focuses first on one variable and then on another, the relationships between all pairs of variables should become more linear. At the conclusion of the process we can judge whether the variables are sufficiently bivariate linear to justify the assumption that they are multivariate normal.

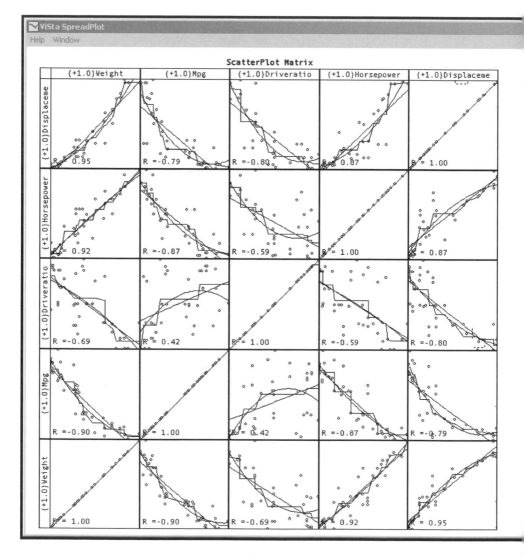

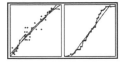

7.7.1 The Transformation Visualization

The spreadplot that implements the visual transformation method that we have just described is shown in Figure 7.17. It is shown as it looks when it first appears. The left-hand portion of the spreadplot is a scatterplot plot matrix that has been modified so that the diagonal shows the transformation for each variable. At first, these transformations are all linear, as shown in the figure. Each off-diagonal plot shows the relationship between a pair of transformed variables. Since the transformations are all initially linear, the initial scatterplot plot matrix is the equivalent of a scatterplot plot matrix of the untransformed variables (compare Figure 7.17 with Figure 7.10). In Figure 7.17 the *MPG × Weight* relationship is the focal relationship, as you can tell by

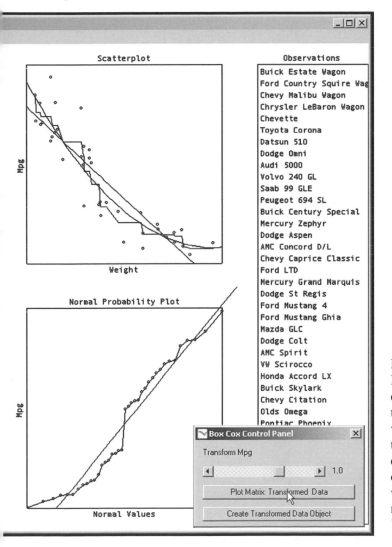

**Figure 7.17
Initial Box–Cox
data
transformation
visualization of
the automobile
data. The focus is
on the *MPG*
variable and its
relationship with
Weight.**

249

looking at the large scatterplot to see which pair of variables is being shown. The *MPG* variable is the focal variable, since it is on the *y*-axis of the focal plot.

Each scatterplot in the scatterplot plot matrix contains a linear, a quadratic, and a monotonic regression guide line (in some plots the linear and quadratic regression lines coincide, making it look as though there are fewer than three lines). When we transform the data, the transformation has linearized the regression as much as possible when the quadratic regression line is identical to the linear regression. Thus, the goal is to make each plot, and each guideline in each plot, as linear as possible.

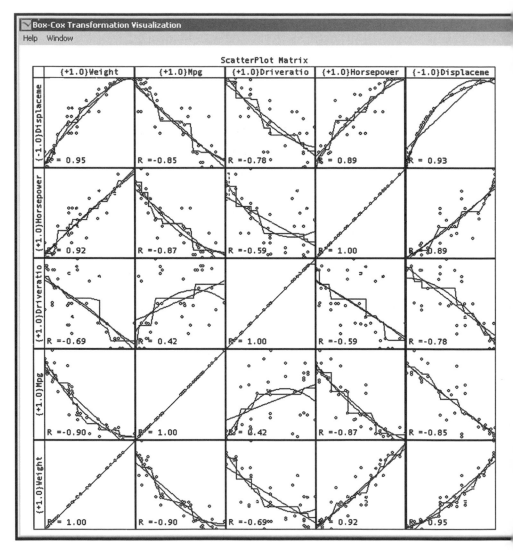

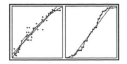

7.7.2 Using Transformation Visualization

The scatterplot plot matrix does more than just show transformations and bivariate relationships:. When you click on a plot in the scatterplot plot matrix, the plot becomes the *focal plot* (the one to be linearized), and its *y* variable becomes the *focal variable* (i.e., the variable to be transformed to make the focal plot linear). The click causes the focal plot to be shown in the top plot to the right of the plot matrix, and causes a normal probability plot of the focal variable to be shown in the lower plot. We see in Figure 7.18 that the focal plot has changed from *Weight × MPG* in Figure 7.17 to *MPG × Displacement* in Figure 7.18. This means that the focal variable has changed from *MPG* to *Displacement*.

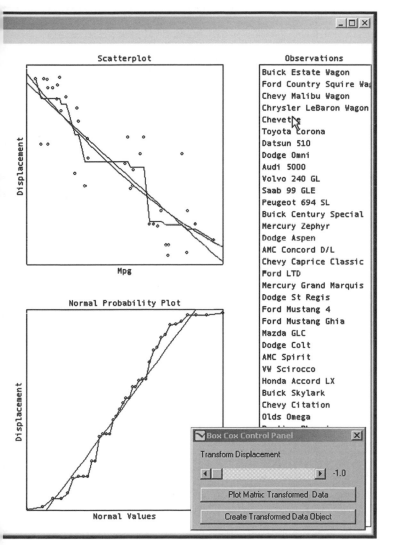

Figure 7.18 Box–Cox data transformation visualization showing the automobile data with the final transformations.

You transform the focal variable by using the slider shown at the bottom right of the spreadplot. Moving the slider changes the transformation of the focal variable, a new transformation being computed every time the slider is moved. Each time a new transformation is computed, the two large plots and all of the plots that are in the focal plot's row and column change. Note that the transformation shown for *Displacement* in the upper-right plot cell is now nonlinear and that the plots in the *Displacement* row and column have changed from the first figure to the second.

The way the visualization has changed from Figure 7.17 to Figure 7.18 shows that the data analyst is searching for transformations that maximize bivariate linearity. If

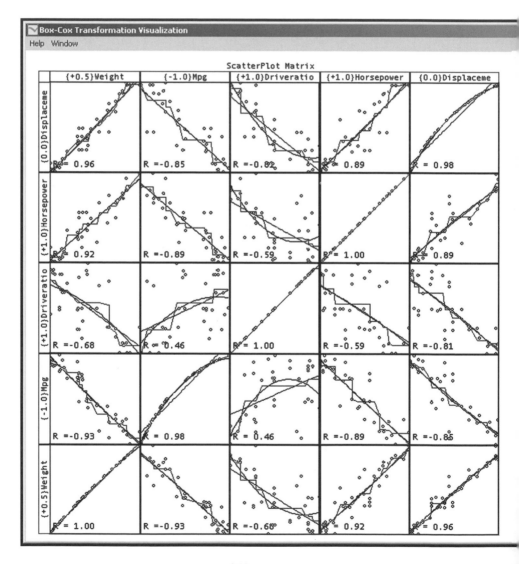

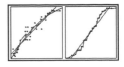

we were the analyst sitting at the computer actively transforming variables rather than reading about it, we would see the transformed variable change dynamically. Because this dynamic feature is the heart of this visualization, we present a thumb-powered version of it in the upper-right corner of the right-hand pages of this chapter. Due to lack of space, we show only the two large plots, which we have arranged horizontally to fit the space on the page. Thumbing through these plots provides a reasonable simulation of what appears on the screen.

In Figure 7.19 we show the transformation visualization as it appears at the end of our search for transformations that create bivariate linearity, at the time that we decided that we had made the bivariate relationships as linear as possible. In fact, we

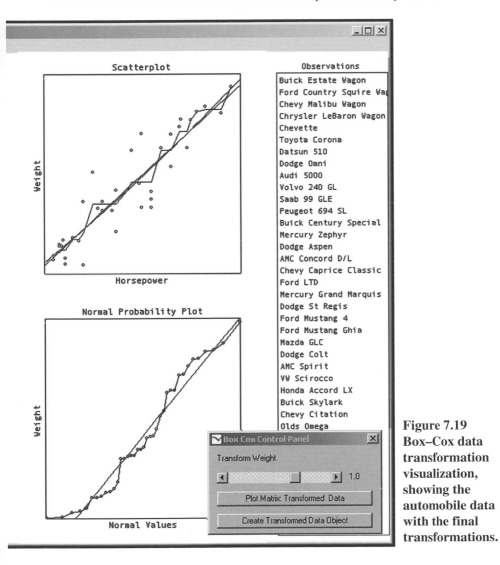

Figure 7.19 Box–Cox data transformation visualization, showing the automobile data with the final transformations.

could not make all of the relationships linear. In particular, the *DriveRatio* variable could not be transformed to have a strongly linear relationship with the other variables. However, the four remaining variables were transformed to be mutually bivariate linear. We show the transformation plot matrix for just these four variables in Figure 7.20, which makes it clear that these four variables are mutually bivariate linear. Thus, it seems that *MPG*, *Weight*, *Horsepower,* and *Displacement* can be transformed to be jointly multivariate normal, whereas *DriveRatio* could not be so transformed. We should keep this in mind when we analyze these data further.

As we explain next, the slider controls a function that computes the transformation. There is a different transformation for each slider value, four of which correspond to well-known transformations (see Table 7.2). We used this information, as well as linearity, to select specific transformations. The final transformations are (see Figure 7.20):

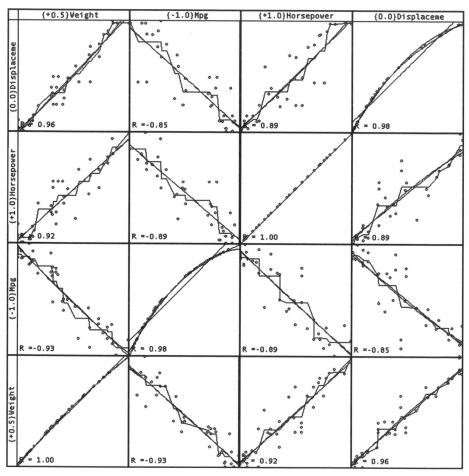

Figure 7.20 Linearized bivariate relationships for four of the five automobile

```
{-1.0}MPG
{+0.5}Weight
{0.0}Displacement
{+1.0}Horsepower
```

The reciprocal of *MPG*
The square root of *Weight*
The log of *Displacement*
Horsepower untransformed

Interestingly, the transformation of *MPG* also linearizes with respect to the L/100km (liters per 100 kilometers) measure used in most of the rest of the world.

Table 7.2 Special Members of the Box–Cox Transformation Family

Parameter Value	Transformation Name	Equation
-1.0	Reciprocal	$-1/y$
0.0	Natural Log	$\log(y)$
0.5	Square Root	$\sqrt{y}$
1.0	Identity (no transformation)	y
2.0	Square	y^2

7.7.3 The Box–Cox Power Transformation

Here is what is going on behind the scenes when we use the slider to search for transformations that maximize bivariate linearity. The slider controls the value of an argument of a function that is used to transform the focal variable. As the user moves the slider, the argument's value is modified, and the function is reapplied to the focal variable using the argument's new value. This results in a revised transformation of the focal variable which is then used to construct revised plots, both large and small. What the analyst sees, if this happens fast enough, is a smoothly and continuously changing transformation that is instantly responsive to the user's actions.

The. mechanism for transforming variables to remove nonlinearities is based on the family of scaled power transformations that is commonly known as the *Box–Cox transformation* (Box and Cox, 1964; Cook and Weisberg, 1999; Tierney, 1990). The transformation family is defined as

$$f(y) = \begin{cases} (y^p - 1)/p & \text{for } p \neq 0 \\ \log(y) & \text{for } p = 0 \end{cases}$$

The slider controls the value of the parameter *p* in this equation. Certain values of the parameter have special meanings, as shown in Table 7.2. Thus, the data analyst changes the value of this parameter dynamically by moving the slider of the dialog box in the figure. When the data analyst moves the slider back and forth, the parameter value changes. All this happens smoothly and instantly, and for datasets that are not too large, can happen many times per second, providing an excellent example of dynamic interactive graphics.

7.8   Visual Fitting: Simple Regression

In this section we demonstrate visual fitting with *simple regression*, the process of fitting the simple regression model to magnitude data. Simple regression analysis is a technique for predicting the observed values of a response variable from the observed values of a predictor variable. Simple regression is a special case of multiple regression, which has multiple predictors, and of multivariate multiple regression, which also has multiple response variables. We fit the regression model using ordinary least squares, finding the strongest linear relationship between predictor and response.

The spreadplot for fitting the data with a simple regression model is shown in Figure 7.21 for the raw (untransformed) automobile data. This spreadplot has a large

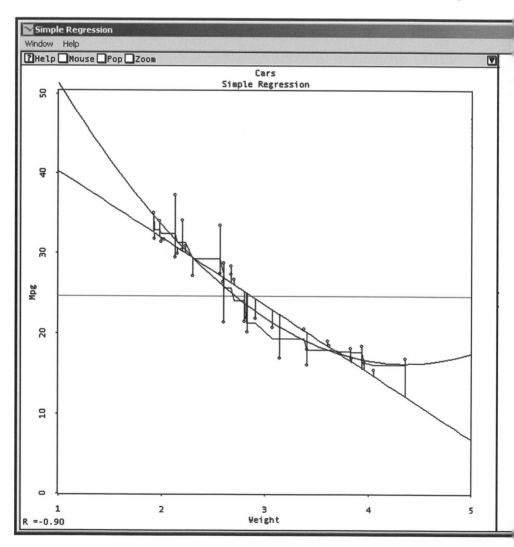

scatterplot, two smaller plots, a namelist, and a dialog box. The large scatterplot is of most interest, since it plots the regression.

We can see that the control panel has been used to display the linear, quadratic, monotone, and residual guidelines in the regression plot. The spreadplot's two smaller plots are a residuals plot and an influence plot. These are specialized scatterplots that help you check on the validity of the OLS assumptions and help you see outliers.

OLS regression assumes that errors are normally distributed and independent, and that the predictor is measured without error. OLS regression is the best method when the assumptions are satisfied. However, when the data fail to meet the assumptions or

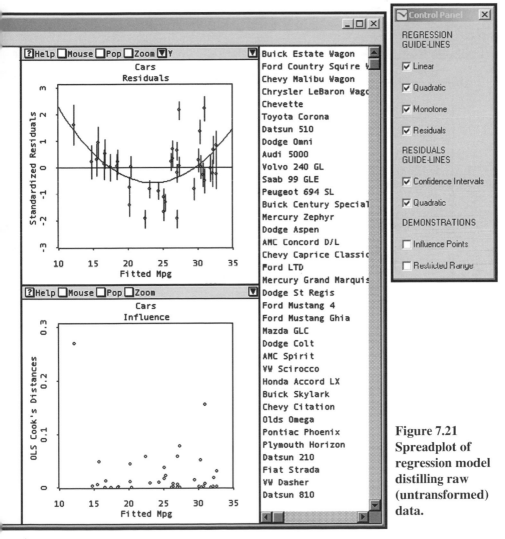

**Figure 7.21
Spreadplot of
regression model
distilling raw
(untransformed)
data.**

when there are problems such as outliers, the OLS coefficients may be inaccurate estimates of the population parameters. If the data fail to meet the assumptions of OLS regression, you can try transforming variables (as we did) or you can try changing fitting techniques, using robust rather than OLS methods.

Residuals plot. The residuals plot is shown as the most-upper of the two smaller plots. It is a plot of the residuals versus the values predicted for the response variable. The residuals plot is a regression diagnostic plot that helps diagnose the suitability of the assumptions underlying regression analysis for the data being analyzed. Residual plots may be used to detect nonnormal error distributions, nonconstant error variance (heteroscedasticity), nonlinearity, and outliers.

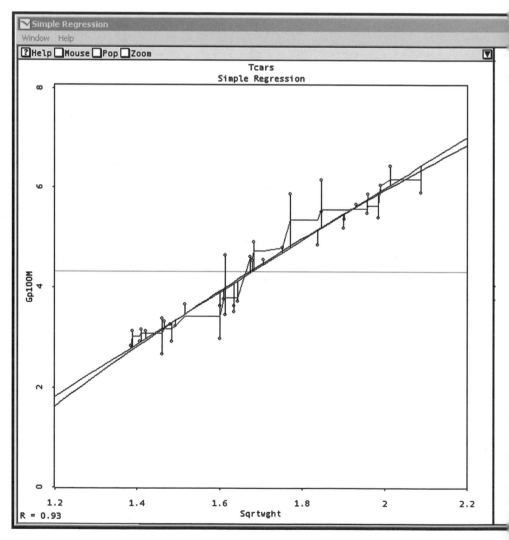

The residuals should show no pattern. To help diagnose whether transformation is needed, we fit a linear and a quadratic guideline to the residuals. Since the linear line is always horizontal and located at zero, it tells us nothing about the nature of the residuals. However, it serves as a reference line for the quadratic guideline: If the quadratic guideline appears to depart from the linear guideline, as it clearly does for the residuals from fitting the model to the raw data shown in Figure 7.21, transformation is called for. Accordingly, we use the data from the visual Box–Cox transformation in Figure 7.7.3 in a second regression analysis. The resulting spreadplot is shown in Figure 7.22, where we see that the regression is much more linear, and that the residuals plot shows much less departure from linearity.

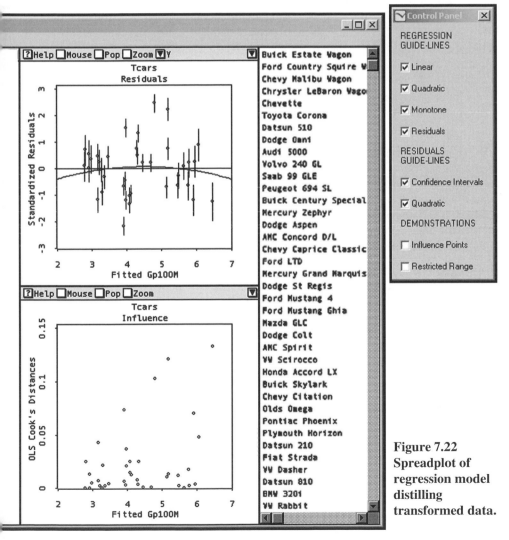

**Figure 7.22
Spreadplot of
regression model
distilling
transformed data.**

259

These two figures also show a vertical line for each residual in the residual plot. These vertical lines represent a confidence interval for the residuals. The plot shows internally studentized residuals with Bayesian error bars. References to the definition of these residuals are given in Tierney (1990, p. 57). The Bayesian error bars represent the mean plus or minus 2 times a Bayesian standard error, defined by Tierney (1990, p. 57) so that they cover a range of values within which we would roughly expect to find the residual 95% of the time.

Influence Plot. An influence plot is a regression diagnostic plot that helps diagnose the stability of the regression analysis. The plot may be used to determine the influence of a particular observation on the regression parameter estimates.

An influence plot shows the effect on the values of the predicted response variable, of removing an individual observation. The plot uses the *Cook's distance* measure, a measure that determines the influence of removing an observation by estimating the difference between the regression coefficients calculated when the observation is included in the analysis and when it is omitted from the analysis.

A large Cook's distance suggests that the observation has a large influence on the calculation of the parameter estimates; small changes in the observation will have relatively large effects on the parameter estimates. If such an observation is not reliable, the model is also not reliable and we do not have stable estimates of the parameters.

7.9   Conclusions

In this chapter we have presented bivariate and multiple bivariate plots. We have used them, along with the univariate plots discussed in Chapter 6, as building blocks to build two bivariate visualizations for exploring data, a multiple bivariate visualization for transforming data, and a bivariate visualization for fitting data with a simple regression model. We illustrated these methods with data about automobiles.

Visual exploration revealed:
- That there is a different linear function relating efficiency and weight of automobiles for light cars than for heavier cars, although we could also conclude that there is one nonlinear function.
- That when you remove differences in weight, American automobiles built in the 1970s were slightly more fuel efficient than all cars of the same weight, a clearly counter-folkloric finding.
- That the variables are not linearly related, and thus not multivariate normal.

Visual Transformation revealed
- That four of the five variables can be transformed to be mutually linear: the fifth variable cannot. We should keep our eye on that variable in future analyses to see whether it should be dropped from further analysis.
- That the transformation of *MPG* is the reciprocal transformation, yielding a measure, such as gallons per 100 miles), which is linearly related to the Liters per 100 kilometers used in much of the world.

Visual fitting of a simple regression model revealed:

- That simple regression analysis using OLS methods on both the original raw data and the transformed data showed that the transformed data are more appropriate for simple regression analysis since they appear to more nearly satisfy the underlying assumptions.

8 Seeing Multivariate Data

8 Seeing Multivariate Data

In this chapter we continue the discussion of methods for seeing magnitude data that we began in Chapter 6 and continued in Chapter 7. In Chapter 6 we discussed univariate methods designed to help us see the distributional characteristics of a single magnitude variable. In Chapter 7 we turned to methods designed to help us understand the relationship between two variables. These plotting methods, collectively known as bivariate plots, enable us to look simultaneously at the relationship between two magnitude variables. In Chapter 7 we also reviewed multiple bivariate plots (plot matrices), methods that show us the bivariate relationship of every pair of variables in a dataset. Here, we discuss graphical methods that help us understand three or more variables simultaneously.

The task faced by the graphical methods presented in the preceding two chapters was quite simple compared to the task facing methods presented in this chapter. After all, up to now, the graphics only had to cope with one or two variables presented in one or two dimensions. Now, however, the task involves visualizing data that have many variables. So we can't construct, say, a 13-dimensional picture of 13 variables and expect a happy viewer! Rather, we need to provide a way of viewing multivariate data that will help us understand their high-dimensional structure even though our vision is limited to three dimensions.

In fact, we identify three distinct ways of viewing multivariate data. Each calls on our cognition to help out our vision, because in our mind we can manipulate five or six, or even eight or nine things simultaneously. This cognitive ability, which is known in cognitive science as the "seven plus or minus two" rule, was originally proposed by Miller (Miller, 1956) and has been well substantiated ever since.

All of the ways that have been proposed for viewing multivariate data begin with the concepts of a point cloud and a data space: Data are represented by a cloud of points that includes a point for every multivariate observation in the data. The cloud of points resides in a high-dimensional space that is analogous to our familiar three-dimensional Euclidean space, but which has many dimensions, one for each variable in the data. The space is called the *data space*.

265

The three distinct ways of viewing multivariate data are distinguished from each other in how they suggest we look at the high-dimensional point cloud. One way suggests that we look at the point cloud through a two- or three-dimensional window to see how the view changes as we maneuver the window in the data space containing the point cloud. Another way suggests that we draw a picture of the point cloud that represents the dimensions of the data space (the variables) by axes that are parallel rather than perpendicular. The third way suggests showing us many two- or three dimensional views of the point cloud simultaneously, each view taken from a different vantage point in the data space. In each case the user's cognitive abilities are enlisted to circumvent visual limitations in an effort to make the right inferences about the structure of the multivariate data.

Before focusing on these three families of multivariate visualization methods, we introduce the multivariate data that will be used throughout the chapter. Since the data that we use are not directly amenable to meaningful visualization, we present a transformation method that uses principal components analysis to improve the visualizability data.

We then address the main topic of this chapter, a detailed explanation of graphics that help us understand multivariate data. We begin by briefly outlining the three families of multivariate plots. Then we present several examples of each plot family, discussing each plot in detail. We then show how these plots can be used to construct visualizations of the data.

8.1   Data: Medical Diagnosis

In Chapter 2 we presented several examples, including one about medical diagnosis. The data for this example were obtained from patients examined by a dermatologist about problems with their skin. The data include information obtained during the office visit and the results of a laboratory analysis of a biopsy taken from the patient. The analysis showed dynamic interactive graphics being used with these data to develop a tool to help the physicians diagnosis their patients. In this chapter we delve into these data more deeply, using them to illustrate the visual analysis of multivariate data.

The data. We introduced these data in Chapter 2, so we only summarize their nature here, recommending that you reread the Section 2.2 for a more complete description. The data, whose variables are listed in Table 2.1, are observations of 34 variables obtained from 366 dermatology patients. Twelve of the variables were measured during the office visit, and 22 were measured in laboratory tests performed on a skin biopsy obtained during the office visit. Of the 34 variables, 32 were measured on a scale running from 0 to 3, with 0 indicating absence of the feature and 3 the largest amount of it. Of the remaining two variables, *Family History* is binary and *Age* is an integer specifying age in years. The data for eight patients who did not state their age were removed. The data are from Nilsel Ilter, the University of Ankara, Turkey (Guvenir et al., 1998).

As we pointed out in Chapter 2, these data are not suitable for visualization, the problem being that the variables provide discrete rather than continuous measurements of magnitude: Of the 34 variables, one is binary and 32 have only four categories of observation. Thus, only one of the 34 variables—the *Age* variable—provides us with a continuous measurement of magnitude.

The problem that arises when we attempt to visualize these data is revealed by the scatterplot matrix shown in Figure 8.1, where we display only three of the 32 variables, just a small portion of the entire data. We see that each plot in the scatterplot matrix has a lattice-like structure. Generally, we are happy to see structure in data, but this lattice-like structure is an uninterpretable artifact of the discrete nature of the variables. Scatterplots of discrete variables always have a lattice structure. Each point in the lattice actually represents many observations, since the discrete data make the points overlap each other.

In essence, the resolution of the data, which is four values per variable, is too low. It is as though we took a picture of our data with a 16-pixel camera that shows only four pieces of information along each of the axes, rather than with a megapixel camera that delivers the fine detail that we need.

Data transformation. Although these data cannot be visualized, they can be transformed into variables that can be visualized, the new set of variables being the continuously measured magnitude variables that are required by the techniques described in this chapter.

Although it may seem a bit strange to use data that are not appropriate for the methods being illustrated, we selected these data specifically because they are indeed a real set of data that present a real set of difficulties in their analysis. We could have selected data that were especially concocted to illustrate our visualization methods,

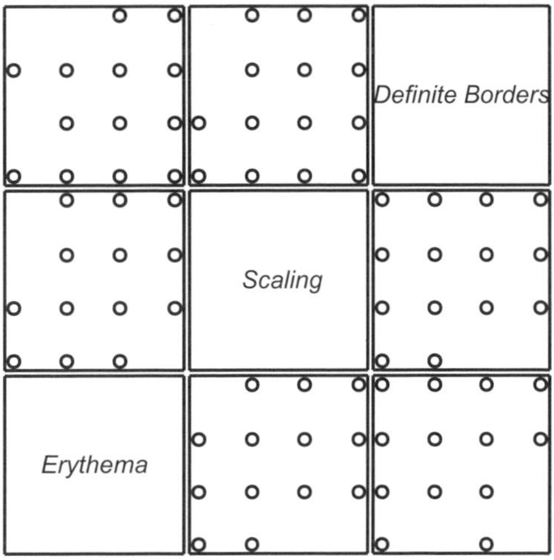

**Figure 8.1 Scatterplot matrix of the first three variables
of the medical diagnosis data.**

but we know that a real set of data collected for their own purposes will, in the long run, be much better at illustrating the power of statistical visualization.

The new set of variables are the principal components of the original set of variables. We use principal components because they are continuous, and because they can be used to reduce the number of variables from the 34 variables of the raw data to a much smaller number of variables that (1) contains the essence of the information in the original data, and (2) are easier to understand because there are fewer variables.

Geometry of the transformation. We illustrate the geometry of the principal components transformation in Figure 8.2. In the top-left figure we see a two-dimensional cloud of points in an arbitrary orientation. The variables whose values form the coordinates of the points in the space are named Var Y and Var X. We have drawn a line through the point cloud representing the first principal component and have labeled it PC 1. We have also drawn short "projection" lines that show the points being projected orthogonally onto the principal component line. We presented all of this in section Section 7.3.1.2, referring to the principal component guideline and calling the short lines residual lines.

These short projection lines have two aspects that are important to understand. One is the location at which the projection line arrives on the component line: This location is the score of the observation on the principal component, the *component score*. The principal component is oriented so that the scores have the largest possible variance, making the principal component the longest direction through the point cloud. The second aspect is their length: The principal component line is oriented so that when the points are projected onto the component orthogonally, the residual lines are,

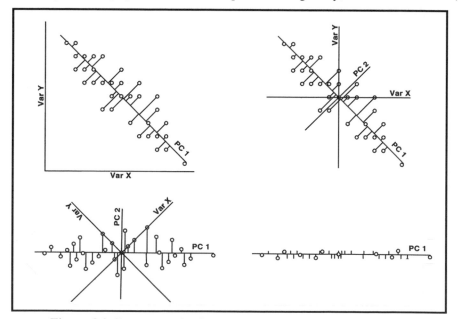

Figure 8.2 Geometry of principal component transformation.

on average, as short as possible. In other words, the line "fits" the cloud better than any other line can, in the sense of minimizing the sum of squared residuals.

In the upper-right pane we have drawn the same space, but with the second principal component line added. Note that it is perpendicular (orthogonal) to the first component line. Note also that it is shorter because there is less variation in its direction. We added the *X* and *Y* lines to emphasize what happens during rotation.

In the lower-left pane we have drawn the same space, but now rotated into its principal component orientation, the orientation in which the horizontal axis of the graphic becomes the first principal component and the vertical axis becomes the second component. Note that the first component is longer than Var *X* and Var *Y*.

Finally, in the lower-right pane we show the one-dimensional approximation of the first component to the original data. Note that we have gained in parsimony (one variable rather than two) but lost in accuracy, although not very much (the first component accounts for 94.22% of the variation in *X* and *Y*).

To summarize, the space formed by the scores on the first *r* principal components is the *r*-dimensional space that best fits the original high-dimensional data space. This characteristic is very useful for exploring multivariate data: The first several principal components provide the most succinct possible visual representation of the data.

Decisions. There are two decisions that the user must make when using principal components: 1) Whether to base the analysis on correlations or covariances, and 2) how many principal components we should use.

The first decision is fairly straightforward: Use correlations unless the variables are all measured in the same units, in which case you should consider whether or not to use covariances. The reasoning is as follows: Computing correlations involves standardizing the variables to have mean equal to zero and variance equal to 1. Normally, this is the most sensible option because otherwise, variables with large variances will contribute more to the analysis. However, if the variables are all measured in the same units, you cannot adjust the units of each variable separately. Since changing a variable's variance involves changing its unit, this can be done only when variables are in different units.

Making the second decision is more difficult. There are several rules of thumb that suggest to us how many principal components are needed to adequately reproduce, and therefore replace, the original data, rules that we discuss at the end of the chapter. These rules suggest from five to seven components. However, with dynamic interactive graphics, it is possible to interact with the graphics to determine exactly which components are needed to understand the data. So it is our approach to use interpretability to determine the number of components that are needed. Note that if a component contributes to our understanding of the data, we not only use that component but must also use all the components before it. That is, all the components that account for more variance must be used. For example, if components 1, 2, 4, and 5 contribute to understanding the data but 3 does not, we must still use number 3, using all of the first five to replace our data (as is the case for our data).

8.2   Three Families of Multivariate Plots

Each of the graphical methods for seeing multivariate data that we discuss in this chapter belongs to one of three families of multivariate graphics. We call these families the orthogonal-axes family, the parallel-axes family, and the paired-axes family. The families are represented by the schematics shown in Figure 8.3, each schematic being labeled to identify the plot family. Each of the three schematics represents a four-dimensional example of the family, the dimensions being *W*, *X*, *Y* and *Z*.

Orthogonal-axes plot family. The family of orthogonal-axes plots includes the spinplot, orbitplot, and multivariate distribution comparison plot. We discuss these plots in this chapter.

The members of the orthogonal-axes family have the strong advantage of being easy to interpret correctly, since we are so familiar with three-dimensional space. However, the members of this family have the strong disadvantage of showing only three variables, even when there are many more.

You will note that the orthogonal-axes schematic, shown in Figure 8.3, is drawn as a 3D cube with three lines meeting in the center of the cube. The lines are labeled *X*, *Y* and *Z*. The cube represents a 3D space, and the three lines represent the three dimensions of the space, which are *X*, *Y* and *Z*. The lines are drawn meeting in the center, which represents the fact that the space can be rotated around its center. The lines are also drawn meeting at right angles to represent the orthogonality of the axes.

Note that the center is labeled *W*, the fourth variable/dimension name, and that there is no line representing the fourth dimension. This is the way the schematic portrays the fact that the dimensions beyond the three that are showing are all orthogonal to the 3D space you see, and therefore are invisible. They do, however, intersect with the three visible dimensions at the center of the space—thus, the *W* in the center.

Parallel-axes plot family. The family of parallel-axes plots includes the parallel coordinates plot, the parallel comparisons plot, and parallel versions of one-dimensional plots that were discussed in Chapter 6, including parallel dotplots, jittered dotplots, boxplots, and diamond plots.

You will note that the parallel-axes schematic in Figure 8.3 displays four vertical lines, labeled *W*, *X*, *Y* and *Z*. These lines represent the four axes as they are depicted

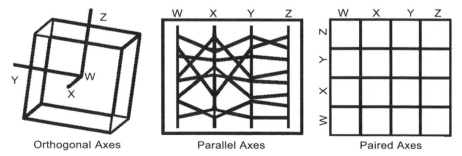

Orthogonal Axes   Parallel Axes   Paired Axes

Figure 8.3 Schematics of the three families of multivariate plots.

by a parallel-axes plot. In addition to the four axes, the schematic has several jagged lines that run horizontally. These lines, called *trace lines*, represent the observations, there being one trace line for each observation. The value of the trace as it crosses an axis corresponds to the observation's value on the variable represented by the axis.

All members of the family of parallel-axes plots have the unique and remarkable characteristic of being able simultaneously to display all of the variables of a multivariate dataset, providing the only way to glimpse all of the dimensions simultaneously. Although they can display all of the dimensions simultaneously, there are two major concerns of which the user should be aware.

The first concern is that even though we see all of the dimensions, we do not see the entire space. If, for example, we look at the parallel-axes schematic in Figure 8.3, we see trace lines between axes *W* and *X*, *X* and *Y*, *Y* and *Z*, but we do not see tracelines between *W* and *Y*, *W* and *Z* or *X* and *Z*. If we reordered the axes, we would have a different view of the structure. Thus, although we see all of the dimensions, we do not see all of the space.

The second problem is that the basic nature of the display is unfamiliar, especially when compared to our extensive familiarity with the orthogonal-axes space. Parallel-axes plots, however, are certainly worth becoming familiar with, given that they can display all of the dimensions simultaneously.

Paired-axes plot family. The family of paired-axes plots includes the plot matrices that we discussed in Section 7.4, including scatterplot plot matrices, quantile plot matrices, numerical plot matrices, and boxplot plot matrices. As we discussed in Chapter 7, a plot matrix is a matrix of plots with a row and column for each of the variables in a multivariate dataset. Each cell of the matrix contains a plot of the relationship of the cell's row and column variables. As such, it is a multiple bivariate plot, being neither truly bivariate nor truly multivariate. Given the choice of where to discuss plot matrices, we decided to include the discussion in the chapter on bivariate methods rather than in this chapter on multivariate methods.

However, the family of paired-axes plots includes additional plot matrices that are truly multivariate, plot matrices where the plots in the cells of the matrix are themselves multivariate. That is, we could have a plot matrix with cells that are orthogonal-axes plots or parallel-axes plots—or, for that matter, paired-axes plots. This kind of plot matrix is a multiple multivariate plot, and discussion of it clearly belongs in this chapter. Thus, our discussion of paired-axes plots is split between two chapters: Multiple bivariate plots were discussed in Section 7.4 and multiple multivariate plots are discussed in this chapter.

The advantages and disadvantages of paired-axes plots really come down to the amount of information presented by the plot and the amount of space it takes to present the information. The user is, in essence, being asked to integrate the information obtained from each of the plots in the plot matrix into a coherent whole. Since the number of plots increases as the square of the number of rows or columns of the plot matrix, the task rapidly becomes impossibly difficult. Also, keep in mind that the physical space occupied by the plot matrix either becomes very large and difficult to manage, or it remains constant and the individual plots become impossibly small. Either way, the number of rows and columns of a paired-axes plot must be rather small, say not more than four, five, or six, for the user to benefit from the presentation.

8.3   Parallel-Axes Plots

In this section we discuss two plots for multivariate data, the parallel-coordinates plot and the parallel-comparisons plot. These two plots are based on a non-Cartesian multidimensional coordinate system that opts to represent the dimensions by axes that are mutually *parallel*. Thus, we identify these plots as members of the family of parallel-axes plots. The material presented in this section owes its existence to the fundamental work on the parallel-coordinates representation of multivariate information by Inselberg (Inselberg, 1985; Wegman, 1990).

8.3.1 Parallel-Coordinates Plot

While the traditional Cartesian coordinate system represents all axes as mutually perpendicular, the parallel-coordinates system represents all axes as mutually parallel. As an example, consider the data on the left of Figure 8.4. We refer to these data as a *data gauge*, a specially constructed artificial dataset that helps us gauge, or demonstrate, a specific plot, in this case the parallel coordinates plot shown in the upper-right panel of the figure.

The parallel coordinates representation involves parallel, equally spaced axes, one axis for each variable. For the data gauge in the left panel of Figure 8.4, there are six variables, so the representation of these data that is shown in the right panel of Figure 8.4 has six parallel, equally spaced axes, labeled X1 through X6. Each observation in the data gauge is represented in the parallel coordinates system by a line called the *trace line*, a jagged line of connected dots, one dot on each axis. A dot is located on an

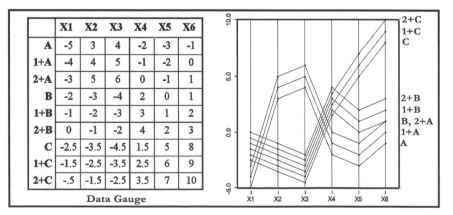

	X1	X2	X3	X4	X5	X6
A	-5	3	4	-2	-3	-1
1+A	-4	4	5	-1	-2	0
2+A	-3	5	6	0	-1	1
B	-2	-3	-4	2	0	1
1+B	-1	-2	-3	3	1	2
2+B	0	-1	-2	4	2	3
C	-2.5	-3.5	-4.5	1.5	5	8
1+C	-1.5	-2.5	-3.5	2.5	6	9
2+C	-.5	-1.5	-2.5	3.5	7	10

Data Gauge

Figure 8.4 Data gauge (left) and its representation by parallel coordinates.

272

axis according to the value that the observation has on the variable plotted on the axis. Furthermore, the *n* observations of a dataset are represented by *n* trace lines, one for each observation, each line connecting *r* dots, one for each of the *r* observed values, for the *r* variables. For the data gauge in Figure 8.4, there are nine observations, so there are nine lines in its representation in the right panel of the figure.

The parallel-coordinates plot is a plot of the data using the parallel-coordinates system of representation. The parallel-coordinates plot displays all of the variables in a simple two-dimensional graphic that portrays the dimensions by axes drawn next to each other. This plot has the distinct advantage that it shows all of the dimensions simultaneously, but it has the disadvantage that parallel-axes plots are not as familiar and comfortable as are orthogonal-axes plots.

If you haven't noticed it yet, a bit more study of the data gauge should reveal that it has structure, and that the observation label tells us about the structure. The first observation row is named A and the next two observation rows are named A+1 and A+2. We then see that the data gauge has values in the A+1 row that equal the values in the A row plus 1, and that the values in the A+2 row are equal to those in the A row plus 2. Note that the observation labels indicate there is an observation B, with two additional observations calculated from it, and that there is a third observation, C, which has two additional observations calculated from it. Thus, the data in our data gauge have three clusters of three points, each cluster consisting of a point with two more points located just one or two units up on each axis.

When a parallel-coordinates plot is brushed, the trace lines flash on and off, with the trace lines of the selected portions of the data being "on" and the trace lines for the unselected portions being "off." When a parallel-coordinates plot is brushed, only the trace lines for the observations selected are shown.

Figure 8.5 shows two different selections of data in the data gauge. The panels in the figure shows a selected subset of data, the members of the subsets being selected by brushing the data. Thus, when we select three observations, as we have done in both of the panels of Figure 8.5, we see three trace lines, since the trace lines for unse-

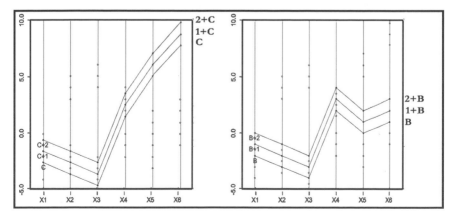

Figure 8.5 Two selections of data gauge clusters.

lected observations are left undrawn. Because we know the structure of our data gauge, we know what to expect when we brush the parallel-coordinates plot. These two panels of Figure 8.5 show two of the three clusters that we built into the data gauge. The left panel reveals cluster C and the right panel reveals cluster B.

Medical diagnosis data. Of course, we do not usually know the structure ahead of time, and the structure is not usually as clear as that which we built into the data gauge. Thus, we return to the medical diagnosis data for a realistic example.

The parallel-coordinates plot of the principal components of the medical diagnosis data is shown in Figure 8.6. Note that principal components (as will be described shortly) create new variables by obtaining variance-maximizing linear combinations of the original data. Thus, in Figure 8.6 we see that the variance of each variable decreases as we move from the left portion of the figure to the right portion.

Actually, Figure 8.6 demonstrates one of the main problems with the parallel-coordinates plot: Even with a fairly small number of observations, the lines overlap so much as to turn the plot into a jumble in which it is difficult to see any structure. This problem can be addressed by repeated application of the following three-step process:

Brush the plot. Brush the parallel-coordinates plot in search of subsets of observations that show similar trace-line profiles.

Change color. Once a coherent subset of observations is found, change the color of its members. This emphasizes the subset's structure.

Hide the subset. Hiding the subset reduces the clutter of lines so that you can more easily see any remaining structure.

We keep cycling through these three steps until no more progress can be made. If we feel that we have hit on a good structure, we can save it; otherwise, we can start over.

We begin by paying particular attention to prominent trace-line features, such as the set of trace lines that are very positive on the first component and those that are very negative on the third component. Brushing each of these prominent features revealed that each feature was part of a pattern shared with numerous other observations, so we

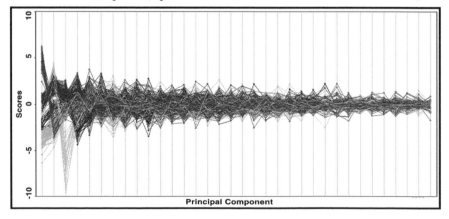

Figure 8.6 Parallel-coordinates plot of the medical diagnosis data.

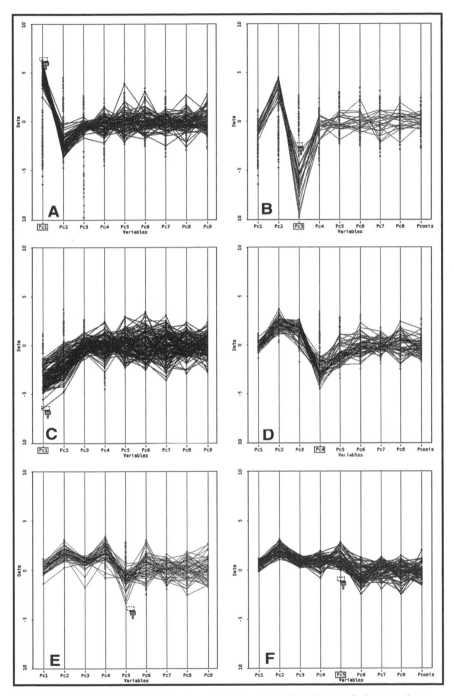

Figure 8.7 Parallel-coordinates plot showing six subsets of observations.

formed two subsets, one for each. Each of these can be seen in the top row of Figure 8.7. Once we had identified a subset, we changed the color of its trace lines and hid the subset. This process was repeated several times until we felt that it was complete, either because a good structure was found or because the process had reached a dead end.

We identified six clusters in five cycles of brushing, coloring, and hiding. These clusters are shown in each of the panels of Figure 8.7, which are arranged and alphabetized according to the order in which there were identified: Subset A was identified, colored, and hidden first; subset B was second; and so forth. It certainly would not have been possible to identify subsets E and F without having identified and hidden the first four. The first five principal components were used to identify the six clusters of data.

8.3.2   Parallel-Comparisons Plot

The parallel-comparison plot is a comparison plot of three or more variables that extends the idea of the bivariate distribution comparison plot discussed in Chapter 7 to any number of variables. This plot provides a simple but powerful way of comparing the distribution shapes of the several variables in your data.

The parallel-comparisons plot takes the fundamental idea of the simple comparison plot (the quantile–quantile plot, or QQplot as it is usually called) and constructs a parallel-coordinates representation of it. The fundamental idea is that of sorting the values of each variable into order, from, say, smallest to largest (or the reverse, it doesn't matter as long as they are all sorted the same way), and then using these sorted values for plotting. This gives us a way of comparing the distribution-generating functions of the variables: If the sorted values are all linearly related, this implies that the distribution-generating functions are all similarly shaped. When this is shown in parallel rather than in orthogonal coordinates, the resulting plot will consist of approximately

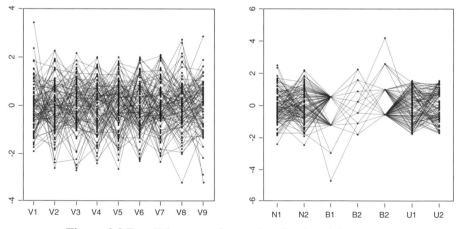

Figure 8.8 Parallel-comparisons plots for two data gauges.

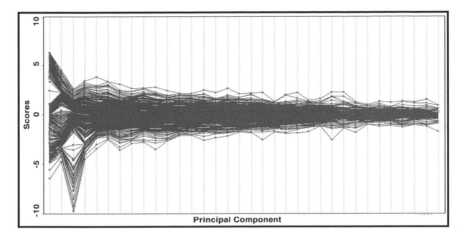

Figure 8.9 Parallel-comparisons plot of the medical diagnosis data.

straight lines. If the observation trace-lines are more or less straight and parallel, the distribution-generating functions for the variables all have similar shapes.

Consider Figure 8.8. The left panels presents a parallel-comparisons plot for nine variables with values sampled from a normal distribution. Our interpretation of this plot is that the trace lines are, at least roughly, straight and parallel, indicating that the generating distributions are all similar. In fact, they are all identical, being the standard normal distribution. The right panel shows a parallel-comparisons plot of seven variables with generating functions that are a mix of different functions, just for reference as an example where the trace lines are not considered to be straight or parallel. A parallel-comparisons plot of the medical diagnosis data is presented in Figure 8.9. We note that the distribution of principal component scores on the first several components is not random and does have structure, but that after the first five or six components the variables all seem to be fairly featureless, although decreasing in variance. This is as we would expect from previous views we have gotten of these data.

8.3.3 Parallel Univariate Plots

Parallel versions of several of the univariate plots presented in Chapter 6 can be quite useful for multivariate data. A parallel version of a univariate plot can be made from any univariate plot that uses only one axis to communicate what it has to say. These include the boxplot, diamond plot, and dot plot. All of the other plots, though unidimensional, have two axes. In fact, the boxplots and diamond plots that we reviewed in Chapter 6, when used with individual observations represented by unjittered dots, are precisely the same as a parallel-coordinates plot ,with a boxplot or diamond plot diagram added for each variable.

To create a parallel version of a univariate plot, you first make several instances of the univariate plot, one for each variable in the multivariate data. Then shape each

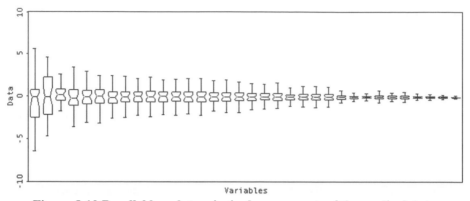

Figure 8.10 Parallel boxplot: principal components of the medical data.

plot so that it is tall and narrow. Finally, arrange these instances next to each other to form a horizontal array of the instances. Now you have a parallel version of the plot. Examples of a parallel boxplot and a parallel diamond plot are shown in Figures 8.10 and 8.11, where the plots are being used to display the principal components of the medical diagnosis data. Applications of parallel univariate plots come to mind, one of which is shown in Section 8.6.2.

As we indicated in Chapter 6, we distinguish between what we call side-by-side univariate plots and parallel univariate plots: Both consist of multiple exemplars of a specific univariate plot drawn in parallel and side by side, so the terminology is a bit arbitrary, but the difference is important. A side-by-side plot (boxplot, dotplot, diamondplot, etc.) consists of two or more plots that represent two or more groups of observations on one variable. On the other hand, parallel plots consist of two or more plots of two or more variables.

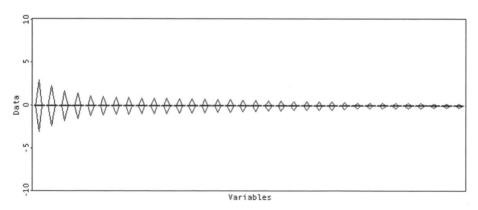

Figure 8.11 Parallel diamond plot: principal components of the medical data.

8.4   Orthogonal-Axes Plots

We begin this section on orthogonal-axes plots for multivariate data with a scatterplot of the medical diagnosis data (see Figure 8.12) even though the scatterplot is a bivariate plot, not a multivariate plot. We begin this way because a scatterplot is an orthogonal-axes plot, and because we believe that starting with the bivariate version of such a plot will make it easier to understand the multivariate versions.

Orthogonality. The scatterplot, of course, is two-dimensional, with the dimensions being drawn at *right angles*, the X-axis corresponding to one of the two variables, and the Y-axis to the other. The two axes are drawn at right angles because this is the geometric rendition of the algebraic notion of orthogonality. Two axes that are algebraically *orthogonal* are geometrically at right angles. They are also said to be mutually *perpendicular*. All three phrases—right angles, perpendicular, orthogonal—are equivalent, all of them referring to the *rendering method* of the plot. We say that a scatterplot is rendered with two axes that are at right angles.

Basis. The basis of an *r*-dimensional space is a set of *r* linearly independent vectors of *r* elements. One set of *r* such vectors, called the *canonical basis*, are the vectors whose elements are all zeros except for a 1 in the *r*-th position. The 2-dimensional space associated with a scatterplot has a basis consisting of two vectors, one that defines each axis, each axis being drawn through the origin and the point with coordinates specified by the basis vector. Thus, the origin [0,0] and the canonical basis vector [1, 0] defines the *x*-axis, and origin [0,0] and the canonical basis vector

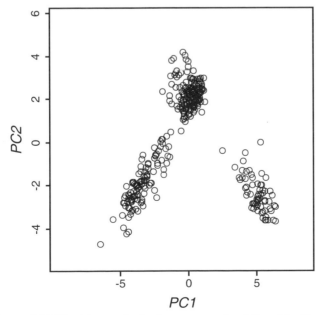

Figure 8.12 First two principal components of the skin disease data.

279

[0, 1] defines the *Y*-axis. The axes of the space are drawn through the origin and the basis points.

A space can have more than one basis, but it has only one canonical basis. Other bases can all be viewed as rigid orthogonal rotations of the canonical basis (and of each other). No basis is "better" than another, although the canonical basis is simpler to use mathematically, since its basis vectors form an identity matrix.

Correlation and orthogonality. Even though a pair of axes are orthogonal, they can represent variables that are correlated, even perfectly (negatively as well as positively) correlated. Orthogonality has nothing to do with the correlation, as these words are generally used for plots. We have just seen how orthogonality is defined: It is a characteristic of the basis of the space. Correlation, on the other hand, is a characteristic of the coordinates of the axes (which are the same as the values observed on the two variables that are being plotted). Thus, in the context of an orthogonal axes graphic, orthogonality and correlation are separate ,unrelated concepts.

8.4.1 Spinplot

A spinplot is a three-variable scatterplot that can be spun by the user or can spin on its own. The spinplot represents the three variables as dimensions of a three-dimensional space that contains points representing the observations. The points are located in the space according to their values on the three variables.

Like a scatterplot, a spinplot can reveal patterns in the relationship between its variables, including the strength and shape of their relationship, and whether there are

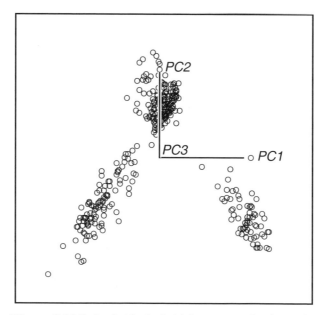

Figure 8.13 Spinplot in its initial unrotated orientation.

clusters of observations. Of course, it can do that for three variables, whereas the scatterplot can do that only for two. Also, like a scatterplot, guidelines can be added to help guide and interpret our exploration. In addition, guide planes and guide surfaces (if you will) can be added.

Figure 8.13 shows an unrotated spinplot, displaying the same view as the view afforded by the scatterplot. Both display the structure as seen in the plane formed by the first two variables being plotted. The two plots differ, however, in that the spinplot has a third dimension, whereas the scatterplot does not. The name of the variable represented by the third dimension of the spinplot is located at the center of the plot shown in Figure 8.13, the name being PC3. The dimension is at right angles to the page, so we do not see it.

The most important feature of a spinplot is that it is spinnable. Rotation is a very powerful tool for understanding relationships among three variables. We typically rotate plots in search of one or more interesting views that cannot be seen in a scatterplot because the interesting view does not align with the plot's axes.

For example, when the spinplot of the medical diagnosis data is spun from the unrotated orientation in Figure 8.13, it only has to rotate a few degrees for it to become obvious that there is a fourth cluster, as we see in Figure 8.14. What looks like a single cluster at the top of both Figures 8.13 and 8.14 is actually two clusters. In Figure 8.14 the space has been spun in a way that reveals the fourth cluster, a long and sinewy distribution of points in the upper-left part of the space.

The rotation can proceed in a noninteractive fashion requiring no input from the user, or it can be tied explicitly to the user's actions. For example, when a spinplot is first shown, it can be shown rotating on its own, or it can be shown without any ongo-

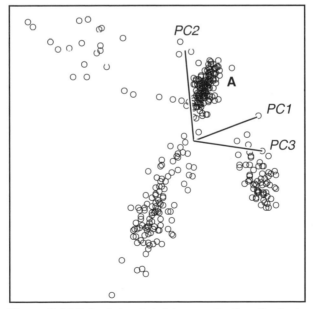

Figure 8.14 Spinplot rotated to reveal a fourth cluster.

281

ing rotation, waiting for the user to spin the space. Usually, when the plot first appears, it is shown rotating, because the rotation shows us different views of the data and the rotation produces a 3D effect while moving, helping us see depth.

The 3D effect is important, as it makes it easier to see the data. In addition to creating the 3D effect by spinning the space, some implementations use point size and/or grayness to enhance the depth effect. Also, "rocking" the space back and forth gives a better feel for depth, and rocking the cloud of points has the same effect as rocking your head back and forth, a movement that one uses to better "pinpoint" just how far away something is. On the other hand, geometries with vanishing points, which are often used in art and architecture to create depth, are not used in statistical visualization because they make judgments of linearity difficult. Also, stereoscopic projections are not used because they require special viewing equipment.

Most implementations provide two ways to control spinning. One way is by clicking buttons on the screen, the other is to use mouse motion. Buttons provide straightforward controls, usually implementing the ability to rotate the space relative to the window's canonical directions—the window's X (horizontal), Y (vertical), and Z (at right angles to the computer screen) axes. There are usually additional buttons to zoom the space in and out and to control the speed of rotation/zooming. See Chapter 4 for details on using buttons to control dynamic graphics such as spinning.

Using the cursor to control movement adds flexibility to the movements that can be created with buttons. The cursor can reorient the point cloud relative to the window, so that spinning the plot along the window's canonical directions provides new views. Also, the direction in which the cursor is moved can be used to create rotation in any arbitrary direction, not just along the window's canonical directions.

Spinplots and high-dimensional space. When a spinplot is used with datasets containing more than three variables the user must determine which three are shown since the spinplot can only show three at a time. Because of this limitation, the plot has the advantage of being easy to interpret correctly, since we are so familiar with three-dimensional space, but it has the disadvantage of showing only three variables, even when there are many more.

Several methods are available to deal with this limitation. One way is to make it easy to switch which of the variables in the dataset are the three that are being shown. Thus, the typical implementation of a spinplot includes easy access to menus with variable names that, when selected, change an axis to display the new variable. A second way is to construct a plot matrix of spinplots, an alternative which we discuss further in the Section 8.5.1. A third way is to extend the concept of a spinplot, which spins in three dimensions, to the concept of an orbitplot, which spins in more than three dimensions. We take this topic up in the next section. Finally, a fourth way is to modify plots (be they scatterplots, spinplots, or orbit plots) to become biplots, an enhancement of an orthogonal axes plot that gives it two (or more) sets of basis dimensions, one of which can be variables and the other, say, principal components. We can then look at the principal components and see all (or many) of the variables simultaneously. We discuss biplots in Section 8.4.3.

8.4.2 Orbitplot

An *orbitplot* is a spinplot that can spin in more than three dimensions simultaneously. Just as a spinplot is designed to help the user visualize structure in three-dimensional data, an orbitplot is designed to help the user visualize structure in high-dimensional data. Figure 8.15 shows the medical data oriented to show five clusters of points, as demarcated by the point symbols. The orientation of the space is indicated by the axes drawn on the plot. The interpretation of the four dimensions is explained in Section 8.4.3.

Like a scatterplot and a spinplot, an orbitplot can reveal patterns in the relationship between its variables, including the strength and shape of their relationship and whether there are clusters of observations. Also, like a spinplot, an orbitplot represents its multiple variables by multiple axes that are algebraically orthogonal, showing the observations as low-dimensional projections of the high-dimensional cloud of data.

Orbitplots were called *tour plots* by those of us who developed them (Asimov, 1985; Buja and Asimov, 1986; Young, 1989; Young et al., 1993). Here we call them orbitplots because when one views points that are spinning in more than three dimensions, one sees the points as orbiting rather than spinning. Of course, points that are spinning are also orbiting, so what's the distinction?

The distinction is as follows: When rotation is in 3D, every point appears to be orbiting in unison with every other point, a kind of movement that we perceive as spinning. On the other hand, when we watch a projection of points that are spinning in more than three dimensions, the points in the cloud appear to be orbiting idiosyncrati-

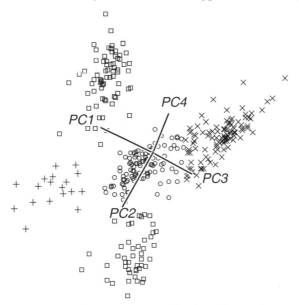

Figure 8.15 Orbit-trails. Orbitplot rotated in four dimensions to reveal five clusters.

cally, each following its own path. What we perceive is that each point has its own orbit.

The orbitplot is particularly good at revealing high-dimensional cluster structure. To find such structure, one watches the orbiting points to see if there are *flocks* of points: groups of points that are always close together as they orbit in the space. The reason we look for flocks is as follows: Points that are truly close together in high-dimensional space will always appear close together in any projection of the high-dimensional space onto a lower-dimensional space. Thus, if we are watching a dynamic projection of the type we are talking about here, and we see a group of points that are moving along similar paths and always staying close together, then it may be the case that they are close together in the high-dimensional space as well. We must say "may be" not "is" because we have not seen, nor will we ever see, all of the projections. But the longer we watch and see no evidence to the contrary, the more confident we can become in our judgment.

Orbit Trails. Because of the importance of orbits, we have developed a way to emphasize them by adding lines to the dynamic graphic that represent the actual orbits. These lines, which we call *orbit trails*, can be used to help identify flocks of points and to explain the static rendition of the plot that one must show in print.

Figure 8.16 shows orbit trails being used with the medical diagnosis data to identify a flock of points. The points selected (the ones with trails) were all in one locale when they were selected. Specifically, the orientation of the space at the time the points were selected was that shown in Figure 8.14, which is the figure showing the spinplot rotated to reveal a fourth cluster. The points in what appears to be a cluster in that fig-

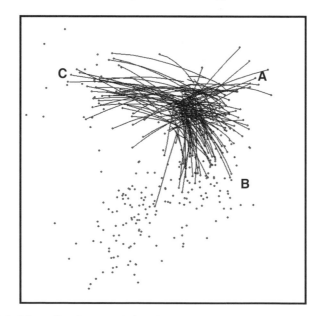

Figure 8.16 Orbit trails. Cluster A in Figure 8.14 looks like three clusters here.

ure, the points labeled A, were all selected. Then, orbiting was started, with orbit trails turned on. Now, at the time the snapshot in Figure 8.16 was taken, we see that what appeared to be one cluster in Figure 8.14 now seems to be three clusters. When added to the three other clusters shown in Figure 8.14 we are up to a total of six clusters.

In Figure 8.17 we present the orbit plot, with tracing on for all points, as it appears when tracing starts from the original principal components orientation shown in Figure 8.13 and proceeds for some relatively short interval (too long an interval creates too much clutter). We clearly see six clusters of trace lines: A and B are the lower two groups of points in Figure 8.13. They continue to look like unitary clusters. C through F started out looking like the one group of points at the top of Figure 8.13, which we see splits into four subgroups.

Finally, notice that Figure 8.17 also has a question mark (located below cluster B). The question mark is at the head of a single orbit line that seems to be headed in a direction that is shared with no other points. Studying this point for awhile reveals that it seems to be an outlier, moving through space in a way that no other point moves.

Orbitplot algorithms. Orbitplots fall into two major classifications, those that are interactive and those that are not. *Interactive orbitplots*, which are called *guided tours* by others, spin as directed by the user, the user creating a guided tour of the data. *Non-interactive orbitplots* (*grand tours*) spin as they wish, taking the viewer on a grand tour of the data.

What you see when you are watching an orbitplot is called the *visible space*. The visible space is a 2D dynamic rendering on the computer screen that is the end result

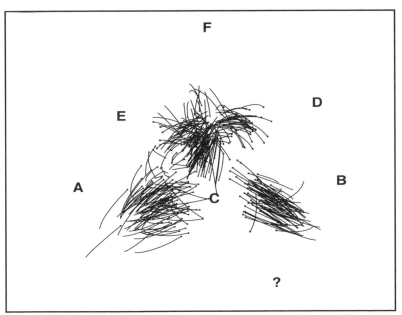

Figure 8.17 Orbit Trails show 6 clusters and an outlier.

of a series of dynamically changing projections of the multivariate (i.e., high-dimensional) data. The details of these projections, which are not presented here, have been discussed by the two groups of researchers working on this technique: namely Buja and Asimov (1988) and their colleagues, and Young (1989) and his colleagues (Young and Rheingans, 1991; Young et al., 1993).

The main problem in designing an orbitplot is how to create the sequence of projections, and how to enable the user to control the sequence in a way that helps understand the structure of the data. This is done by creating a series of "target" spaces, along with the tools needed for smoothly interpolating between the target spaces.

The various methods all begin by defining two target spaces and setting the visible space to the first target. An interpolation is then created to generate a sequence of interpolates, with the visible space always being equal to the interpolate. The interpolation is a trigonometric interpolation, so that every interpolation rotates a bit around the two targets. The interpolations are arranged such that there is a 90° rotation from target 1 to target 2, followed by a 270° rotation back to the original target. Each step of the interpolation is small enough, and the calculations are quick enough so that the resulting display shows smooth and continuous real-time change

Some aspects of the process described in the preceding paragraph can be placed under the user's control so that the interpolation process becomes a dynamic graphic that is interactive. For example, the user can control the speed and angle of the rotation process. This can be done, depending on the implementation, for the axes of the screen or for the dimensions of the space.

The final aspect of the orbitplot algorithm is called *residualization* (Young, 1989) This is a feature that allows the user to redefine the targets, providing a way for the user to see a new path of interpolations through the data's high-dimensional space One way of doing this is to provide the user with a button which, when clicked, creates a new first target that is equal to the visible space (i.e., the space displayed on the screen at the time the button was clicked), and a new second target that is the largest invisible space (i.e., the first three principal components of the residuals that remain after fitting the visible space to the data). The user can residualize repeatedly, providing a way of traversing many paths throughout the data space.

8.4.3 BiPlot

A *biplot* (Gabriel, 1971; Gower and Hand, 1996) is an enhanced scatterplot of the first two principal components of a set of multivariate data. The points of the plot represent the scores that the observations have on these two components. The plot is enhanced by the addition of vectors representing the variables of the original data. Hence the name biplot, as two aspects of the data are shown on the plot, the observations (as points) and the variables (as lines). The biplot for the medical diagnosis data is presented in Figure 8.18.

One may legitimately wonder if this is a multivariate plot —isn't it just a scatterplot with a few bells and whistles? Isn't it a bivariate plot? Well, when looked at from the point of view of how many axes are being plotted, yes, it is a bivariate plot. But there is a more fundamental characteristic of the biplot that makes it truly multivariate—the information being plotted. Since the axes of the plot are the first two principal components, the plot itself is showing us the structure of our data as it appears in the principal plane—the plane formed by the first two principal components, and the plane that shows more variation than any other plane in the entire data space. Thus, the biplot presents the data as seen from a viewpoint that maximizes the variation of the data, and in that sense is certainly multivariate.

Interpreting Points. The points of a biplot are interpreted in the same way as the points of a scatterplot: Points that are close together correspond to observations that have similar scores on the components displayed in the plot. When these components fit the data well, the points also correspond to observations that have similar values on the variables.

Interpreting Vectors. The vectors are drawn from the origin of the plot in a way that makes their direction and length interpretable. Specifically, the vector is drawn so that it is aligned with the direction that is most strongly related to the vector's varia-

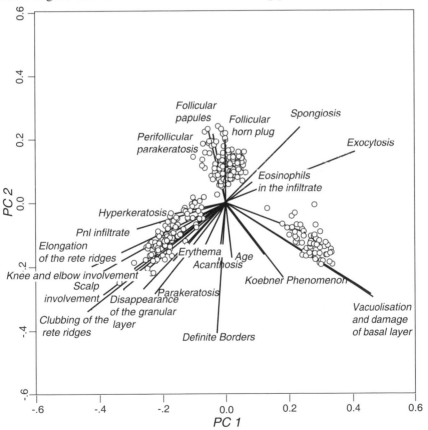

Figure 8.18 Biplot of the skin diagnosis data.

ble, with the length of the vector portraying the strength of that relationship. Thus, as one moves from one side of the plot to the other in the direction given by the vector, one moves from a side of the plot that is typified by little of whatever the vector represents, to the opposite side where there is a lot of whatever the vector represents. This interpretation of the direction holds no matter how long the vector is, but is more accurate for long vectors than for short ones. It follows that vectors pointing in the same direction correspond to variables that have similar response profiles, and, therefore, similar meaning in the context set by the data. Vectors pointing in opposite directions correspond to variables with similar but reversed response profiles, implying there is a negative relationship between the two variables. Long vectors are more strongly related to the components being displayed than are short vectors. Long vectors are more important in interpreting the meaning of the components.

We note that there are several long vectors for which labels are showing in the plot. in Figure 8.18. These variable-vectors have labels such as *Exocytosis*, *Spongiosis*, *Definite borders*, *Band like infiltrate*, etc. Since these vectors are long, their associated variables contribute strongly to the first principal component plane.

Note that some of the long vectors tend to point right through the main body of a cluster of observations. These variable-vectors are particularly important for that specific cluster of observation-points, since they all have a lot of whatever the vectors' variables are measuring. Thus, for the cluster of patients at the lower left side of the space their lab work revealed a lot of *Clubbing of the Rete Ridges*, and quite a lot of *Parakeratosis*. Patients who are not in that cluster have little of these.

The opposite is true for variables that are long but point away from a cluster of points (such as *Spongiosis* and *Exocytosis*). For them, the patients represented by the lower-left cluster of points are patients that show an absence of *Spongiosis* and *Exocytosis*. Regardless, for those variables that are represented by long vectors that are aligned with a particular cluster of patients or are pointing away from a cluster of points, these are the variables that are the best at differentiating the patients in the cluster from those that are not in the cluster.

Orthogonal projection of points onto vectors. There is an additional aspect of interpretation that involves the orthogonal projection of points onto vectors, which shows how much of the vector's variable the point's observation is expected to have. The schematic shown in Figure 8.19 presents details about this aspect of the interpretation of biplots. The schematic displays five points, labeled with the capitol letters A through E, and one vector, labeled X_i. The points are represented by the dots beside the labels. The schematic also shows each point's projection onto the vector. The projection is portrayed by the dashed line, and the projected point is shown by the arrow-tip at the end of the projection path.

If a point projects onto the vector near the head of the vector, then the observation is likely to have a lot of whatever the vector's variable is measuring. Point B in the schematic is such a point. On the other hand, if the point projects near the tail of the vector, then the opposite is true. Point D is a point like this.

Finally, you will note that along the vector are some tic-marks located beside lower-case letters a through e. These lower case letters represent five values observed of variable X_i. Note that the projection of point A falls near a on the vector, as is true for each of the other observations as well. The distance along the vector between the arrow-tip of the projected A and the observed a is the *residual difference* between the observed a and the fitted A that is minimized by principal components analysis.

We can interpret the locations where the arrow-tips of the projections fall along the vector for variable X_i. The locations specify the best estimate, for each observation, of the amount of whatever variable X_i is measuring that is associated with each observation. Thus, as we move along the vector from the tail towards the head we come across projected observations (the arrow-tips) that are estimated to have increasingly greater amounts of the information measured by variable X_i.

Finally, we discuss the so-called *iso-value bars* identified in the schematic. These bars emphasize that several observations that are located in very different parts of the principal plane can be estimated to have the same or very similar amounts of a variable, despite the fact that they are not closely located in the principal plane. In fact, all of the observation points that are located along a given iso-value bar are estimated to have the same amount of the variable.

Non PCA based biplots. Note that biplots can be constructed by using a wide variety of linear combinations other than those determined by principal components

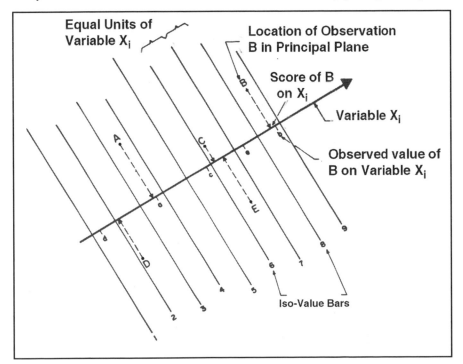

Figure 8.19 Interpreting biplots with point projection.

analysis, the major restriction being that the two sets of linear combinations that are used to construct the horizontal and vertical axes of the plot must be orthogonal to each other. That is, their basis vectors must be orthogonal. Developments have been reported that construct biplots on the basis of canonical discriminant analysis (Young and Sarle, 1982), redundancy analysis (Muller, 1981; Young & Sarle, 1982), Correspondence Analysis (Gifi, 1991; Greenacre, 1993). Any source can be used is the new basis is a set of orthogonal coefficients applied to the variables.

When we say that any way of generating orthogonal linear combinations is acceptable, we do in fact mean just that. In particular, we have no problem with the data analyst simply constructing linear combinations on the basis of experience or theory, without using an analysis method to determine the specific linear combinations. Since the linear combinations must be orthogonal, it behooves the software designer to provide access to routines that orthogonalize a set of coefficients. Alternatively, the coefficients created by a researcher on the basis of knowledge or theory can be treated as data for a principal components analysis, which will not only orthogonalize them but also rotate them into maximum variance orientation. An example appears in the very last step of the Section 8.6.5.

Triplots, and so on. One way of looking at biplots is in terms of the basis of a space. In Section 8.4 we introduced the concept of an orthogonal basis of a space, mentioning that a given set of data always has a canonical basis—the identity matrix that underlies the basic representation of the variables of multivariate data by the axes of the space.

We also mentioned that any given set of data can have other bases in addition to the fundamental canonical basis. In fact, we can look at a biplot as a plot that incorporates two bases: One of the bases is the canonical basis for which the axes are the variables. The second is the principal component basis for which the axes are the principal components. Being able to switch the viewing reference frame is sometimes a powerful tool for looking at structure.

There is nothing to stop us from defining additional bases of the space, bases that provide other kinds of glimpses of the data: Such additional bases could be, say, a basis defined by discriminant analysis, which would allow us to look at the data from a viewpoint that emphasizes the cluster structure of the data, a viewpoint that provides a quite different view than is afforded by the principal components-based basis commonly used with biplots. Having multiple bases allows us to define TriPlots or what might be called QuadraPlots, QuintaPlots, and so on.

Spinning and Orbiting BiPlots, TriPlots, and so on. There is also nothing to prevent us from defining spinning and orbiting versions of biplots (or, for that matter, of triplots, quadraplots, and so on.). An example of such is presented in Figure 8.20. Note that each group of points has it's own group of variables, but that many fewer variables need to be observed.

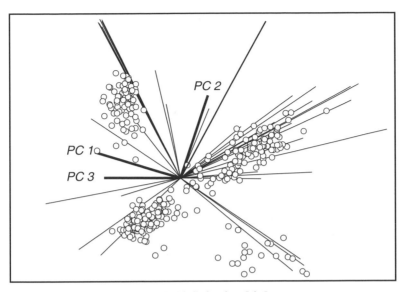

Figure 8.20 Spinning biplot.

8.4.4 Wiggle-Worm (Multivariable Comparison) Plot

The *multivariable comparison plot*, which every viewer inevitably wishes to rename the *wiggle-worm plot*, is a multi-variable generalization of the two-variable distribution comparison plot discussed in Chapter 7. An example is shown in Figure 8.21. This plot, which is a dynamic high-dimensional generalization of the well-known quantile–quantile plot, can suggest whether the several variables come from similarly shaped distributions.

We call this plot the wiggle-worm plot because it consists of a jagged line that spins and changes shape, looking like a wiggle worm, if you will. In its simplest form, the plot is noninteractive. The data analyst simply looks at the plot to see if the wiggle-worm line is usually more or less straight. If so, the several variables all tend to have similar distribution shapes. If it is not straight, but kinks or bends, as is the case here, some of the variables do not have the same distribution shape as some of the other variables. The argument that leads to this interpretation is the same as that presented for the bivariate version of this plot (see Section 7.3.2).

These plots have three or more dimensions, one for each of the three or more variables that are being plotted. The dimensions can be shown, but they add little to the display, it being more effective to have just the wiggle worm showing. All of the plot controls associated with the spinplot and orbitplot can be added to create an interactive version of the dynamic graphic, but there seems to be little to recommend doing this, since the noninteractive version is effective enough.

As with the two-variable comparison plot (the QQplot), the wiggle-worm plot is constructed from the values of the variables after they have been sorted into order: Each variable is sorted separately from smallest to largest (or largest to smallest, it

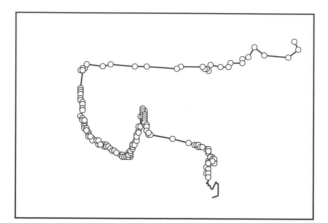

Figure 8.21 Six-dimensional wiggle-worm plot.

doesn't matter as long as they are all sorted in the same direction). Thus, the first observation becomes the one that has the smallest value observed on each variable, the second becomes the one that has the next smallest value, and so on.

A connected-point spinplot or orbitplot is constructed from the sorted values, and the space is spun in however many dimensions it has. If the several variables are all from similarly shaped distributions, the line connecting the points will tend to be a straight line; otherwise, it will not be straight.

8.5   Paired-Axes Plots

A paired-axes plot is an $[r \times r]$ plot matrix of multivariate plots, each showing the same r-dimensional subspace of the multivariate data, but each with its own unique permutation of the dimensions formed from all combinations of r dimensions taken two at a time. Thus, each cell of the plot matrix presents a unique view of the r-dimensional space, and taken together they present all possible two-dimensional views of the r dimensional space. This arrangement of dimensions provides the viewer access to all possible two-dimensional views simultaneously, thereby improving the chances of finding interesting structure. Note that the correspondence between the plot's first two dimensions and the plot matrix row and column numbering is straightforward: Plot-matrix cell $[i, j]$ contains the plot whose first two dimensions are i and j. We have already discussed plot matrices in Chapter 7. The plots we discussed there were multiple bivariate plots, since the plots in the matrix were bivariate. Here the plots in the plot matrix are multivariate, so the overall plot is multiple multivariate.

One of the primary difficulties with plot matrices is that the images in the cells get very tiny when the number of variables being plotted goes up. Thus, while the plot can show us many different views of the same data, these views get uselessly small when the number of variables being plotted exceeds 7 or 8. One solution to this problem is to treat the plot matrix as though it is a control-panel, so that a click on a cell produces a larger version of the cells' plot. One then has many views of the data without being restricted to tiny views.

We do not present an interpretation of the medical diagnosis data with a paired-axes plot because the process is essentially identical to the interpretation using the scatterplot matrix, a process that was shown in sufficient detail in the Chapter 2.

8.5.1 Spinplot Plot Matrix

The spinplot plot matrix for the first four principal components of the medical diagnosis data appears in Figure 8.22. Since we have already discussed spinplots and plot matrices, we do not go into detail here. Note that when all of its plots are in their initial canonical orientation, the spinplot matrix provides the same view of the data as is provided by the scatterplotmatrix, an example of which is presented for these data in Figure 2.6 (allowing for differences in the number of variables being plotted and in the orientation of the diagonal).

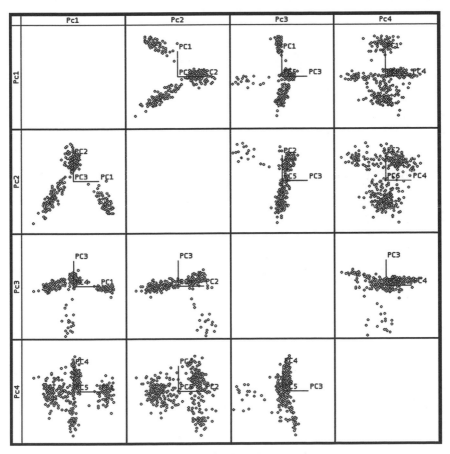

Figure 8.22 Spinplot plot matrix.

Of course, the matrix of spinplots is more flexible than a scatterplot matrix, permitting many views of the data that are not possible with a scatterplot matrix. Note, however, that although we see many of the possible three-dimensional subspaces, we do not see all of them, and while the three-dimensional plots can be spun, they still do not have the potential of showing us all of the three-dimensional subspaces that exist within the data space. The rotations that we can create are all functions of three variables, and there is no way to perform rotations involving more than three variables. Naturally, we could solve this problem partially by creating an orbitplot-plot-matrix, and while we have the potential of seeing all six-dimensional subspaces, the complexity of the task is too great. It is difficult enough to manipulate one orbitplot, let alone an entire collection.

8.5.2 Parallel-Coordinates Plot Matrix

The parallel-coordinates plot-matrix for the principal components of the medical diagnosis data appears in Figure 8.23. The cells of this plot all contain an *r*-dimen-

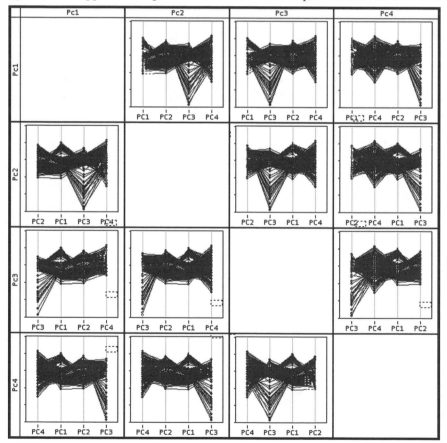

Figure 8.23 The Parallel Coordinates Plot-Matrix - A Paired Axes Plot

sional parallel-coordinates plot, but each cell presents a unique permutation of the dimensions. As you can see by looking at the plot, the permutation has a marked effect on the patterns that you see, suggesting that this plot can augment our ability to see structure in the data. This is likely to be particularly true when the information being plotting is the principal components of a dataset, since there are likely to be just a few components to understand, and seeing many permutations of them will improve our ability to understand our data.

8.6 Multivariate Visualization

We have now finished presenting the individual multivariate graphical methods, having discussed all of the plots in the paired axes, parallel axes and orthogonal axes families. Consequently, since these plots are the constituent parts of the multivariate visualization, we are now ready to visualize the data.

But, of course, you will recall that we have actually been looking at the principal components of the data, because the data themselves cannot be visualized, due to their discrete nature. Consequently, we will be using the visualization of the principal components analysis of the data rather than of the data themselves.

Interestingly enough, the principal component visualization itself consists of two separate spreadplots, a fit visualization and the multivariate data visualization. So, rather than losing out on being able to use the multivariate visualization, we gain, since we can use it plus another. This will provide us with a more complete understanding of our data than we would have gotten had we simply proceeded forward with the multivariate visualization of the raw data, since the data are discrete.

But note: Even for continuous data we recommend using principal components analysis to visualize and explore the data. The visualization not only provides dynamic interactive graphics to help clarify your data, it also provides visualizations about the nature of the data and about the fit of the data by the PCA model, and often, it seems, visualizing the first several components of one's data is much more informative than visualizing the data themselves.

8.6.1 Variable Visualization

We begin with the variable visualization (see Figure 8.24), a visualization that helps us see if the variance of the variables is homogeneous. This in turn helps us decide whether the principal components analysis that we will be performing should be based on the correlations or on the covariances of the variables. The visualization provides information about the variances, medians, quartiles, and other quantiles of the variables. It also provides information about the similarity of the shape of the variable distributions across variables.

The variance of each variable is portrayed in the top panel of the spreadplot. This helps the user of principal components analysis to decide whether the analysis should be performed on correlations or covariances. If the variables all have the same variance, it makes no difference which choice is made. However, if there is a large difference from one variable to the next, the decision may make a big difference. In the upper panel of the three plots in Figure 8.24 we see that the variation in the age variable (the last one on the right) is very much greater than that of the other variables, indicating that the variation is sufficiently different between it and the other variables that correlations should be used rather than covariances.

The medians, quartiles, and other quantiles of the variables are shown by the parallel boxplot that appears in the middle of Figure 8.24, which gives us an impression of the shape of each variable's distribution. This plot is created from standardized variables, so that variation in scale is removed, providing us a view of each variable's distributional shape. Looking at the boxplots for the raw skin disease data, we see that most of the variables are somewhat positively skewed—tail pointing upward. Some are extremely skewed. Although this is not a problem for PCA, it could be a problem if other analyses are contemplated.

The parallel comparisons plot shown at the bottom of the visualization addresses whether the variables have similarly shaped distributions, a requirement for some analyses but not for principal components analysis. If the lines in the plot were more-or-less horizontal, the variables would have similarly shaped sample distributions. They do not, however; thus, once again we stand warned, although principal components is quite immune to these problems.

8.6.2 Principal Components Analysis

It is not crucial that you understand the mathematics in this short section, and for those who wish to do so, skip past it to the next section. However, for those who have had linear algebra or matrix algebra, this section gives a succinct algebraic statement of the principal components model. For more information, see Jollife (2002).

The Model. The fundamental equation for principal components is

$$X = SC' \tag{8.1}$$

Figure 8.24 Multivariate data variable spreadplot.

where X is the data matrix, an $[n \times h]$ matrix of data from n observations of h variables; S is the $[n \times h]$ matrix of the scores of the n observations on the h principal components, and C is an $[h \times h]$ transformation matrix that describes how the original variables are transformed into the principal component variables, and whose transpose describes linear combinations of the principal components that reconstitute the original data. The principal components model imposes orthogonality restrictions:

$$S'S = D^2$$
$$C'C = CC' = I$$

(8.2)

where D^2 is a diagonal matrix, and I is an identity matrix (except for some zeros).

Finally, the maximum variance criterion of the principal components model is achieved by obtaining the set of solutions to the eigenequations

$$\left[(n-1)^{-1} X'X \right] C = CD^2$$

(8.3)

where n is the number of observations. The eigensolution defines the diagonal of D^2 as the eigenvalues and the columns of C as the eigenvectors. We can then solve for S by substitution and algebraic manipulation: The eigenvalues (diagonal of D^2) equal the variance of their corresponding principal components and are often used as a measure of the "summary power" of the component, or of the "importance" of the contribution of the component in summarizing the data.

The entire set of principal components are (1) mutually orthogonal and (2) variance optimal. Because of these two properties, the first several principal components provide the most parsimonious and most accurate representation of your data. There is no set of r linear combinations of the variables that can account for more variance in the variables than is accounted for by the first r principal components. That is, the first principal component is the linear combination of the original variables that retains the greatest amount of information; the second principal component retains the most information possible from what is left after the first, and so on.

Eigenvalues. Each eigenvalue specifies the amount of variation in the data that is fit by its principal component. Thus, if we divide a given principal component's eigenvalue by the sum of all of the eigenvalues (which equals the total variation in the data), we know what proportion of the total variation in the data is accounted for by the specific principal component. That is,

$$p_i = \frac{d_{ii}^2}{\sum_j d_{jj}^2}$$

(8.4)

specifies the proportion of variance fit by each component. We can then take these proportions and for each component obtain the cumulative sum (the sum of all the proportions for the components up to and including the current component). That is,

$$f_i = \sum_{j=1}^{i} p_j, \ 0 < i < n \tag{8.5}$$

is the proportion of variance fit by the i-dimensional principal component space.

Scores. The space formed by the scores S on the first r principal components is the r-dimensional space that best fits the original high-dimensional data space. This characteristic is very useful for exploring multivariate data because the first several principal components provide the most succinct visual representation of the data.

Another way of looking at the principal component scores S is that they are the orthogonal rotation of the original data to the biggest part of the original data space; the first principal component is the longest direction through the original space. The second principal component is the longest direction through the space that is orthogonal to the first component; and so on. The matrix C is the coefficients matrix that orthogonally rotates X to S. It is also the basis of the space that expresses X as S.

Assumptions. The only assumption underlying principal components analysis is that the data are magnitude data. The analysis does not involve hypothesis testing, so we do not need to assume that the data are sampled from a normally distributed population or any other theoretical probability distribution. However, if you analyze data with variables that are asymmetrically distributed, have heteroskedastic variances or show nonlinear relationships among themselves, the results may be distorted by the influence of a few observations, and therefore their interpretation will be uncertain.

Decisions. As was mentioned earlier, the data analyst must make two decisions when using principal components analysis: (1) whether to base the analysis on correlations or covariances, and (2) how many components are enough to fit and describe our data well but still be parsimonious. The variable visualization, presented just above, helps with the first decision. It is obvious that the analysis must be performed on correlations, because the age variable is so very clearly on a different scale than all the other variables. The fit visualization, discussed next, helps with making the second decision.

8.6.3 Fit Visualization

The fit visualization, presented in Figure 8.25, provides two plots that show information about the fit of the model to the data. The plots are a scree plot and a slope ratio plot. This visualization is used after the analysis has been done, but before the model visualization is shown. It is designed to help you decide how many principal components should be retained for further investigation. The decision rests in large part on the nature of the eigenvalues, since they tell us about how much of the variance in the original data is accounted for by each principal component. Therefore, a plot of the eigenvalues is often recommended as an aid to making the decision. The plot—called a Scree Plot after the shape of the rubble at the bottom of a cliff—shows the relative fit of each principal component. It does this by plotting the proportion of the variance of the data that is fit by each component versus the number of components. The plot shows the relative importance of each component in fitting the data. The components will always be sorted according to their relative importance, so initial components will always explain more variance than will those placed in subsequent positions.

Two rules of thumb. There are two rules of thumb that one uses with a scree plot when deciding how many principal components may be interpretable. One is to keep those whose eigenvalues are greater than 1.0, and the other is to look for an elbow in the plot, leaving those which are at or larger than the value at the elbow.

The *eigenvalues greater than 1* rule is suggested because when correlations have been computed the average variance of a variable is 1.0 (indeed, the variance of every variable is 1), so those principal components with eigenvalues greater than 1 account for more than average variance, and those with less than 1 for less than average variance. (Clearly, this is modified from 1.0 to the average variance, whatever it is, when covariances are used as the basis of the components analysis.) For our data, as one can see in Figure 8.25, this rule of thumb tells us that as many as seven components may be meaningful. Experience suggests that this may be taken as a rough "upper limit" on the number that will prove to be useful.

The second rule of thumb says that we should look for an "elbow" in the plot—in other words, for a place where the plot bends more tightly, with the decision as to the number of components that may be useful corresponding to the number at the location of the elbow. For our data the elbow seems to suggest that we interpret five components. Although experience suggests that this is often a good guess as to the number that will prove to be useful, experience also suggests that it is usually pretty difficult to decide where the elbow is. Since what one is doing when one is looking for the elusive elbow is making a judgment about the slope of the curve of the scree before the point in question as compared to the slope afterward, a curve that specifies exactly that is shown in the bottom panel of Figure 8.25. This curve, which is a function of the ratio of the slope of the curve before and after the point, clearly shows that the maximum change in slope is at five components. As we now know, that number proved to be a good estimate.

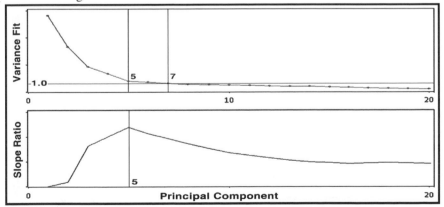

Figure 8.25 Criteria for Selecting the Number of Components

8.6.4 Principal Components Visualization

The PCA model visualization spreadplot, which is shown in Figure 8.26, is constructed from six graphics (plus three namelists that are not shown). The graphics are a scatterplot matrix in the upper left, with a parallel-coordinates plot next to it, and an

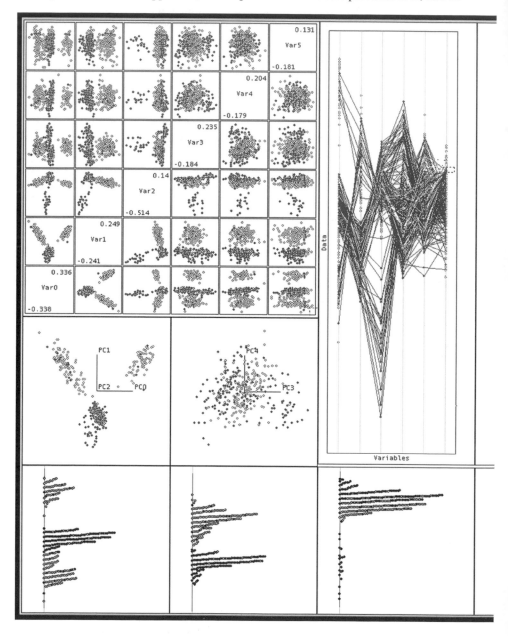

orbit plot completing the top row. Below the scatterplot matrix are two spinplots, the left-hand one showing dimensions 1 to 3 and the right-hand one showing dimensions 4 to 6. In the bottom row is a parallel jittered dotplot.

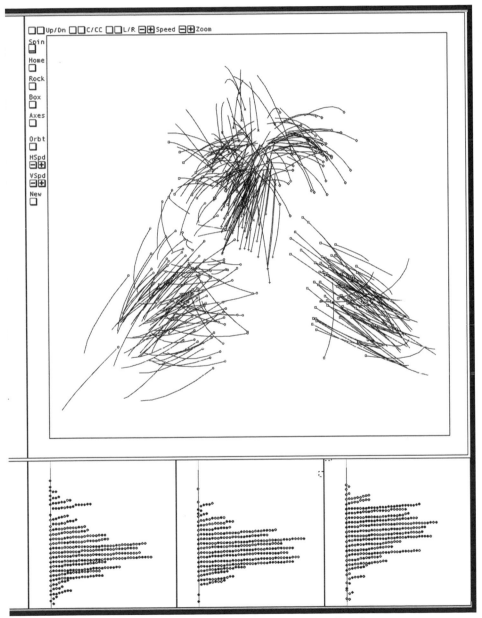

Figure 8.26 Principal components visualization

This spreadplot is specialized to communicate structure whose dimensionality is known approximately. Thus, with these data, the graphics are constructed to provide information that is six-dimensional. Each plot in the figure is showing the first six variables (which are the first six principal component score variables of the medical data). The namelists that are not shown include a list of all the observations (for identifying the points), a list of all of the variables, and a list of all of the principal components. By selecting six variables the user can change the variables that are assigned to each plot in the spreadplot, the order of selection being retained in the presentation. Thus, any permutation of any combination of six variables can be inspected. Also, clicking on the scatterplots in the scatterplot matrix changes the order of the dimensions in all of the other windows—the two variables of the scatterplot that was clicked on becoming the first and second variables of the other plots (counting the two 3D spinplots as one plot). We do not have much to say about the visualization, because we have already said what needs to be said while we were using these data with the graphics we were introducing.

8.6.5 One More Step - Discriminant Analysis

When the data include grouping information, such as the clusters that we have formed during the course of our analysis, we can use that information as part of the visualization process. Specifically, if we are really interested in seeing the cluster structure, we should no longer use principal components analysis to look at the data. Afterall, it is showing us the data as they look from a viewpoint that shows the maximum variation in the data, not one that shows the cluster structure as well as possible. What we want to do is to take a look at the data from a viewpoint that best shows the group structure. We would be particularly happy with a view that maximizes the variation between groups relative to the variation within groups. Such a view would separate the groups as much as possible, while at the same time it would make the groups as compact as possible. The analysis that does this is called *discriminant analysis*. We turn to a discussion of it next.

When we perform discriminant analysis we obtain orthogonal linear combinations of the magnitude variables such that the fitted values that result from the linear combinations fit the grouping information as well as possible (in the specific sense of maximizing the ratio of the between-group variance to the within-group variance).

Geometrically, discriminant analysis rotates the data space into an orientation such that the resulting "first" dimension (the first linear combination) is the one that gives us the view of the data that makes the groups look as compact as possible and at the same time look as well separated as possible. The second dimension does the same thing, given that it must be orthogonal to the first dimension.

We carried such analysis in the following way. For each variable in the raw data, we calculated the mean of the values of the variable for each of the groups that we had formed. We then formed a matrix of means. This matrix had the same number of variables and observations as the raw data, but every raw observation of a variable was replaced by the mean of the variable for the group it was in.

We then did a principal components analysis of this matrix—the matrix with the individual values observed replaced with their group means—providing us with a space such that the first dimension is the one which has the greatest variance in the means—equivalent to the most widely separated group means, weighted by the number of patients receiving the diagnosis.

We then used the first few principal components of the means—we used six—as a target to which we rotated the raw data, the thought being that if we could orient the raw data so that they were as much like the principal components of their group means, that when we looked at the resulting plane or 3d space it would show us the best view of the raw data's group structure.

The conclusion of these analyses is shown in Figure 8.27, where we clearly see the six group structure, although two of the groups, while clearly separated, do show some overlap. The figure is drawn with each member of a group connected by a line to the group's mean, to emphasize the group structure. So, for the first time, we can see the six groups without the need for movement related features. That is, for the first time we have a picture that not only convinces us that we can see six groups in the data, but is also capable of convincing readers of a printed article as well.

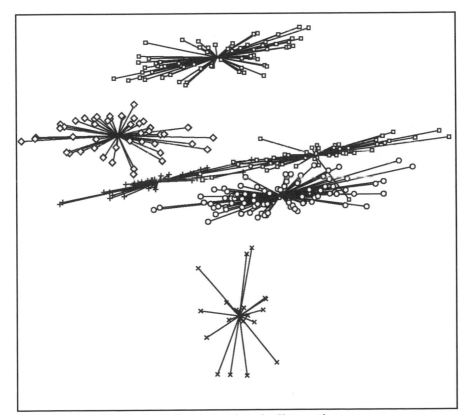

Figure 8.27 Plane showing six diagnostic groups.

8.7   Summary

8.7.1 What Did We See? Clusters!

Let's pause for a moment and summarize what we have done with the medical diagnosis data—and what we have found out about the data.

8.7.2 How Did We See It?

We have looked at the data extensively, using a wide variety of graphical methods designed to help us find structure in multivariate magnitude data. We have, in fact, used at least one example of each family of methods. Using one trick or another, each graphing method attempts to show us structure that has more dimensions than we can comprehend—parallel-axes plots show us many axes at the same time, albeit in an unfamiliar parallel orientation—orthogonal-axes plots show us 6 or 8 dimensions at the same time, attempting to communicate that elusive high-dimensional structure via motion—paired-axes plots more or less take the easy way out, by showing us all combinations of two or three dimensional plots, leaving us with the problem of trying to comprehend them all together.

Orthogonal-axes plots. The orthogonal-axes plots clearly show the clustery nature of these data. The scatterplot of the first two principal components (Figure 8.12) clearly shows there are at least three clusters, easily passing the IOI (interocular impact) test. The spinplot of the first three components, when rotated (Figure 8.14), shows that one of the three clusters we saw in the scatterplot splits into two, giving four clusters, also breezing through the IOI test. The orbitplot of the first six components also quite easily shows that there are five clusters, once again satisfying the IOI test. Although it was easy to see this with a dynamic graphic, it was a struggle to find a view that revealed the five groups when displayed as a static graphic on the page, although such a view was found (Figure 8.16). Finally, adding orbit-trails to the orbitplot revealed the sixth group of points, with the orbits quite clearly showing the additional structure in the dynamic plot. They also aided in preparing a static printed image (Figure 8.17) showing all six groups.

Parallel-axes plots. The parallel-axes plot did not clearly display the cluster structure, and it was difficult to uncover the structure. The main problem with the plot is that the lines obscured whatever structure there might be to see. To cope with this problem, one has to identify and hide whatever clusters can be seen, a process that makes the process of discovering structure much more problematic, since the prior history of steps that you have taken is shaped by the decisions that are available for you to make at any given time.

There is also the problem that, as argued above, although you see all dimensions, you don't see the entire space. What you see is a series of planes through the space, the first one being formed by axes 1 and 2, the next by axes 2 and 3, the next by 3 and

4, and so on. So if the interesting structure is in planes 3 and 5, you won't see it unless you change the order of the axes.

There is one aspect of the parallel-coordinates plot that at least partially compensates for the problems just identified, and that is that we can see structure in more dimensions simultaneously than can be seen with other methods. Perhaps this means that the best use of a parallel-axes plot is in conjunction with a linked orthogonal-axes plot, where the main decisions are based on the orthogonal-axes plots, with the reasonableness of all decisions being double-checked using parallel plots.

Paired-axes plots. In Chapter 2 we used a scatterplot matrix in association with enlarged versions of the scatterplot to investigate these same data, coming to remarkably similar conclusions. Note, however, that we followed the usual way of carrying out this task, which is to use the paired-axes plot as a control panel. The small plot cell scatterplots are just too small to see, so when one is clicked on a plotcell, a large version of the plot in the plotcell is created for the user to use.

This is, in a way, an admission that paired-axes plots (i.e., plot matrices) are not as effective a tool for seeing data as are orthogonal-axes plots and parallel-axes plots, the problem being that the plotcells are simply too small to discern fine structure. They are very good at communicating large structure (see, e.g., the frontish piece of Chapter 2), and very good as a control panel to manage access to more powerful graphics.

Since clicking on a cell of the paired-axes plot produces a large orthogonal-axes plot, most of the work reported in Chapter 2 is based on orthogonal-axes plots, not on paired-axes plots. However, since the paired-axes plot provides simultaneous access to many views of the data, we are justified in identifying the methods as being different from the usual single-view orthogonal plot.

Principal components visualization. While we didn't dwell very long on the principal components visualization, it was in fact the vehicle by which we were shown the various plots that we discussed in earlier sections of the chapter. Thus, it was of major importance.

Discriminant analysis. Once we discovered that we had a clustery structure in our data, we were ready to use some form of discriminant analysis to look more carefully at the cluster structure. It was the only approach that clearly revealed the six clusters in a static plane, a very useful view for scientific communication.

8.7.3 How Do We Interpret It? With Diagnostic Groups!

You may recall that at the end of the presentation of these data in the Chapter 2 we mentioned that the data also included the doctor's diagnosis of each patient. It is interesting to compare our visual classifications, each of which was done without using the knowledge of the diagnoses, with the diagnostic classification. This comparison results in what is called a confusion matrix, a table that shows the frequency with which members of each of our visual classes had already, without our knowledge, been assigned to each of the diagnostic classes. If our visual classification agrees exactly with the diagnostic classification, the confusion matrix will have the same number of classes as diagnoses, and the frequencies will fall along the diagonal, there being zeros in all off-diagonal cells. To the extent that our visual classification is "confused," there will be nonzero frequencies off the diagonal.

The confusion matrices all appear in Figure 8.28. We see that our visual classes correspond very closely to the diagnostic classes: All of the patients diagnosed with either psoriasis, lichen Planus, or pityriasis rubra pilaris were visually classified correctly by each of the three plot families, as was also the case with chronic dermatitis for two of the three plot families. Of the 250 patients diagnosed with one of these four diseases, only 2 of the 750 visual classifications disagreed with the diagnostic classifications. The only misclassification problems of concern occurred between seborrheic dermatitis and pityriasis rosae, where 16 or 18 of the 108 patients with one of these two diagnoses were misclassified visually. Of course, there could also be misdiagnoses by the doctor—we can't tell.

Overall, the plots allowed us to classify the data quite well visually: With paired-axes plots we agreed with the doctor's diagnosis 336 out of 358 times (93.85%); with parallel-axes plots we agreed 338 of 358 times (94.41%); and for orthogonal-axes plots we were in agreement with the doctors 342 of the 358 times (95.53%). Our visual accuracy also compares favorably with that of a special algorithm named VFI (Guvenir et al., 1998) which was in agreement with the doctors 345 of the 358 times.

8.8   Conclusion

In this chapter we presented multivariate plots, noting that they fall into three families, according to the way the variables are represented by the axes: orthogonally, in parallel, or in pairs.

We have used them with a set of data concerning medical diagnosis of skin disease data to demonstrate how these plots are used. We discovered that they are considerably more difficult to use than the univariate and bivariate plots presented in the Chapters 6 and 7, but then the problem of trying to uncover structure that may be more than three-dimensional, in high-dimensional discrete data is a much more difficult problem than we were pursuing in the earlier chapters. We also performed a principal components analysis and discussed the relationship between its visualizations and the visualization for multivariate data. We also compared it with a discriminant analysis that clearly showed the data structure. Remarkably stable views of the data structure were obtained with several different graphing techniques, suggesting that the data are strongly structured and that when one has strongly structured data the graphical techniques are capable of helping you discover it.

Paired-Axes Plots
Visual Clusters

Diagnosed Disease	A	B	C_1	C_2	C_3	C_4	Sum	Errors
Psoriasis	111						111	
Lichen planus		71					71	
Pityriasis rubra pilaris			19		1		20	1
Pityriasis rosae				39	8	1	48	9
Seborrheic dermatitis		1		10	48	1	60	12
Chronic dermatitis						48	48	
Sum	112	71	19	49	57	50	358	22

Key: The letters A, B, etc., refer to the letters in Figure 2.7.

Parallel-Axes Plots
Visual Clusters

Diagnosed Disease	C	A	B	F	E	D	Sum	Errors
Psoriasis	111						111	
Lichen planus		71					71	
Pityriasis rubra pilaris			20				20	
Pityriasis rosae				46	2		48	2
Seborrheic dermatitis				16	44		60	16
Chronic dermatitis					2	46	48	2
Sum	112	71	20	62	48	46	358	20

Key: The column captions C, A, etc., refer to the letters in Figure 8.7.

Orthogonal-Axes Plots
Visual Clusters

Diagnosed Disease	A	B	F	C	E	D	Sum	Errors
Psoriasis	111						111	
Lichen planus		71					71	
Pityriasis rubra pilaris			20				20	
Pityriasis rosae				45	3		48	3
Seborrheic rermatitis				13	47		60	13
Chronic dermatitis						48	48	
Sum	111	71	20	58	50	48	358	16

Key: The column captions refer to the symbols in Figure 8.17.

Figure 8.28 Confusion tables for three visual clusterings vs. diagnosis.

9   **Seeing Missing Values**

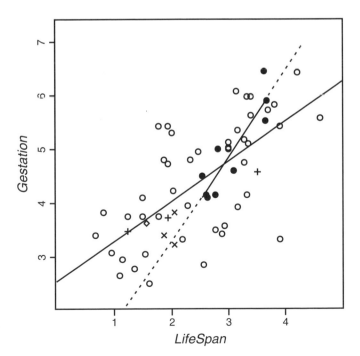

9 Seeing Missing Values

Missing values are values that the researcher had planned to collect, but for some reason beyond the control of the researcher, did not collect. Data that have missing values have "holes" in the dataset like those schematized in Figure 9.1. Data with missing values are very common. In fact, in any given empirical situation, it is probably more likely than not that there will be missing data.

There are many reasons why there can be missing data. Here are some real-world examples:

- People usually do not like to provide information about their income, so some people may refuse to answer some questions.

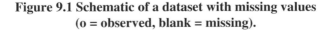

**Figure 9.1 Schematic of a dataset with missing values
(o = observed, blank = missing).**

311

- Manufacturers may fail to give details about their products for various reasons (trade secrets, concern about competition).
- Students do not attend all of their exams, and we wonder if the missing values on test exams would have been very low.
- Some samples can be destroyed due to mistakes in the manipulation process, and we ask ourselves if this is truly random.
- Data are unexpectedly not available, due to ethical, organizational, or practical issues, and there is just nothing that we can do.
- The data are impossible, so the research records them as missing, as when supposedly male rats become pregnant, even in the most controlled situation.
- The very nature of the subject makes it difficult or dangerous to get the data. It may be dangerous to find out how much time some mammals sleep (the example we use throughout this chapter).

9.1 Introduction

Although missing data are very common, discussion of how to analyze data with missing values has been often, well, missing! It is especially noticeable that most, if not all, introductory and middle-level textbooks on statistics ignore the possibility of missing data. In these books, the examples are overly immaculate, with all the rows of the data tables complete. However, close examination of these examples often reveals that they have been manipulated to produce such spotless tables of data. Furthermore, when methods are discussed for coping with missing data, the procedure known as casewise deletion is often recommended, despite the fact that it can produce biased results based on severely reduced sample sizes.

So why is it that a problem of such importance is not taught routinely in basic material on statistics? Perhaps the fundamental reason that most books on statistics ignore the problem of missing values is the complexities that result when we do not ignore the problem.

Consider what happens when we wish to compute the mean of each of the variables in a dataset. When there are missing values, we can no longer divide the sum of the values of a variable by the number of cases in the sample because we must recognize the fact that the number of cases varies for each variable. Two alternative ways of computing the mean when there are missing values come to mind: One way is to use all of the nonmissing values for a given variable, summing them all and then dividing by the number of non-missing values for the variable. This approach has the desirable feature of using all of the available cases for each variable, but it has the undesirable characteristic that each mean is based on a different subsample of the data. The second way that comes to mind is to delete all the observations that have one or more missing values and then to compute the means of each variable. This has the nice characteristic that each mean is based on the same cases, but it suffers from the fact that we have to throw away a number of non missing observations. The problem is still worse for covariances, since each pair of variables may have a different combination of observed and missing values.

We begin to appreciate the extent of the difficulties associated with missing values when we realize that many of the standard data analysis techniques (i.e., regression analysis, principal components analysis, analysis of variance, etc.) start by computing means and covariances. If some of the values in the data matrix are missing, computing the means and the covariances becomes more difficult. This problem has been handled traditionally by simple such as casewise deletion of data or formulae that use only the cases available. Better alternatives have been developed in recent years. In particular, the estimation-maximization (EM) algorithm (Little and Rubin, 1987; Schafer, 1997) provides maximum likelihood estimates for the means and the covariances that are preferred to the means and covariances provided by traditional approaches.

Although missing values complicate the computational methods, there is a much more fundamental difficulty: The pattern of missing values can be much more of a problem than the number of missing values. Of course, the greater the number of incomplete data, the greater the loss of information, and hence the lower the statistical precision. However, if the missing values are scattered randomly throughout the data matrix, most methods for dealing with missing values will produce similar results, whereas when the probability of obtaining a missing value depends on other variables in the analysis (or, even worse, on unknown or unmeasured variables), the statistical results can be biased.

Missing values are also a problem for statistical visualization. The usual way of coping with missing values is simply to delete the observations that are incomplete and then to visualize the remaining observations. Of course, this is unsatisfactory for the same reasons that this approach is unsatisfactory when it is used with any other statistical method. But there are additional reasons why this approach is unsatisfctory for dynamic interactive graphics. In particular, when there are several linked plots, the linking methods no longer work correctly when each plot drops those observations that are incomplete for the variables that the plot is using at the time.

On the other hand, missing data should not be regarded as a nuisance that can be ignored. On the contrary, with the right tools, missing data can be a valuable source of insight for data analysis. So, although visualization of missing data is, in a sense, a contradiction, as you can not see what you do not have, this chapter will show that, there are information in the missing values that can come to the surface using the appropriate tools. Thus, there may be outliers that would otherwise remain hidden, imputed values that fall in interesting locations, and patterns of missing values may tell something about the propensity of certain subgroups of observations to generate them.

Chapter outline. In this chapter we focus our attention on the visual exploration of missing data. We do not consider issues related to modeling and inference for data with missing values. See Schafer (1997) or Little and Rubin (1987) for more on this topic. Given that our primary interest is in visual exploration of data with missing values, we pay particular attention to:

1.   What the obstacles are when we wish to explore data with missing values, and how to avoid these obstacles as much as possible.

2.   How to visualize the data after plausible values have been imputed from information in the observed parts of the data matrix.

3.   How to display patterns in the grouping of cases that have missing values in the same variables. Patterns that stand out from the rest of the patterns may constitute subgroups of observations deserving special attention.

This chapter starts by introducing a dataset about the sleeping behavior of various mammals. These data have missing values as a consequence of the problems of determining the sleep behavior of certain animals. We then explore this dataset, paying attention to the limitations of ordinary visualization tools and suggesting modifications that make explicit the missing values in the data. These modifications are designed so that the researcher will always have a graphical impression of the existence of the missing values and of their relation to the observed values.

We then turn to a discussion of the different imputation techniques that can be used to partially recover the missing information, and to the methods for displaying these imputed values in plots that include values both observed and imputed. We point out that the process of imputing missing values can generate useful substitutes for unobserved values, emphasizing that when the imputed values are used to replace the missing values, the visualizations will work as they normally do. We point out that the imputed values should always be marked in some way so that it is obvious which elements of a plot are imputed and which are not.

Finally, we discuss several plots designed to visualize patterns of missing values. These plots display those variables that have missing values, showing them as well as their summary statistics (means, variances, and covariances). Visualization of these patterns are a good companion to standard inferential techniques designed to test the mechanisms of missing data, as they can provide insight about the significance or lack of significance of the results attained with them. In addition, patterns help us check assumptions related to the mechanisms that may have generated the missing values. These assumptions can be assessed using inferential techniques, but to provide a new example of the old remark "numerical summaries are not enough" (Chambers et al., 1983), plots that display the elements that are combined to produce these summaries help us see important features that otherwise would remain unseen.

9.2   Data: Sleep in Mammals

The function of REM (rapid eye movement) sleep is one of the great mysteries in the field of sleep and wakefulness. It seems that REM sleep must have an important function because almost all mammals have it. One piece of evidence that has been used to investigate its function is the correlation of characteristics of animals with the amount of their REM sleep. Variables that have been found to be correlated with REM sleep in mammals are: measures of the amount of non-REM sleep, safe sleep conditions,

Table 9.1 Variables in the Sleep in Mammals Data[a]

Variables	Description	Number of Missing Values
NonDreaming	Amount of non-REM sleep (hours/day)	13
Dreaming	Amount of REM sleep	11
TotalSleep	Total of sleep (dreaming + nondreaming)	3
BodyWeight	Logarithm of body weight (g)	0
BrainWeight	Logarithm of brain weight (g)	0
LifeSpan	Logarithm of maximun life span (years)	4
Gestation	Logarithm of the gestation time (days)	4
Danger	0 = least danger; 1 = most danger (from other animals)	0

[a] Note that TotalSleep = Dreaming+NonDreaming. Thus, it is always possible to compute the value of an observation for any of these three variables if the value of the other two is known.

and inmaturity at birth (Siegel, 1995). We will use a dataset collected by Allison and Cicchetti (1976) that has variables related to these measures. There are 61 cases in this dataset. The variables are described in Table 9.1.

This dataset has been used to illustrate the technique of multiple regression analysis, where any of the three first variables is taken as the response variable and all or a subset of the other five are used as predictor variables. These data have missing values in some of the variables, but the published analysis usually excludes missing cases. This will result in a loss of cases ranging from 19 (models that use *NonDreaming* as dependent variable and include *LifeSpan* and *Gestation*) to only three (models that use *TotalSleep* as the dependent variable and *BodyWeight* and *BrainWeight*).

9.3 Missing Data Visualization Tools

We have mentioned previously that conventional graphical techniques often simply ignore missing values. This is unfortunate, as the analyst is left without the opportunity of checking a part of the data that may be informative. In the special case of interactive plots, the consequences of this approach are even more important, as many of the basic techniques may not work properly.

Consider linking. When several linked plots display different variables, each variable having its own unique pattern of missing values, an observation selected in one of the plots will not necessarily link to the same observation in another plot. The problem stems from the lack of response in some of the plots, which have a missing value in some of the variables displayed, but not in others. Also, other interactive actions, such as selecting the missing values in a given variable and seeing their values observed in other variables, are simply not possible.

One solution to these problems is used by the Manet (Missings Are Now Equally Treated) program (Unwin et al., 1996). This program incorporates modifications to basic plots and introduces new plots that can use missing values, thereby ensuring that missing values will always be represented. The specific way that each plot represents the values depends on the characteristics of that type of plot.

9.3.1   Missing Values Bar Charts

Manet introduces this plot as a way of displaying the proportion of missing values in a variable. This plot is a rectangle split in two, with black and white parts representing, respectively, the proportion of values observed and missing in the data.

9.3.2   Histograms and Bar Charts

Manet incorporates missings in histograms by adding a bar of a different color that counts the number of missing values in the variable displayed. This bar is active, so that it can be used to select cases. The cases in the bar can be linked to other plots. As the bar for missing values could be confused with the bars for values observed, a small gap separates it from the rest of the bars. Also, plots specialized to categorical data, such as mosaic and barcharts plots, include the missing values as an additional category automatically.

9.3.3   Boxplots

Boxplots are not easily modified to show missing values. Manet solves this problem by showing automatically in the boxplot window a missing values chart.

9.3.4   Scatterplots

Scatterplots show the observations with missing values in one of the represented variables as points on the axis of the plot. The scatterplots also have boxes representing the proportion of missing x values, the proportion of missing y values, and the proportion of missing x and y values.

Figure 9.2 shows a scatterplot inspired by Manet's proposals. Two variables from the mammals data are displayed in this figure: *LifeSpan* and *NonDreaming*. values observed are represented by circles. Missing values are represented by filled circles. Cases that are missing in both variables simultaneously are displayed in the plot by a separate white rectangle that indicates the proportion of cases with regard to the total. This plot helps us see that missing values for the variable *NonDreaming* happen more often for large values of the variable *LifeSpan*. On the other hand, missing values for the variable *LifeSpan* are located at about the center of the distribution for the variable *NonDreaming*. Finally, we also see that the proportion of missing values occurring simultaneously in both variables is very low because the missing bar chart placed in the lower part of the plot is predominantly white. Manet does not include other plots

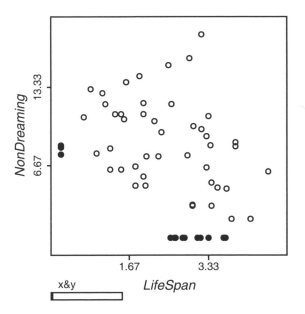

Figure 9.2 Scatterplot with missing values displayed on each axis as filled points and in a separate bar.

for numerical data such as 3D plots or scatterplot matrices, but it seems possible to extend the same ideas to these other displays quite easily.

Compared with the traditional way of treating missing values in many statistical systems, Manet's approach is unquestionably sensible. However, using imputation techniques, as we discuss in the following section, we can further improve the plots. Imputation provides us with reasonable values that are computed taking advantage of the observed part of the data. This way, the missing values can be displayed almost as if they were values observed.

9.4 Visualizing Imputed Values

Ad hoc methods for imputation of missing data have been around for many years. However, over the last decade theoretically sound statistical methods for imputing missing values have become available. Furthermore, these methods offer promising approaches for adapting visualization methods to the realities of missing data.

There are a number of issues involved in using imputation methods for statistical visualization:

- Imputed data must be marked uniquely to differentiate them from observed data, and the resulting visualizations must include this marking in order to reflect the special status of the values (Swayne and Buja, 1998) .
- Although there are several imputation methods, not all of them can be recommended. In particular, it is well known that the "quick and dirty methods" can-

317

not be recommended because they are simplistic methods that distort the data. These methods include replacing missing values with unconditional means or removing cases with missing values. These difficulties apply whether the data are imputed for inference or for visualization. Figure 9.3 and the discussion associated with it is an example of the danger of using inappropriate imputation metods.

- Even though good imputations will be less harmful than bad imputations, there is still the issue that the values obtained do not include any indication of the uncertainty of their estimates. Multiple imputation (Rubin, 1987; Schafer, 1997) is a technique that improves over single imputation because it provides information about the variability of imputations. Plots that include this information are of interest because the analysts can obtain insight about the variability associated with the imputed observations.

We discuss these issues in the remainder of this chapter. We use the sleep in mammals data as an example.

9.4.1 Marking the Imputed Values

Before we start discussing methods for imputing reasonable values, we will introduce the way of marking the values in the plots so we do not forget that they are not observed values.

Imputed values can be considered to be reasonable approximations of the unvalues observed. However, it is important not to confuse them with observed values. Therefore, we include in the plot reminders that they are not actually observed and that they should be interpreted with care. The way to remind us of this fact usually consists of marking the imputed values differently from the rest of the values in the plots. This way, the analyst can, on the one hand, obtain general overviews of all the data, imputed and not imputed, and, on the other hand, appreciate the areas where the imputed values tend to behave in an interesting way. Thus, in this section we discuss some ways in that the imputed values can be marked in plots.

The strategies for marking the imputed values in plots generally will differ depending on the types of plots considered. In this section we focus on the plots that use elements for representing each individual observation, such as dotplots, scatterplots, and so on. Other types of plots may require different strategies than used for these plots, but they are not considered here for reasons of space.

Plots based on points can be modified in three ways to indicate that some of them do not represent values observed but that they have been imputed. The modifications consist in using different symbols, colors, and/or sizes. Assuming that we have started with a multivariate data matrix where some of the variables had missing values that have been imputed, these modifications will apply to different plots in the following way:

1. *Dotplots/boxplots/parallel plots.* In this case the imputed values can be marked with a different feature, such as a different color/symbol/size. The

same feature can be used in several plots. Histograms that have individual bricks for each observation can use different colors for displaying imputed values.

2. *Scatterplots.* In bivariate data, missing values can occur in any of the variables separately and also in both simultaneously. Figure 9.3 shows a scatterplot of the variables *LifeSpan* and time *NonDreaming*. The method used to impute values is the unconditional means method discussed in section Section 9.4.2. Observed points are displayed as hollow circles. Missing values in the variable *LifeSpan* are displayed with an ×, and missing values in the variable *Non-Dreaming* use a +. Finally, observations with missing values in both variables are displayed using a ◊. This plot is in many aspects very similar to the plot in Figure 9.2 but with the difference that it does need any special addendum for missing values in both variables.

3. *Spinplots.* The same strategy as that used for scatterplots can be applied to spinplots of three variables. However, if all combinations are to be displayed in the plot, eight (rather than four) symbols or colors are needed to represent all types of missing values. In practice, the number of patterns of missing values is probably not going to be so large, so the number of symbols/colors would be lower.

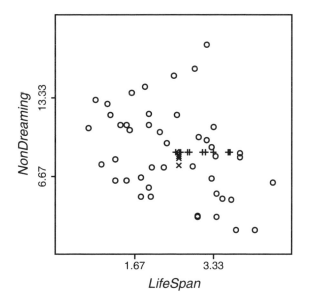

Figure 9.3 Scatterplot with imputation of unconditional means. Imputed values for variables are marked with the symbols: *LifeSpan* **(×),** *NonDreaming* **(+), and both (◊).**

4. *Scatterplot matrices.* Using different symbols in scatterplot matrices for each
    pattern of missing values generates too many symbols. A more convenient
    approach consists of treating each scatterplot individually as described before.
    For example, we would know that a point marked with an × in one of the scat-
    terplots would correspond to an observation that has a missing value in the
    variable displayed on the row of the scatterplot matrix in which this scatterplot
    is located. Notice that this observation would also be marked on the scatter-
    plots along that row with an ×, except if the variable on the vertical axis is also
    missing, in which case it would be marked with a ◊ (see Figure 9.5). This
    marking allows identifying the other variables in which an observation would
    be missing simultaneously. Interactively, this would be accomplished by
    selecting an observation and looking along the corresponding row or column
    of the scatterplot matrix.

Marking the values in the plots is important to avoid the pitfall of interpreting an
imputed value as an observed value, as we shall see in the next section.

Even though it is convenient to emphasize that imputed values are not *real* values,
we expect them to be credible values, so that we will be able to use them for obtaining
insight about our data. However, not all the imputation methods are equally appropriate for producing reasonable values. In the following two sections we review methods
appropriate for this endeavor.

9.4.2   Single Imputation

In this section, we introduce single imputation methods and discuss its application to
the sleep in mammals example. The first method of imputation that we explain, *single
imputation* (Schafer, 1999), estimates a single value for each missing value. Single
mputation is in contrast with *multiple imputation*, a method whereby, using simulation, each missing value receives several imputed values. Multiple imputation has the
advantage over single imputation that uncertainty about the missing values can be
introduced in the statistical model such that intervals of confidence around the estimations can be computed. On the other hand, single imputation requires less effort and
can be also acceptable when the number of missing values is small. Also, as multiple
imputation methods can be understood as extensions of single imputation methods, it
is useful to review them before discussing the others. The methods are:

1. *Unconditional means.* This method consists of substituting the mean of the
    variable for the missing values. This method is available in some statistical
    packages even though it presents considerable disadvantages and should, in
    general, be avoided. The reason is that this method inflates the number of val-
    ues in the center of the distribution and understimates the variance of the vari-
    able (see Little and Rubin, 1987, p. 44). The covariances among the variables
    are also underestimated. From the point of view of statistical visualization, this
    method has the disadvantage that the imputed values are laid out in lines paral-

lel to the axis as shown in the Figure 9.3 on page 319 for the variables *LifeSpan* and *Dreaming*.

2. *Conditional means.* With conditional means, the values used for the imputation are obtained by conditioning on the observed part of the incomplete cases. The method was proposed by Buck (1960, cited in Little and Rubin, 1987) and it consists basically of predicting each missing value using the values observed in a particular pattern of missing values. This method underestimates the variance of the completed data (but less than the unconditional means) and inflates the covariance of the variables. More serious problem is that the imputed data are predicted values without any component of error and give an overoptimistic sense of precision. This suggests inmediately the strategy of adding a suitable random component to the data that is function of the residual variance. Multiple imputation methods (discussed in the next section) can be considered as extending this strategy and combining it with the method in the following paragraph.

3. *Estimation-maximization (EM) algorithm.* EM can be understood as an extension of the conditional means method. In EM, after the missing values have been substituted by a first round of imputed values, the parameters of the data are recomputed, and a new imputation is carried out. The same process is repeated until convergence. The advantage of the method is that the imputed values are not only a function of parameters of the observed part of the data, but also use the information contained in the nonobserved part *given* the observed part of the data. The EM algorithm is a general strategy apropriate for different types of data (Little and Rubin, 1987) but the one that has been implemented most often corresponds to incomplete multivariate normal samples. Notice that the estimation-maximization algorithm has as its main purpose the computation of maximum likelihood estimates for the parameters of the data (e.g., means, variances, covariances), and not of imputations of values. However, the parameters so computed can be used to compute estimations of missing values similar to the conditional means method, but the imputed values still lack a component of error.

We will show an example of single imputation using the EM algorithm applied to the mammals data. As mentioned in the preceding paragraph, the most common implemented versions of the EM method assume that the data follow approximately the normal multivariate distribution. A way to check that assumption approximately is the scatterplot matrix shown in Figure 9.4. Scatterplot matrices are limited to the univariate and bivariate distributions and do not allow direct inspection of more than two dimensions. However, as discussed in Section Section 7.4, this can be a reasonable approach to the problem in many cases.

Figure 9.4 shows that all the bivariate plots look well except for the dummy-coded categorical variable *Danger*. This would be a problem if this variable had missing values, because some of the imputed values would probably fall out of the 0 to 1 range. However, *Danger* does not have any missing values, and consequently, this problem does not arise here. Note that as indicated in Table 9.1, some of the variables have been transformed using logarithms in order to improve the symmetry and linearity of the data. Also, it is important to mention that we assume that the mechanism of missing data is *ignorable*. (See Section 9.5.2 to learn more about assumptions with missing data.)

The scatterplot matrix in Figure 9.4 suggests that it is reasonable to use the EM algorithm to compute the maximum likelihood parameters for the data. Using these parameters, imputed values for the missing data use most of the information in the data observed. Also, once the data have been imputed, it is possible to visualize them using the marking strategies discussed in Section 9.4.1. In our case we examined several scatterplots for pairs of variables until we found one that seemed to us to be of special interest. This scatterplot is shown in Figure 9.5 and corresponds to the vari-

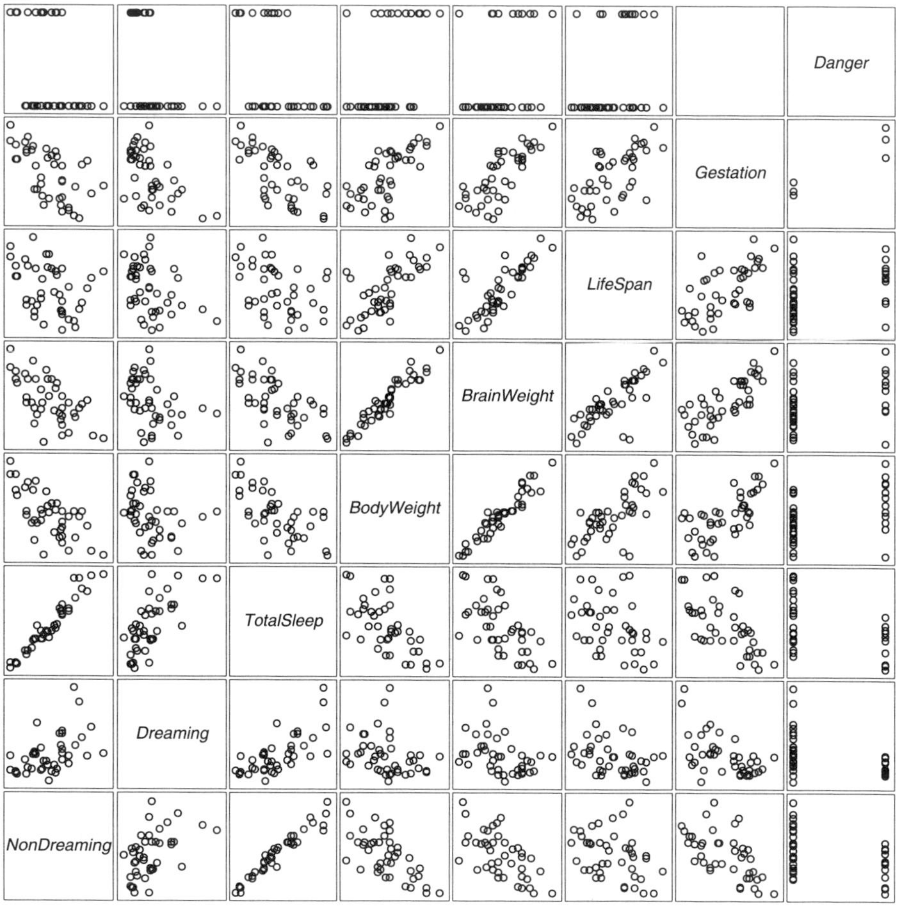

Figure 9.4 Scatterplot matrix for the sleep in mammals data.

ables *LifeSpan* and *NonDreaming*. Note that the imputed values in Figure 9.5 do not lie at the means of the variables as they did in Figure 9.3, which used the unconditional means method. So, although the imputed values actually fall on fitted hyperplanes, they do not portray evidence of regularity when they are displayed in lower dimensionality views.

The special feature we saw in Figure 9.5 is the group of imputed points located at the lowest right side of the plot. These points are marked with a +, denoting that they were observed in the variable *LifeSpan* but not on *NonDreaming*. Selecting and labeling the points in this area provided some additional insight on these observations. So the labels of observations with missing values are the African Elephant (placed very close to the observed Asian Elephant*)*, the Roedder, the Donkey, the Giraffe, and the Okapi. Also, there are five animals selected in Figure 9.5 that have no missing values for the variable *NonDreaming* (Asian Elephant, Goat, Sheep, Horse, and Cow). Interestingly, these mammals are all ruminants, a group of animals that generally have low *TotalSleep;* they are also in the *Danger* species and presumably share other features. So the parallel-boxplots plot in Figure 9.6 displays the profiles of the mammals selected previously along all the variables in the dataset. These profiles reveal that the mammals selected feature low sleep time (low values at *Dreaming, NonDreaming,* and *TotalSleep*), large *BodyWeight* and *BrainWeight*, and long *LifeSpan* and *Gestation* time. They also are species in *Danger* (except the African Elephant).

The plot in Figure 9.6 can be used for exploring the missingness of the observations selected because they are marked according to the strategies for boxplots outlined in

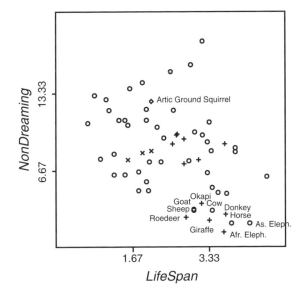

Figure 9.5 Scatterplot with values imputed using the estimation-maximization method.

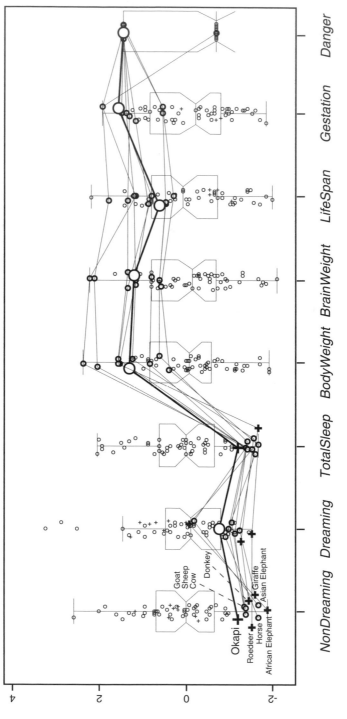

Figure 9.6 Paralell boxplots with values imputed using estimation maximization.

Section 9.4.1. However, this black-and-white printed, static figure suffers from over-lapping, a problem that would be solved in the interactive version by selecting each of the observations individually. Also, colors on the computer screen stand out more clearly than do black-and-white symbols. Hence, to improve the effect of this plot, we have enlarged the size of the symbols (the Okapi has a symbol larger than the others).

Note that missing values in the first three variables (*Dreaming, NonDreaming* and *TotalSleep*) must happen in at least two of them at the same time. This is a conse-quence of the variable *TotalSleep,* being the sum of the variables *Dreaming* and *Non-Dreaming*. In Figure 9.6 we have selected observations that have missing values in any of these three variables. So the most common pattern of missing values among the selected observations is one in which the variable *TotalSleep* is known, but not its split into time *Dreaming* and *NonDreaming*. Two typical examples of observations in this pattern are the African Elephant and the Donkey. For these animals, it may be dif-ficult to differentiate time *Dreaming* from time *NonDreaming*.

Two of observations selected in Figure 9.6 keep a pattern different from the typical pattern described in the preceding paragraph. The pattern consists of knowing the time *Dreaming* but not the *TotalSleep* or their time *NonDreaming*. Two mammals are in this situation: the Giraffe and the Okapi. This pattern is striking because it seems to us easier to record the variable *TotalSleep* than the variables *Dreaming or Non-Dreaming* separately. However, this does not seem to be true for Giraffes and Okapis. An explanation of this pattern is that Giraffes spend most of the night standing up, so it is difficult to know if they are sleeping, except in the short periods of time that they place their heads on the ground. This time seems to consist of REM sleep and so can be observed more easily than the total sleep and nondreaming times. Okapis the clos-est relative of giraffes, are also difficult to observe.

In summary, the plots in Figures 9.5 and 9.6 are useful because they offer a more complete picture of the dataset analyzed than do equivalent plots that exclude the missing values. The advantage stems from two sources:

1. Cases with missing values are not completely removed from plots.

2. Although the imputed values are only a reasonable approximation to the "true" values, and their interpretation must be done very cautiously, they can be of enormous help in understanding the missingness and its causes.

Single imputation is an interesting strategy for exploration of data with missing val-ues. However, there is not information about the confidence that can be placed in the imputed values. The following section, which describes multiple imputation, offers a way to go beyond this limitation.

9.4.3 Multiple Imputation

As indicated in the Section 9.4.2, single imputation does not provide any measure of the uncertainty of the quantities estimated. Rubin (1987) suggested multiple imputa-tion as a way of overcome this problem. The idea of multiple imputation is to compute several simulated versions of the missing values that incorporate a random component. The random component reflects uncertainty about the missing values in such a way that the various draws would cover the space of the possible results. These results can later be combined to produce estimates of the parameters and, more important, confidence intervals of the estimates. Imputations can be computed using different methods but one that has been implemented in several packages can be regarded simply as a stochastic version of the EM algorithm (Schafer, 1997).

If several imputations are computed for a missing value, and assuming a normal distribution of the imputations, we can estimate the standard error of the mean of the imputations for the value using the formula

$$e_i = \frac{\sigma_i}{m-1}$$

where e_i is the error estimated and σ_i is the standard deviation of the m imputed values for the missing observation i.

The error associated with each imputation can be used for adding intervals of confidence to the plots as shown in Figure 9.7. The scatterplots in these figures use horizontal lines for displaying 95% intervals of confidence around values that are missing in the variable in the abscissa and vertical lines for the variable in the ordinate. When the value is missing in both axes, the symbol turns into a diamond that displays the intervals for both variables simultaneously. The interactive version of this plot only shows the symbols for the intervals of confidence when the points are selected, so the graphical impact of the lines can be controlled by the user.

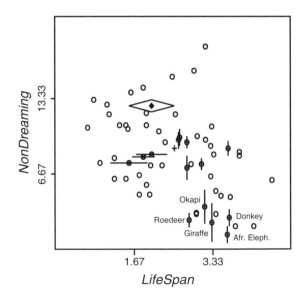

Figure 9.7 Scatterplot with intervals of confidence computed using 10 multiple imputed datasets.

The imputed values for the ruminants in Figure 9.7, which have low *NonDreaming* and high *LifeSpan*, have in general short intervals of uncertainty. Therefore, we can trust that the real values for those animals will not differ much from the estimations.

Note that the intervals for some observations are larger than for others. Focusing on the labeled observations, the Roedeer, Donkey and African Elephant have small intervals and those around the Okapi and the Giraffe are somewhat larger. Of course, the differences in size depend on having good predictors for the missing values. Thus, in the sleep in mammals example, the three variables related to sleep are highly related among themselves —e.g. the correlation of the variable *TotalSleep* with *NonDreaming* has a value of 0.97—and consequently, observations with *NonDreaming* missing and *TotalSleep* observed will be predicted very accurately. This is what happens with the Roedeer, Donkey and African Elephant. On the other hand, the Okapi and Giraffe have only the value for the variable *Dreaming*. As the correlation of this variable with *NonDreaming* is of 0.53, the confidence intervals are larger for these two observations.

9.4.4 Summary of Imputation

In this section we have shown that imputation can be a powerful tool for exploring data with missing values. The data imputed allow us to build visualizations that are similar to those for complete datasets but that include all the cases in our sample. Of course, the imputed data must not be interpreted as if they were values observed. However, by marking them, the analyst can always see if a given value is imputed or observed and in this way carry out interpretations of individual values as appropriate. Also, using multiple imputation, it is possible to add intervals of confidence to the imputed values, such that the analyst can have a visual impression of the trust that he/she can put on them.

Notice that this section has focused on the exploration of the individual imputations. This is an unusual approach to the topic of missing data, which generally is more concerned with the problems of inference of parameters under the presence of missing data. For a more complete treatment of these issues, the reader can consult, for example, Schafer (1997).

A theme that has been lurking behind the previous discussion is of patterns of missing data. Patterns of missing data are defined by the variables that have or do not have missing values for a group of observations. If the pattern has as observed variables those that are good predictors of the variables with missing values, their estimations are more accurate. This is an example of the conclusions that can be derived from patterns of missing values. Hence, in the next section we show methods and possible derivations that can be obtained from the exploration of patterns of missing data.

9.5 Missing Data Patterns

A pattern of missing data indicates the variables in a group of observations that have missing or values observed. We have seen in Section 9.4 that patterns of missing data can be a source of valuable information about data. Thus, for example, two observations in Section 9.7 had values observed in a variable—*Dreaming*—that seemed more difficult to record than the variables that had not actually been observed (*TotalSleep* and *NonDreaming*). This strange pattern was eventually explained by the fact that two mammals, Giraffes and Okapis, are able to sleep while they stand up, but they lay down when in REM sleep. Hence, there are more evident cues associated with the state of *Dreaming* than for *TotalSleep*, as it can be difficult to know whether or not an animal is actually sleeping when it stands up.

Exploration of the patterns of missing values in our data may inform about:

- The patterns themselves and the number of cases in each of them.
- The mechanisms that may have produced the missing data.
- The summaries of the values observed given the missing data patterns.

Notice that the number of different patterns of missing data can be large for multivariate datasets and that it is convenient to visualize the observed and imputed part of the data simultaneously. Therefore, to obtain a complete account of the missing data patterns, the analyst can make good use of special tools tuned to this problem. In the following subsections we review some visualizations that have been designed specifically for the exploration of missing data patterns.

9.5.1 Patterns and Number of Cases

The first thing to do with missing data patterns is to list them. Also, the count of cases in each of them is important. A bar chart or a table can be used for this purpose. Table 9.2 shows the patterns for the mammals data. The table has been sorted accord-

Table 9.2 Patterns of missing data in the Mammals Data[a]

NonDr	Dream	TotalS	Body	Brain	LifeSp	Gestati	Dange	**Count**
×	×	×	×	×	×	×	×	42
×	×	•	•	•	•	•	•	9
•	•	•	•	•	•	×	•	3
•	•	•	•	•	×	•	•	2
×	•	×	•	•	•	•	•	2
×	×	•	•	•	×	•	•	1
•	•	•	•	•	×	×	•	1
×	×	×	•	•	•	•	•	1

[a](×=observed; •=missing)

ing to the number of cases in the patterns. Thus, it is posible to see that the pattern for *Complete Data* (no missing values) is the largest in our data, followed by the pattern with missing values in the variables *NonDreaming* and *Dreaming*. Other patterns with fewer cases are listed below.

As patterns with a high number of missing values are probably affecting more the estimations of parametters than those with fewer, it is important to check if the means, variances, and so on, of the observations in those patterns differ much from the observations in other patterns (especially from the *Complete Data* pattern). Therefore, an important use of Table 9.2 is to identify patterns with many cases.

However, the possibilities of Table 9.2 for data analysis are not limited to show patterns that are large. Thus, we can see in it that there is a mammal so mysterious that researchers have not been able to determine its *LifeSpan* or its *Gestation* time (Desert Hedgehog) and also that there is a mammal whose sleep behaviour is completely unknown (Kangaroo). Therefore, Table 9.2 is a source of valuable insight for data analysis, especially if linked to plots and labels of the observations in the missing data patterns.

Table 9.2 can be re-sorted in other ways, so it is possible to put together patterns with similar variables. Interactive capabilities such as displacing rows manually, or sorting according to one or several variables, make possible explorations of this type.

The mammals data have only eight different patterns of missing data, so their study is easy using a table. However, datasets with many variables may have many more patterns of missing data. Also, to know the count of cases in each pattern is normally only the first step to be undertaken in the exploration of this aspect of the data. Thus, statistics such as means, variances, and covariances displayed by pattern are usually required also. We will see in the following sections visualizations especially tuned to this problem, but first we discuss some preliminary theory that will be of help in that endeavor.

9.5.2 The Mechanisms Leading to Missing Data

One of the main uses of the analysis of the patterns of missing data is to shed light on the mechanisms producing the missingness. Knowledge of those mechanisms is of considerable interest because it may guide the strategy to follow in the analysis of data. In this section we first describe a widely accepted classification of the mechanisms of missing data, and thereafter, a statistical test that has been proposed for checking the mechanisms that underlie the data at hand. As patterns of missing values play a central role in this test, it is interesting to provide visualizations that provide insight about the various factors that take part in the computation of such single summary value. These visualizations are the goal of the Section 9.5.3.

The mechanisms that may have produced the missing data may be classified as (Dempster et al., 1977; Little and Rubin, 1987):

- *Missing completely at random* (MCAR). This mechanism implies that the missing values are a random subsample of the data. This case happens when the probability of response is independent of the values of the variables. Therefore, summary statistics like means or variances should be the same for the data observed and for the non-observed. Likewise, the matrix of variances-covariances for the whole data-matrix should be equal for the complete and for the incomplete data.

- *Missing at Random* (MAR). This MAR mechanism is less restrictive that MCAR. It assumes that the probability of response on a variable *Y* with missing values may be related to other variables *X*, but not with itself. Despite its name, MAR does not mean that missing values are a simple random subsample of all values. MAR is less restrictive than MCAR because it requires only that the missing values behave like a random sample of all values within subclasses defined by data observed (Schafer, 1997). In this case, incomplete data have different summary statistics than those of complete data but it is assumed that the data observed possess enough information to recover the lost information, helping to estimate the correct summary statistics. MAR and MCAR are said to be *ignorable* missing data mechanisms.
- *Non Missing at Random* (NMAR). Sometimes, researchers will have additional information on the data that lead them to be suspicious of data being MAR or MCAR. In this case, it is supposed that there is a nonmeasured variable that could be related to the missing values. This is called a *nonignorable missing data mechanism*. This situation is particularly problematic because there is no test that that can be used to evaluate whether or not the missing data mechanism is random.

Most of the literature concerning missing data is based on the ignorable mechanisms of MAR or MCAR. However, it is important to mention that *ignorability* can be regarded as relative. Although a missing data mechanism may not be known by the researcher in full, there can be variables that can explain the *missingness* to a lesser or greater extent. Including a number of the variables in the analysis that explain the missingness will make the assumption of MAR much more plausible (Graham et al., 1996; Schafer, 1997). Although there are no tests for NMAR, it is possible to check if data are MCAR versus the less restrictive situation of MAR.

One of the most intuitive analyses for MCAR is computing the differences in variables between observed and missing values for the rest of variables. Tests for the differences of means are also sometimes reported. Significant differences for the tests can be taken as evidence that data are not MCAR. However, even though this procedure can be considered informative, it produces a large number of *t*-tests that are difficult to summarize, because they are correlated with a complex structure, depending on the patterns of missing data and the correlation matrix (Little, 1988).

The previous strategy could also be applied to visualizations of data. Plots could be constructed for all the variables for the missing and nonmissing parts of each variable. However, due to the presence of missing values in the variable displayed, the size of the groups would not always be homogeneous. Also, exploring bivariate relationships among variables for missing and nonmissing values of other variables would be more complex, as the plots for different variables might vary widely in the observations displayed.

A test that evaluates if the mechanism of missing values in a dataset is MCAR or MAR was suggested by Little (1988). Little's test evaluates the differences between the means observed for patterns of missing data and the maximum likelihood estimations obtained using the EM algorithm and has the following formula:

$$d^2 = \sum_{j=1}^{J} d_j^2 = \sum_{j=1}^{J} m_j (\bar{y}_{\text{obs},j} - \hat{\mu}_{\text{obs},j}) \hat{\Sigma}_{\text{obs},j}^{-1} (\bar{y}_{\text{obs},j} - \hat{\mu}_{\text{obs},j})^{\text{T}} \tag{9.1}$$

where there are p variables, J patterns of missing values, m_j observations in pattern j, μ and Σ are the maximum likelihood estimates of the vector of means and the matrix of covariances obtained using the EM algorithm, and obs, j are the subsets of the parameters corresponding to nonmissing observations for pattern j. Finally, $\bar{y}_{\text{obs},j}$ is the p-dimensional vector for the sample average of data observed in pattern j.

This test has a χ^2 distribution with degrees of freedom equal to

$$\sum_{j=1}^{J} p_j - p$$

where p_j is the number of variables observed in pattern j. Notice that the test above can be regarded as a sum of normalized Mahalanobis distances of each pattern with respect to the maximum likelihood means. Examination of individual contributions to the test can be used to find out which patterns contributed most to the test. However, straightforward comparisons of the sizes of the terms can be misleading, and corrections are necessary to explore these contributions (Hesterberg, 1999).

The test assumes that the matrix of covariances is the same for all patterns. Little (1988) and Kim and Bentler (2002) provide tests for homogeneous means and covariances. However, this test is quite limited, due to patterns with few cases and is not reported os often as the test for means.

Although Little's test offers a convenient summary of the deviations of the parameters from the maximum likelihood estimators, it is always interesting to check the patterns individually. For example, in the mammals data, Little's test for means returns 39.80 with 42 degrees of freedom (p = 0.57) so we woud not reject the MCAR hypothesis. However, looking at Table 9.2, we can see that the variable *LifeSpan* is involved in several patterns with few cases each. But as examination of equation (9.1) makes clear, patterns with few cases can not contribute much to increasing the value of the test, and also,the existence of more patterns increases the degrees of freedom of the test. Therefore, if the variable *LifeSpan* is excluded from the test, we may expect that the MCAR hypothesis will be rejected more easily. Indeed, the output of Little´s test after excluding this variable is 30.54 with 20 degrees of freedom (p = 0.06), which is marginally significant. This result alerts us to the dangers of accepting the results of the test without examining the contributions of the individual patterns.

9.5.3   Visualizing Dynamically the Patterns of Missing Data

We saw in section Section 9.5.2 that the MCAR mechanism involves homogeneity of means and covariances and that a statistical test exists for checking this assumption. However, visualizations of the components of these formula may provide insight into the characteristics of each pattern that would otherwise remain hidden.

In this section we discuss visualizations that address, on the one hand, means and variances, and on the other hand, covariances and correlations of the data. DIG techniques are useful for these visualizations because they allow the user to explore step by step the information in the patterns and focus on those that present more interesting features. The first visualization to be described uses raw data and does not involve imputation. The other two are focused on summaries of the data computed by patterns, such that the summaries can be compared easily among themselves and with the maximum likelihood estimates of the statistics. These two plots are the most closely related to Little's test and hence can be a useful companion to it.

Parallel diamond plots for observations in missing data patterns. Figure 9.8 a shows diamond plots for the observations in two of the missing data patterns of the mammals data. The plots are made in the following way: First, the mean and the standard deviation of each of the variables is computed using the available observations. Second, the variables are standardized using these means and standard deviations to

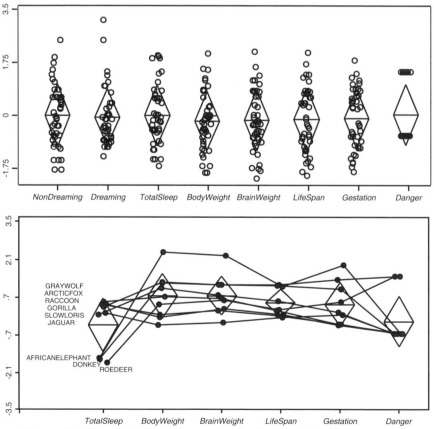

Figure 9.8 Diamonds for observations in the *Complete Data* (top) and *Dreaming/Non Dreaming* (bottom) missing data patterns.

have mean 0 and standard deviation 1. Third, observations in each of the pattern are selected and plotted separately. Fourth, diamond plots for the means and standard deviations of the observations in the pattern are superimposed over the observations in the pattern.

Figure 9.8a shows the observations in the *Complete Data* pattern (no missing values in any of the variables). Figure 9.8b shows the observations in the pattern with variables *Dreaming* and *NonDreaming* missing. The plots can be interpreted as showing if the observations for each variable in a pattern are different from the remainder of the observations. If the observations for a variable in a given pattern are not different, the points will have mean zero and unit standard deviation, and consequently, the diamond for these observations will be centered in the zero line and will have a height of 2. On the contrary, if the observations are different, the diamonds will be uncentered in the zero line or will not have a height of 2. This plot is meant to be used interactively, by selecting one pattern after the other from a list to see the changes in the plot.

Figure 9.8 top shows that the means and the standard deviations for the variables in the *Complete Data* pattern are similar to those for all the available cases. On the other hand, Figure 9.8 bottom, reveals that the same statistics for the pattern of missing values *NonDreaming/Dreaming* differ from those for the available cases. The differences are that the mean of the observations for this pattern is lower for the variables *TotalSleep* and *Danger* and higher for the rest of variables than the means for the available data.

The plots discussed here have the appeal of being based on raw data, without resorting to any special procedure or computation. Furthermore, using selection and linking, they can be implemented using the techniques described in Chapter 4.

In the next section we review plots that follow a different strategy. Using imputation to fill the gaps, all the variables and all the patterns are displayed in the plot. This results in a very complete representation of the data that is amenable to more sophisticated exploratory data analysis.

Diamond plots for missing data patterns. The display discussed in the preceding subsection has the disadvantage that it requires a different plot for each missing data pattern and comparisons among patterns are complex because the variables represented at each plot are different. One way of reducing the number of displays consists of using imputation of the missing values in order to obtain completed representations, similar to those discussed in Section 9.6 but modified to take into account patterns of missing values.

Figure 9.9 is an adaptation of diamond plots designed to represent the summary information in a dataset after it has been filled using imputed data. This plot is meant to be used interactively to view the distribution of values for a given pattern. The different parts of the plot refer to the following:

- Points stand for the means of observations in a missing data pattern for a variable. The symbol $\Diamond$ stands for cases observed. If the symbol for the point is an $\times$, the cases used to compute the value of the point have been imputed.
- Horizontal lines placed over each point indicate the relative size (the number

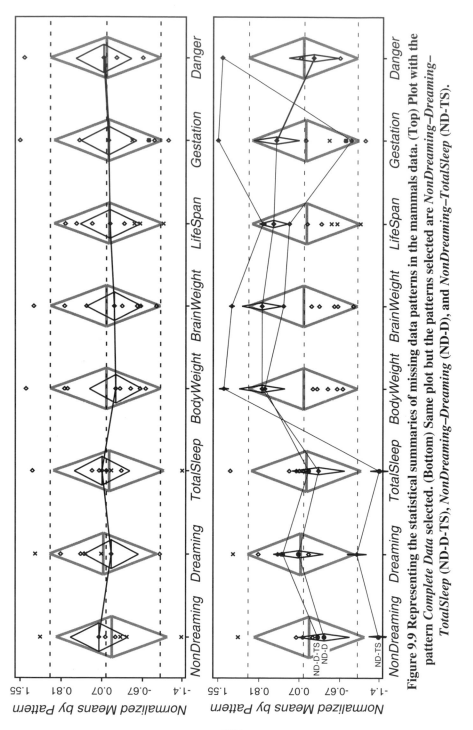

Figure 9.9 Representing the statistical summaries of missing data patterns in the mammals data. (Top) Plot with the pattern *Complete Data* selected. (Bottom) Same plot but the patterns selected are *NonDreaming–Dreaming–TotalSleep* (ND-D-TS), *NonDreaming–Dreaming* (ND-D), and *NonDreaming–TotalSleep* (ND-TS).

of cases) of each pattern. Larger lines correspond to patterns with more cases.

- Large diamonds drawn using gray lines are located at the maximum likelihood means for the variables. The height of the diamonds corresponds to their standard deviations (in standard scores). Note that these means and standard deviations can differ from 0 and 1. The reason is that the algorithm for computing the maximum likelihood estimates starts by standardizing the available data to mean 0 and standard deviation 1, but, the final estimates can vary from these initial values.

- Selecting a point in this plot shows a diamond that informs about the standard deviations of the observations in the pattern for each variable.

Note that Figure 9.9 top is similar to Figure 9.6 but that instead of showing individual observations, it shows the schematics for the summaries of the observations by pattern.

We will now make some comments about Figure 9.9 with regard to our specific example on sleep in mammals. A pattern of special relevance is the *Complete Data* pattern because it refers to observations without missing values. This pattern is the one selected in Figure 9.9 top. The diamonds for this pattern are centered in the plot, but we see that they do not coincide completely with the maximum likelihood means. For example, the mean time of the mammals in this pattern spent *NonDreaming* is higher than the one computed using the EM algorithm. Also, these mammals seem to have means of *BodyWeight* and *BrainWeight* lower than the ones computed using EM. Therefore, it looks like observations without missing values correspond to relatively small animals that spend a short time *NonDreaming*.

The schematics in Figure 9.9 can be linked with a plot of the individual values of equation (9.1). These individual values represent contributions of each of the patterns to Little's MCAR test and can be used to evaluate the distance of the patterns to the mean.

The patterns with the highest values of Little's test are:

- *NonDreaming, Dreaming,* and *TotalSleep (ND-D-TS)*. This pattern has only one observation (Kangaroo) and is displayed using the smallest diamond in Figure 9.9 bottom.

- *NonDreaming* and *Dreaming (ND-D)*. This pattern is the second largest after the *Complete Data* pattern. It has the largest diamonds in Figure 9.9 bottom. The means for the observations in this pattern for the variables *Body* and *Brain Weight,* the *LifeSpan,* and the time of *Gestation* are larger than for the *Complete Data* pattern and the EM means. This pattern seems characteristic of big mammals, where, for some reason, it is difficult to know how *TotalSleep* time is split into time *Dreaming* and *NonDreaming*.

- *NonDreaming* and *TotalSleep (ND-TS)*: This pattern was discussed in Section 9.4. It corresponds to only two observations, Giraffes and Okapis.

As we have seen, interactive tools make the plot in Figure 9.9 more useful. In our case, clicking on a point selects all the points in the pattern. Ctrl-click selects additional patterns. A common situation with this plot is when there are small patterns (few cases) that are very extreme and the analysis wants to remove them to focus on the larger patterns. This action is carried out with a menu item that also rescales the plot. Other actions of interest are the combination of very similar patterns in order to reduce the complexity of the plot or linking it to plots for the components of Little's MCAR test.

Scatterplot matrices for data with missing values. In Section 9.5.2 we discussed visualizations intended to explore the homogeneity of means and variances by patterns of missing data. In this section we explore visualizations of covariances (or relationships) among variables.

Little (1988) expressed several concerns about the power of the version of the MCAR test if used for checking the homogeneity of both means and covariances among patterns of missing data. In particular, he pointed out that patterns with fewer cases than observed variables cannot be used for this purpose, and that the test is probably too sensitive to departures from the normality assumption to be of practical use. As a consequence, this test is probably not used as often as the test for homogeneity of means. Nevertheless, it seems interesting, from an exploratory point of view, to have the capability of checking the relationships among variables of the observations falling into patterns of missing data.

Figure 9.10 shows a scatterplot of the variables *Gestation* and *LifeSpan* for the mammals data. This scatterplot includes the imputations obtained from the EM algo-

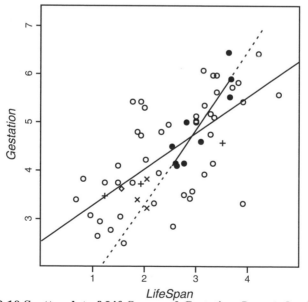

Figure 9.10 Scatterplot of *LifeSpan* and *Gestation*. Imputed values have the following symbols: *LifeSpan* (×), *Gestation* (+) and both (◊). Points shown with a filled circle are in the missing data pattern *NonDreaming–Dreaming* and have been used to compute the regression line with dashed parts. The other regression line has been computed using the parameters estimated using EM.

rithm using the symbols described in Section 9.4.1. This scatterplot could be seen as being part of a scatterplot matrix for all the variables in the dataset.

The scatterplot in Figure 9.10 differs from the scatterplot in Figure 9.5 in that it displays a regression line indicating the relationship among the observations *by patterns of missing values*. The solid lines in the plot refer to the regression lines estimated using the covariances computed with the EM algorithm; the dashed lines represents the regression line computed using only the subset of values in the pattern of missing values currently selected (in this case, the pattern is *NonDreaming* and *Dreaming*). An examination of Figure 9.10 shows that, in general, the line for the pattern selected does not differ too much from the maximum likelihood line. Using a scatterplot matrix of all the pairs of variables would allow us to observe whether the MCAR assumption for covariances holds in general for this pattern.

9.6   Conclusions

The main theme of this chapter is the use of imputation as a way of recovering the missing data and, more important, to illustrate interactive displays that include both imputed and values observed. This strategy requires marking the values imputed differently from the values observed. The visualizations so created can be very useful for the exploration of data. Thus, for the mammals data, a number of features have been revealed that otherwise would have remained unobserved. These features have lead us to understand the singularities of our data in relation with the missing values. This happened, for example, with the two observations with missing values in the variable *TotalSleep* but values observed in the variable *Dreaming*. It may seem that knowing simply whether or not an animal is sleeping should be easier than detecting whether it is dreaming. However, for mammals that do not necessarily lie down when sleeping but do when dreaming, this seems to be true.

References

Agresti, A. (1990). *Categorical Data Analysis*. Wiley, New York.

Alba, R. D. (1987). Interpreting the parameters of log-linear models. *Sociological Methods and Research*, 16(1):45–77.

Allison, T. and Cicchetti, D. V. (1976). Sleep in mammals: ecological and constitutional correlates. *Science*, 194(12):732–734.

Andersen, E. B. (1996). *Introduction to the Statistical Analysis of Categorical Data*. Springer-Verlag, New York.

Asimov, D. (1985). The Grand Tour: a tool for viewing multidimensional data. *SIAM Journal on Scientific and Statistical Computing*, 6(1):128–143.

Ato, M. and Lopez, J. J. (1996). *Análisis estadísticos para datos categóricos*. Síntesis, Madrid.

Barnett, V. and Lewis, T. (1995). *Outliers in Statistical Data*. Wiley, New York.

Becker, R. (1994). A brief history of S. Bell Laboratories.

Becker, R. A. and Chambers, J. M. (1981). *S: A Language and System for Data Analysis*. AT&T Bell Laboratories, Murray Hill, NJ.

Becker, R. A. and Cleveland, W. S. (1987). Brushing scatterplots. *Technometrics*, 29:127–142.

Becker, R. A. and Cleveland, W. S. (1988). The use of brushing and rotation for data analysis. In Cleveland, W. S. and McGill, M. E., editors, *Dynamic Graphics for Statistics*, pages 247–276. Brooks/Cole, Pacific Grove, CA.

Becker, R. A., Chambers, J. M., and Wilks, A. R. (1988a). *The New S Language*. Brooks/Cole, Pacific Grove, CA.

Becker, R. A., Cleveland, W. S., and Wilks, A. R. (1988b). Dynamic graphics for data analysis. In Cleveland, W. S. and McGill, M. E., editors, *Dynamic Graphics for Statistics*, pages 1–50. Brooks/Cole, Pacific Grove, CA.

Becker, R. A., Cleveland, W. S., and Shyu, M. J. (1996). The design and control of trellis display. *Journal of Computational and Graphical Statistics*, 5:123–155.

Bertin, J. (1967). *Semiologie graphique: les diagrammes, les reseaux, les cartes.* Gauthier-Villars, Paris.

Bertin, J. (1983). *Semiology of Graphics*. University of Wisconsin Press, Madison, WI. (English translation by W. Berg).

Betz, D. (1985). An XLISP tutorial. *Byte*, 10(3):211–236.

Bickel, P. J., Hammel, J. W., and O'Connell, J. W. (1975). Sex bias in graduate admissions: data from Berkeley. *Science*, 187:398–403.

Bishop, Y. M. M., Fienberg, S. E., and Holland, P. W. (1988). *Discrete Multivariate Analysis: Theory and Practice.* MIT Press, Cambridge, MA.

Box, G. E. P. and Cox, D. R. (1964). An analysis of transformations. *Journal of the Royal Statistical Society, Series B,* 26:296–311.

Brooks, F. P. J. (1975). *The Mythical Man-Month: Essays on Software Engineering.* Addison Wesley, Reading, MA.

Brooks, F. P. J. (1995). *The Mythical Man-month: Essays on Software Engineering* (20th anniversary edition). Addison-Wesley, Reading, MA.

Buck, S. F. (1960). A method of estimation of missing values in multivariate data suitable for use with an electronic computer. *Journal of the Royal Statistical Society*, 22(2):67–81.

Buja, A. and Asimov, D. (1986). Grand tour methods: An outline. *Computing Science and Statistics*, 17:63–67.

Buja, A., Hurley, D., and McDonald, J. A. (1988). Elements of a viewing pipeline for data analysis. In Cleveland, W. S. and McGill, M. E., editors, *Dynamic Graphics for Statistics*, pages 277–308. Brooks/Cole, Pacific Grove, CA.

Chambers, J. M., Cleveland, W. S., Kleiner, B., and Tukey, P. A. (1983). *Graphical Methods for Data Analysis.* Wadsworth International Group, Belmont, CA.

Christensen, J., Marks, J., and Shieber, S. (1992). Labeling point features on maps and diagrams. Technical Report TR-25-92, Harvard University, Cambridge, MA.

Christensen, J., Marks, J., and Shieber, S. (1995). An empirical study of algorithms for point-feature label placement. *ACM Transactions on Graphics*, 14(3):203–232.

Christensen, R. (1990). *Log-Linear Models*. Springer-Verlag, New York.

Cleveland, W. S. (1994a). *The Elements of Graphing Data*, revised edition. Hobart Press, Summit, NJ.

Cleveland, W. S. (1994b). *Visualizing Data*. Hobart Press, Summit, NJ.

Cleveland, W. C. and McGill, M. E. (1988). *Dynamic Graphics for Statistics*. CRC Press, Boca Raton, FL.

Cook, R. D. and Weisberg, S. (1994). *An Introduction to Regression Graphics*. Wiley, New York.

Cook, R. D. and Weisberg, S. (1999). *Applied Regression Including Computing and Graphics*. Wiley, New York.

Cruz-Neira, C., Sandin, D. J., DeFanti, T. A., Kenyon, R. V., and Hart, J. C. (1992). The CAVE: audiovisual experience automatic virtual environment. *Communications of the ACM*, 35:67–72.

Cutting, J. E. (2002). Representing motion in a static image: constraints and parallels in art, science, and popular culture. *Perception*, 31:1165–1193.

de Leeuw, J. (2004). Personal communication.

de Leeuw, J. (2005). On abandoning Xlisp-Stat. *Journal of Statistical Software*, 13(7):1–5.

Dempster, A. P., Laird, N. M., and Rubin, D. B. (1977). Maximum likelihood from incomplete data via the EM algorithm. *Journal of the Royal Statistical Society*, 39:1–37.

Donoho, A. W., Donoho, D. L., and Gasko, M. (1988). MacSpin: dynamic graphics on a desktop computer. *IEEE Computer Graphics and Applications*, 8(4):51–58.

Eddy, W. F., Howe, S. E., Teitel, B. F., and Young, F. W. (1991). The future of statistical software. In *National Research Council Forum*. National Academies Press, Washington, DC.

Eick, S. G. (1994). Data visualization sliders. In *ACM Symposium on User Interface Software and Technology*, pages 119–120, Monterey, CA.

Fisherkeller, M. A., Friedman, J. H., and Tukey, J. W. (1975). Prim-9: a data display and analysis system. In *Pacific Regional Conference of the Association for Computing Machinery*, San Francisco, CA.

Fowlkes, E. B. (1971). User's manual for an on-line interactive system for probability plotting on the ddp-224 computer. Technical memorandum, AT&T Bell Laboratories, Murray Hill, NJ.

Friendly, M. (1992). User's guide for MOSAICS. Technical Report 206, Psychology Department, York University, 4700 Keele Street, Toronto, Canada.

Friendly, M. (1994). Mosaic displays for multi-way contingency tables. *Journal of the American Statistical Association*, 89:190–200.

Friendly, M. (1995). Conceptual and visual models for categorical data. *American Statistician*, 49:153–160.

Friendly, M. (1999). Extending mosaic displays: marginal, partial, and conditional views of categorical data. *Journal of Computational and Statistical Graphics*, 8:373–395.

Friendly, M. (2000a). Re-visions of Minard. *Statistical Computing and Statistical Graphics Newsletter*, 12(1):13–19.

Friendly, M. (2000b). *Visualizing Categorical Data*. SAS Institute, Cary, NC.

Friendly, M. (2004). Milestones in the history of data visualization: a case study in statistical historiography. In Gaul, W. and Weihs, C., editors, *Studies in Classification, Data Analysis, and Knowledge Organization*. Springer-Verlag, New York.

Friendly, M. and Denis, D. J. (2001). The roots and branches of statistical graphics. *Journal de la Societe Francaise de Statistique*, 141(4):51–60.

Friendly, M. and Denis, D. J. (2004). Milestones in the history of thematic cartography, statistical graphics, and data visualization. www.math.yorku.ca/SCS/Gallery/milestone/.

Friendly, M. and Kwan, E. (2003). Effect ordering for data displays. *Computational Statistics and Data Analysis*, 43(4):509–539.

Fuchs, H., Poulton, J., Eyles, J., Greer, T., Goldfeather, J., Ellsworth, D., Molnar, S., Turk, G., Tebbs, B., and Israel, L. (1989). Pixel-planes 5: a heterogeneous multiprocessor graphics system using processor-enhanced memories. In *SIGGRAPH '89: Proceedings of the 16th ACM Annual Conference on Computer graphics and Interactive Techniques*, New York, pages 79–88.

Gabriel, K. R. (1971). The biplot graphical display of matrices with applications to principal component analysis. *Biometrika*, (58):453–467.

Gane, C. and Sarson, T. (1979). *Structured Systems Analysis: Tools and Techniques*. Prentice-Hall, Englewood Cliffs, NJ.

Gifi, A. (1990). *Nonlinear Multivariate Analysis*. Wiley, New York.

Gower, J. C. and Hand, D. J. (1996). *Biplots*. Chapman & Hall.

Graham, J., Hofer, S., and MacKinnon, D. (1996). Maximizing the usefulness of data obtained with planned missing value patterns: an application of maximum likelihood procedures. *Multivariate Behavioral Research*, 31(2):197–218.

Greenacre, M. J. (1993). *Correspondence Analysis in Practice*. Academic Press, London.

Guerry, A. M. (1833). *Essai sur la statistique morale de la france*. Crochard, Paris. English translation: Whitt, H. P. and Reinking, V. W. Edwin Mellen Press, Lewiston, NY, (2002).

Guvenir, H. A., Demiroz, G., and Ilter, N. (1998). Learning differential diagnosis of erythemato-squamous diseases using voting feature intervals. *Artificial Intelligence in Medicine*, 13:147–165.

Haberman, S. J. (1973). The analysis of residuals in cross-classification tables. *Biometrics*, 29:205–220.

Hartigan, J. A. and Kleiner, B. (1981). Mosaics for contingency tables. In Eddy, E. F., editor, *Computer Science and Statistics: Proceedings of the 13th Symposium on the Interface*, Springer-Verlag, New York.

Hartigan, J. A. and Kleiner, B. (1984). A mosaic of television ratings. *American Statistician*, 38:153–160.

Henderson, H. V. and Velleman, P. F. (1981). Building regression models interactively:data originally collected from consumer reports. *Biometrics*, 37:391–411.

Hesterberg, T. (1999). A graphical representation of Little's test for MCAR. Technical Report 94. MathSoft, Seattle, WA.

Hofman, H. (2003). Constructing and reading mosaic plots. *Computational Statistics and Data Analysis*, 43(4):565–580.

Hutchins, E. L., Hollan, J. D., and Norman, D. A. (1986). Direct manipulation interfaces. In Norman, D.A. and Draper, S. W., editors, *User Centered System Design: New Perspectives on Human–Computer Interaction*, pages 87–124. Lawrence Erlbaum Associates, Hillsdale, NJ.

Ihaka, R. and Gentleman, R. (1996). R: a language for data analysis and graphics. *Journal of Computational and Graphical Statistics*, 5(3):299–314.

Inselberg, A. (1985). The plane with parallel coordinates. *The Visual Computer*, 1:69–97.

JMP (1989–2002). *Version 5*. SAS Institute. Cary, NC.

Jollife, I. T. (2002). *Principal Components Analysis*. Springer-Verlag, New York.

Kim, K. H. and Bentler, P. M. (2002). Tests of homogeneity of means and covariance matrixes for multivariate incomplete data. *Psychometrika*, 67(4):609–624.

Little, R. J. A. (1988). A test of missing completely at random for multivariate data with missing values. *Journal of the American Statistical Association*, 83(4):1198–1202.

Little, R. J. A. and Rubin, D. B. (1987). *Statistical Analysis with Missing Data*. Wiley, New York.

Little, R. J. A. and Rubin, D. B. (2002). *Statistical Analysis with Missing Data-second edition*. Wiley, New York.

Marsaglia, G. (1968). Random numbers fall mainly in the plane. *Proceedings of the National Academy of Sciences USA*, 61:25–28.

McCloud, S. (1994). *Understanding Comics*. HarperCollins, New York.

McCullagh, P. and Nelder, J. A. (1989). *Generalized Linear Models*. Chapman & Hall, London.

McDonald, J. A. (1982). Interactive graphics for data analysis. Ph.d. dissertation, Stanford University, Stanford, CA.

McDonald, J. A. (1988). Orion I: interactive graphics for data analysis. In Cleveland, W. S. and McGill, M. E., editors, *Dynamic Graphics for Statistics*, pages 179–200, Brooks/Cole, Pacific Grove, CA.

McGill, R., Tukey, J. W., and Larsen, W. A. (1978). Variations of boxplots. *American Statistician*, 32:12–16.

Miller, G. (1956). The magical number seven, plus or minus two: Some limits on our capacity for processing information. *Psycological Review*, 63:81–97.

Minard, C. J. (1844). *Tableaux figuratifs de la circulation de quelques chemins de fer. lith. (n.s.)*. ENPC, 5860/C351, 5299/C307.

Molina, J. G., Ledesma, R., Valero-Mora, P. M., and Young, F. W. (2005). A video tour through ViSta 6.4. *Journal of Statistical Software*, 13(8):1–13.

Mosteller, F., Fienberg, S. E., and Rourke, R. E. K. (1983). *Beginning Statistics with Data Analysis*. Addison-Wesley, Reading, M.A.

Muller, K. E. (1981). Relationships between redundancy analysis, canonical correlation, and multivariate regression. *Psychometrika*, 46:139–142.

Nagel, H. and Granum, E. (2004). Explorative and dynamic visualization of data in virtual reality. *Computational Statistics*, 19(1):55–73.

Nelson, L., Cook, D., and Cruz-Neira, C. (1999). XGobi vs. the C2: results of an experiment comparing data visualization in a 3D inmmersive virtual reality environment with a 2D workstation display. *Computational Statistics: Special Issue on Interactive Graphical Data Analysis*, 14(1):39–51.

NIST/SEMATECH (2004). e-handbook of statistical methods. www.itl.nist.gov/div898/handbook.

Noma, E. (1987). Heuristic method for label placement in scatterplots. *Psychometrika*, 52(3):463–468.

North, C. and Shneiderman, B. (1997). A taxonomy of multiple window coordinations. Technical Report 3854, Computer Science Department, University of Maryland, College Park, MD.

North, C. and Shneiderman, B. (2000). Snap-Together Visualization: A User Interface for Coordinating Visualizations via Relational Schemata. *Advanced Visual Interfaces:* 128-135.

Playfair, W. (1786). *Commercial and Political Atlas: Representing, by Copper-Plate Charts, the Progress of the Commerce, Revenues, Expenditure, and Debts of England, during the Whole of the Eighteenth Century*. Corry, London (3rd edition, Stockdale, London, 1801); French edition, Tableaux d'arithmétique linéaire, du commerce, des finances, et de la dette nationale de l'Angleterre, Chez Barrois l'Aine, Paris, 1789.

Playfair, W. (1801). *Statistical Breviary; Shewing, on a Principle Entirely New, the Resources of Every State and Kingdom in Europe*. Wallis, London.

Pattison, T. and Phillips, M. (2001). View coordination architecture for information visualisation. In *Australian symposium on Information visualisation*, Sidney, Australia.

Preece, J., Rogers, I., Sharp, H., Benyon, D., Holland, S., and Carey, T. (1994). *Human–Computer Interaction*. Addison-Wesley, Wokingham, Berks, England.

Rindskopf, D. (1990). Non-standard log-linear models. *Psychological Bulletin*, 108(1):150–162.

Rubin, D. (1987). *Multiple Imputation for Nonresponse in Surveys*. Wiley, New York.

Savage, S. (2002). *Decision making with insight*. Duxbury Press, Belmont, CA.

Schafer, J. L. (1997). *Analysis of Incomplete Multivariate Data*. Chapman & Hall, London.

Schafer, J. L. (1999). Multiple imputation: a primer. *Statistical Methods in Medical Research*, 8(1):3–15.

Scott, D. W. (1992). *Multivariate Density Estimation*. Wiley, New York.

Siegel, J. M. (1995). Phylogeny and the function of REM sleep. *Behavioural Brain Research*, 69:29–34.

Silverman, B. W. (1986). *Density Estimation for Statistics and Data Analysis*. Chapman & Hall, London.

St. Amant, R. (1997). Navigation for data analysis systems. *Lecture Notes in Computer Science*, 1280:101–109.

Stevens, W. P., Myers, G. J., and Constantine, L. L. (1974). Structured design. *IBM Systems Journal*, 13(2):115–139.

Stine, R. and Fox, J. (1996). *Statistical Computing Environments for Social Research*. Sage, Thousand Oaks, CA.

Sturges, H. A. (1926). The choice of a class interval. *Journal of the American Statistical Association*, 21:65–66.

Stuetzle, W. (1987). Plot windows. *Journal of the American Statistical Association*,82(398): 466-475-

Swayne, D. F. and Buja, A. (1998). Missing data in interactive high-dimensional data visualization. *Computational Statistics*, 13(1):15–26.

Swayne, D., Cook, D., and Buja, A. (1998). XGobi: interactive dynamic data visualization in the X window system. *Journal of Computational and Graphical Statistics* 7(1):113-130.

Swayne, D., Lang, D., Buja, A., and Cook, D. (2003). GGobi: evolving from XGobi into an extensible framework for interactive data visualization. *Computational Statistics and Data Analysis*, 43(4):423.

Symanzik, J., Cook, D., Kohlmeyer, B., and Cruz-Neira, C. (1996). Dynamic statistical graphics in the cave virtual reality environment. In *Dynamic Statistical Graphics Workshop*, Sidney, Australia.

Symanzik, J., Cook, D., Kohlmeyer, B. D., Lechner, U., and Cruz-Neira, C. (1997). Dynamic statistical graphics in the C2 virtual reality environment. *Computing Science and Statistics*, 29:41–47.

Theus, M. (2003). Interactive data visualization using Mondrian. *Journal of Statistical Software*, 7(11):1–9.

Thornes, B. and Collard, J. (1979). *Who Divorces?* Routledge & Kegan, Paul, London.

Tierney, L. (1988). Xlisp-Stat: a statistical environment based on the Xlisp language. Technical Report 528, School of Statistics, University of Minnesota, Minneapolis, MN.

Tierney, L. (1990). *Lisp-Stat: An Object Oriented Environment for Statistical Computing and Dynamic Graphics*. Wiley-Interscience, New York.

Tierney, L. (2005). Some notes on the past and future of Lisp-Stat. *Journal of Statistical Software*, 13(9):1–15.

Tufte, E. R. (1983). *The Visual Display of Quantitative Information*. Graphics Press, Cheshire, England.

Tufte, E. R. (1997). *Visual Explanations: Images and Quantities, Evidence and Narrative*. Graphics Press, Cheshire, England.

Tukey, J. W. (1962). The future of data analysis. *Annals of Mathematical Statistics*, 33:1-67.

Tukey, J. W. (1965). The technical tools of statistics. *American Statistician*, 19:23–28.

Tukey, J. W. (1977). *Exploratory Data Analysis*. Addison-Wesley, Reading, MA.

Turlach, B. A. (1993). Bandwidth selection in kernel density estimation: a review. Discussion paper 9232 Institut of Statistique, Louvain-la-Neuve, Belgium.

Udina, F. (1999). Implementing interactive computing in an object-oriented environment. Economics Working Papers 419. Department of Economics and Business, Universitat Pompeu Fabra, Barcelona, Spain.

Unwin, A. R., Hawkins, G., Hofman, H., and Siegl, B. (1996). Interactive graphics for data sets with missing values: Manet. *Journal of Computational and Graphical Statistics*, 5(2):113–122.

Valero-Mora, P. M. and Udina, F. (2005). The health of Lisp-Stat. *Journal of Statistical Software*, 13(10):1–5.

Valero-Mora, P. M., Young, F. W., and Friendly, M. (2003). Visualizing categorical data in ViSta. *Computational Statistics and Data Analysis*, 43(4):495–508.

Valero-Mora, P. M., Rodrigo, M. F., and Young, F. W. (2004). Visualizing parameters from log-linear models. *Computational Statistics*, 19(1):113–133.

Velleman, P. F. (1997). *DataDesk Version 6.0: Statistics Guide*. Data Description, Ithaka, NY.

Velleman, P. F. and Velleman, A. Y. (1985). *Data Desk*. Data Description, Ithaka, NY.

Velleman, P. F. and Wilkinson, L. (1993). Nominal, ordinal, interval, and ratio typologies are misleading. *American Statistician*, 47:65–72.

Vermunt, J. K. (1997). *Log-Linear Models for Event Histories*. Sage, Thousand Oaks, CA.

Wainer, H. (1988). *Dynamic Graphics for Statistics*, Comment on "Dynamic Graphics for Data analysis" by Becker, Cleveland and Wilks, pages 60–62. Brooks/Cole, Pacific Grove, CA.

Wainer, H. and Velleman, P. F. (2001). Statistical graphics: mapping the pathways of science. *Annual Review of Psychology*, 52:305–335.

Wand, M. P. (1996). Data-based choice of histogram bin width. *Statistical Computing and Graphics*, 51(1):59–64.

Wegman, E. J. (1990). Hyperdimensional data analysis using parallel coordinates. *Journal of the American Statistical Association*, 85(411):664–675.

Wegman, E. J. and Symanzik, J. (2002). Immersive projection technology for visual data mining. *Journal of Computational and Graphical Statistics*, 11(1):163–188.

Wegman, E. J., Symanzik, J., Vandersluis, J., Luo, Q., Camelli, F., Dzubay, A., Fu, X., Khumbah, N.-A., Moustafa, R., Wall, R., and Zhu, Y. (1999). The Mini-CAVE: a voice-controlled IPT environment. In Bulling, H. J. and Riedel, O., editors, *Proceedings of the 3rd International Immersive Projection Technology Workshop*, pages 179–190, Springer-Verlag, Berlin.

Wilhelm, A. F. X. (1990). A data model for interactive Statistical packages. In *Proceedings of the Section on Statistical Graphics, pages 61–70*, Baltimore, ASA.

Weisberg, S. (2005). Lost opportunities: Why we need a variety of statistical languages. *Journal of Statistical Software*, 13(1):1–12.

Wilkinson, L. (1999). *The Grammar of Graphics*. Springer, New York.

Young, F. W. (1994). ViSta: The visual statistics system. Research Memorandum 94-1 (revised 1996). L.L. Thurstone Psychometric Laboratory, Chapel Hill, NC.

Young, F. W. (1989). Visualizing six-dimensional structure with dynamic statistical graphics. *Chance*, 2:22–30.

Young, F., Faldowski, R., and McFarlane, M. (1993). Multivariate statistical visualization. In Rao, C. R., editor, *Handbook of Statistics*, volume 9, Computational Statistics, pages 959–998. North Holland, New York.

Young, F. W. and Lubinsky, D. J. (1995). Guiding data analysts with visual statistics strategies. *Journal of Computational and Graphical Statistics*, 4(4):229–250.

Young, F. W. and Rheingans, P. (1991). Visualizing structure in high-dimensional multivariate data. *IBM Journal of Research and Development*, 35(1–2):97–107.

Young, F. W. and Sarle, W. S. (1982). *Exploratory multivariate data analysis*. SAS Institute, Cary, NC.

Young, F. W. and Smith, J. B. (1991). Towards a structured data analysis environment: a cognition based design. In Buja, A. and Tukey, P. A., editors, *Computing and Graphics in Statistics*, pages 255–279. Springer-Verlag, New York.

Young, F., Valero-Mora, P., Faldowski, R., and Bann, C. M. (2003). Gossip: The arquitecture of spreadplots. *Journal of Computational and Graphical Statistics*, 12(1):80–100.

Author Index

Subject Index

WILEY SERIES IN PROBABILITY AND STATISTICS
ESTABLISHED BY WALTER A. SHEWHART AND SAMUEL S. WILKS

The *Wiley Series in Probability and Statistics* is well established and authoritative. It covers many topics of current research interest in both pure and applied statistics and probability theory. Written by leading statisticians and institutions, the titles span both state-of-the-art developments in the field and classical methods.

Reflecting the wide range of current research in statistics, the series encompasses applied, methodological and theoretical statistics, ranging from applications and new techniques made possible by advances in computerized practice to rigorous treatment of theoretical approaches.

This series provides essential and invaluable reading for all statisticians, whether in academia, industry, government, or research.

† ABRAHAM and LEDOLTER · Statistical Methods for Forecasting
  AGRESTI · Analysis of Ordinal Categorical Data
  AGRESTI · An Introduction to Categorical Data Analysis
  AGRESTI · Categorical Data Analysis, *Second Edition*
  ALTMAN, GILL, and McDONALD · Numerical Issues in Statistical Computing for the
    Social Scientist
  AMARATUNGA and CABRERA · Exploration and Analysis of DNA Microarray and
    Protein Array Data
  ANDĚL · Mathematics of Chance
  ANDERSON · An Introduction to Multivariate Statistical Analysis, *Third Edition*
* ANDERSON · The Statistical Analysis of Time Series
  ANDERSON, AUQUIER, HAUCK, OAKES, VANDAELE, and WEISBERG ·
    Statistical Methods for Comparative Studies
  ANDERSON and LOYNES · The Teaching of Practical Statistics
  ARMITAGE and DAVID (editors) · Advances in Biometry
  ARNOLD, BALAKRISHNAN, and NAGARAJA · Records
* ARTHANARI and DODGE · Mathematical Programming in Statistics
* BAILEY · The Elements of Stochastic Processes with Applications to the Natural
    Sciences
  BALAKRISHNAN and KOUTRAS · Runs and Scans with Applications
  BALAKRISHNAN and NG · Precedence-Type Tests and Applications
  BARNETT · Comparative Statistical Inference, *Third Edition*
  BARNETT · Environmental Statistics
  BARNETT and LEWIS · Outliers in Statistical Data, *Third Edition*
  BARTOSZYNSKI and NIEWIADOMSKA-BUGAJ · Probability and Statistical Inference
  BASILEVSKY · Statistical Factor Analysis and Related Methods: Theory and
    Applications
  BASU and RIGDON · Statistical Methods for the Reliability of Repairable Systems
  BATES and WATTS · Nonlinear Regression Analysis and Its Applications

*Now available in a lower priced paperback edition in the Wiley Classics Library.
†Now available in a lower priced paperback edition in the Wiley–Interscience Paperback Series.

BECHHOFER, SANTNER, and GOLDSMAN · Design and Analysis of Experiments for Statistical Selection, Screening, and Multiple Comparisons

BELSLEY · Conditioning Diagnostics: Collinearity and Weak Data in Regression

† BELSLEY, KUH, and WELSCH · Regression Diagnostics: Identifying Influential Data and Sources of Collinearity

BENDAT and PIERSOL · Random Data: Analysis and Measurement Procedures, *Third Edition*

BERRY, CHALONER, and GEWEKE · Bayesian Analysis in Statistics and Econometrics: Essays in Honor of Arnold Zellner

BERNARDO and SMITH · Bayesian Theory

BHAT and MILLER · Elements of Applied Stochastic Processes, *Third Edition*

BHATTACHARYA and WAYMIRE · Stochastic Processes with Applications

† BIEMER, GROVES, LYBERG, MATHIOWETZ, and SUDMAN · Measurement Errors in Surveys

BILLINGSLEY · Convergence of Probability Measures, *Second Edition*

BILLINGSLEY · Probability and Measure, *Third Edition*

BIRKES and DODGE · Alternative Methods of Regression

BLISCHKE AND MURTHY (editors) · Case Studies in Reliability and Maintenance

BLISCHKE AND MURTHY · Reliability: Modeling, Prediction, and Optimization

BLOOMFIELD · Fourier Analysis of Time Series: An Introduction, *Second Edition*

BOLLEN · Structural Equations with Latent Variables

BOLLEN and CURRAN · Latent Curve Models: A Structural Equation Perspective

BOROVKOV · Ergodicity and Stability of Stochastic Processes

BOULEAU · Numerical Methods for Stochastic Processes

BOX · Bayesian Inference in Statistical Analysis

BOX · R. A. Fisher, the Life of a Scientist

BOX and DRAPER · Empirical Model-Building and Response Surfaces

* BOX and DRAPER · Evolutionary Operation: A Statistical Method for Process Improvement

BOX and FRIENDS · Improving Almost Anything, *Revised Edition*

BOX, HUNTER, and HUNTER · Statistics for Experimenters: Design, Innovation, and Discovery, *Second Editon*

BOX and LUCEÑO · Statistical Control by Monitoring and Feedback Adjustment

BRANDIMARTE · Numerical Methods in Finance: A MATLAB-Based Introduction

BROWN and HOLLANDER · Statistics: A Biomedical Introduction

BRUNNER, DOMHOF, and LANGER · Nonparametric Analysis of Longitudinal Data in Factorial Experiments

BUCKLEW · Large Deviation Techniques in Decision, Simulation, and Estimation

CAIROLI and DALANG · Sequential Stochastic Optimization

CASTILLO, HADI, BALAKRISHNAN, and SARABIA · Extreme Value and Related Models with Applications in Engineering and Science

CHAN · Time Series: Applications to Finance

CHARALAMBIDES · Combinatorial Methods in Discrete Distributions

CHATTERJEE and HADI · Regression Analysis by Example, *Fourth Edition*

CHATTERJEE and HADI · Sensitivity Analysis in Linear Regression

CHERNICK · Bootstrap Methods: A Practitioner's Guide

CHERNICK and FRIIS · Introductory Biostatistics for the Health Sciences

CHILÈS and DELFINER · Geostatistics: Modeling Spatial Uncertainty

CHOW and LIU · Design and Analysis of Clinical Trials: Concepts and Methodologies, *Second Edition*

CLARKE and DISNEY · Probability and Random Processes: A First Course with Applications, *Second Edition*

* COCHRAN and COX · Experimental Designs, *Second Edition*

CONGDON · Applied Bayesian Modelling

*Now available in a lower priced paperback edition in the Wiley Classics Library.
†Now available in a lower priced paperback edition in the Wiley–Interscience Paperback Series.

*Now available in a lower priced paperback edition in the Wiley Classics Library.
†Now available in a lower priced paperback edition in the Wiley–Interscience Paperback Series.

---

*Now available in a lower priced paperback edition in the Wiley Classics Library.
†Now available in a lower priced paperback edition in the Wiley–Interscience Paperback Series.

HUZURBAZAR · Flowgraph Models for Multistate Time-to-Event Data

IMAN and CONOVER · A Modern Approach to Statistics

† JACKSON · A User's Guide to Principle Components

JOHN · Statistical Methods in Engineering and Quality Assurance

JOHNSON · Multivariate Statistical Simulation

JOHNSON and BALAKRISHNAN · Advances in the Theory and Practice of Statistics: A Volume in Honor of Samuel Kotz

JOHNSON and BHATTACHARYYA · Statistics: Principles and Methods, *Fifth Edition*

JOHNSON and KOTZ · Distributions in Statistics

JOHNSON and KOTZ (editors) · Leading Personalities in Statistical Sciences: From the Seventeenth Century to the Present

JOHNSON, KOTZ, and BALAKRISHNAN · Continuous Univariate Distributions, Volume 1, *Second Edition*

JOHNSON, KOTZ, and BALAKRISHNAN · Continuous Univariate Distributions, Volume 2, *Second Edition*

JOHNSON, KOTZ, and BALAKRISHNAN · Discrete Multivariate Distributions

JOHNSON, KEMP, and KOTZ · Univariate Discrete Distributions, *Third Edition*

JUDGE, GRIFFITHS, HILL, LÜTKEPOHL, and LEE · The Theory and Practice of Econometrics, *Second Edition*

JUREČKOVÁ and SEN · Robust Statistical Procedures: Aymptotics and Interrelations

JUREK and MASON · Operator-Limit Distributions in Probability Theory

KADANE · Bayesian Methods and Ethics in a Clinical Trial Design

KADANE AND SCHUM · A Probabilistic Analysis of the Sacco and Vanzetti Evidence

KALBFLEISCH and PRENTICE · The Statistical Analysis of Failure Time Data, *Second Edition*

KARIYA and KURATA · Generalized Least Squares

KASS and VOS · Geometrical Foundations of Asymptotic Inference

† KAUFMAN and ROUSSEEUW · Finding Groups in Data: An Introduction to Cluster Analysis

KEDEM and FOKIANOS · Regression Models for Time Series Analysis

KENDALL, BARDEN, CARNE, and LE · Shape and Shape Theory

KHURI · Advanced Calculus with Applications in Statistics, *Second Edition*

KHURI, MATHEW, and SINHA · Statistical Tests for Mixed Linear Models

* KISH · Statistical Design for Research

KLEIBER and KOTZ · Statistical Size Distributions in Economics and Actuarial Sciences

KLUGMAN, PANJER, and WILLMOT · Loss Models: From Data to Decisions, *Second Edition*

KLUGMAN, PANJER, and WILLMOT · Solutions Manual to Accompany Loss Models: From Data to Decisions, *Second Edition*

KOTZ, BALAKRISHNAN, and JOHNSON · Continuous Multivariate Distributions, Volume 1, *Second Edition*

KOTZ and JOHNSON (editors) · Encyclopedia of Statistical Sciences: Volumes 1 to 9 with Index

KOTZ and JOHNSON (editors) · Encyclopedia of Statistical Sciences: Supplement Volume

KOTZ, READ, and BANKS (editors) · Encyclopedia of Statistical Sciences: Update Volume 1

KOTZ, READ, and BANKS (editors) · Encyclopedia of Statistical Sciences: Update Volume 2

KOVALENKO, KUZNETZOV, and PEGG · Mathematical Theory of Reliability of Time-Dependent Systems with Practical Applications

LACHIN · Biostatistical Methods: The Assessment of Relative Risks

LAD · Operational Subjective Statistical Methods: A Mathematical, Philosophical, and Historical Introduction

*Now available in a lower priced paperback edition in the Wiley Classics Library.

†Now available in a lower priced paperback edition in the Wiley–Interscience Paperback Series.

LAMPERTI · Probability: A Survey of the Mathematical Theory, *Second Edition*
LANGE, RYAN, BILLARD, BRILLINGER, CONQUEST, and GREENHOUSE ·
Case Studies in Biometry
LARSON · Introduction to Probability Theory and Statistical Inference, *Third Edition*
LAWLESS · Statistical Models and Methods for Lifetime Data, *Second Edition*
LAWSON · Statistical Methods in Spatial Epidemiology
LE · Applied Categorical Data Analysis
LE · Applied Survival Analysis
LEE and WANG · Statistical Methods for Survival Data Analysis, *Third Edition*
LePAGE and BILLARD · Exploring the Limits of Bootstrap
LEYLAND and GOLDSTEIN (editors) · Multilevel Modelling of Health Statistics
LIAO · Statistical Group Comparison
LINDVALL · Lectures on the Coupling Method
LIN · Introductory Stochastic Analysis for Finance and Insurance
LINHART and ZUCCHINI · Model Selection
LITTLE and RUBIN · Statistical Analysis with Missing Data, *Second Edition*
LLOYD · The Statistical Analysis of Categorical Data
LOWEN and TEICH · Fractal-Based Point Processes
MAGNUS and NEUDECKER · Matrix Differential Calculus with Applications in
Statistics and Econometrics, *Revised Edition*
MALLER and ZHOU · Survival Analysis with Long Term Survivors
MALLOWS · Design, Data, and Analysis by Some Friends of Cuthbert Daniel
MANN, SCHAFER, and SINGPURWALLA · Methods for Statistical Analysis of
Reliability and Life Data
MANTON, WOODBURY, and TOLLEY · Statistical Applications Using Fuzzy Sets
MARCHETTE · Random Graphs for Statistical Pattern Recognition
MARDIA and JUPP · Directional Statistics
MASON, GUNST, and HESS · Statistical Design and Analysis of Experiments with
Applications to Engineering and Science, *Second Edition*
McCULLOCH and SEARLE · Generalized, Linear, and Mixed Models
McFADDEN · Management of Data in Clinical Trials
* McLACHLAN · Discriminant Analysis and Statistical Pattern Recognition
McLACHLAN, DO, and AMBROISE · Analyzing Microarray Gene Expression Data
McLACHLAN and KRISHNAN · The EM Algorithm and Extensions
McLACHLAN and PEEL · Finite Mixture Models
McNEIL · Epidemiological Research Methods
MEEKER and ESCOBAR · Statistical Methods for Reliability Data
MEERSCHAERT and SCHEFFLER · Limit Distributions for Sums of Independent
Random Vectors: Heavy Tails in Theory and Practice
MICKEY, DUNN, and CLARK · Applied Statistics: Analysis of Variance and
Regression, *Third Edition*
* MILLER · Survival Analysis, *Second Edition*
MONTGOMERY, PECK, and VINING · Introduction to Linear Regression Analysis,
Fourth Edition
MORGENTHALER and TUKEY · Configural Polysampling: A Route to Practical
Robustness
MUIRHEAD · Aspects of Multivariate Statistical Theory
MULLER and STOYAN · Comparison Methods for Stochastic Models and Risks
MURRAY · X-STAT 2.0 Statistical Experimentation, Design Data Analysis, and
Nonlinear Optimization
MURTHY, XIE, and JIANG · Weibull Models
MYERS and MONTGOMERY · Response Surface Methodology: Process and Product
Optimization Using Designed Experiments, *Second Edition*

*Now available in a lower priced paperback edition in the Wiley Classics Library.
†Now available in a lower priced paperback edition in the Wiley–Interscience Paperback Series.

MYERS, MONTGOMERY, and VINING · Generalized Linear Models. With Applications in Engineering and the Sciences
† NELSON · Accelerated Testing, Statistical Models, Test Plans, and Data Analyses
† NELSON · Applied Life Data Analysis
NEWMAN · Biostatistical Methods in Epidemiology
OCHI · Applied Probability and Stochastic Processes in Engineering and Physical Sciences
OKABE, BOOTS, SUGIHARA, and CHIU · Spatial Tesselations: Concepts and Applications of Voronoi Diagrams, *Second Edition*
OLIVER and SMITH · Influence Diagrams, Belief Nets and Decision Analysis
PALTA · Quantitative Methods in Population Health: Extensions of Ordinary Regressions
PANJER · Operational Risk: Modeling and Analysis
PANKRATZ · Forecasting with Dynamic Regression Models
PANKRATZ · Forecasting with Univariate Box-Jenkins Models: Concepts and Cases
* PARZEN · Modern Probability Theory and Its Applications
PEÑA, TIAO, and TSAY · A Course in Time Series Analysis
PIANTADOSI · Clinical Trials: A Methodologic Perspective
PORT · Theoretical Probability for Applications
POURAHMADI · Foundations of Time Series Analysis and Prediction Theory
PRESS · Bayesian Statistics: Principles, Models, and Applications
PRESS · Subjective and Objective Bayesian Statistics, *Second Edition*
PRESS and TANUR · The Subjectivity of Scientists and the Bayesian Approach
PUKELSHEIM · Optimal Experimental Design
PURI, VILAPLANA, and WERTZ · New Perspectives in Theoretical and Applied Statistics
† PUTERMAN · Markov Decision Processes: Discrete Stochastic Dynamic Programming
QIU · Image Processing and Jump Regression Analysis
* RAO · Linear Statistical Inference and Its Applications, *Second Edition*
RAUSAND and HØYLAND · System Reliability Theory: Models, Statistical Methods, and Applications, *Second Edition*
RENCHER · Linear Models in Statistics
RENCHER · Methods of Multivariate Analysis, *Second Edition*
RENCHER · Multivariate Statistical Inference with Applications
* RIPLEY · Spatial Statistics
* RIPLEY · Stochastic Simulation
ROBINSON · Practical Strategies for Experimenting
ROHATGI and SALEH · An Introduction to Probability and Statistics, *Second Edition*
ROLSKI, SCHMIDLI, SCHMIDT, and TEUGELS · Stochastic Processes for Insurance and Finance
ROSENBERGER and LACHIN · Randomization in Clinical Trials: Theory and Practice
ROSS · Introduction to Probability and Statistics for Engineers and Scientists
ROSSI, ALLENBY, and McCULLOCH · Bayesian Statistics and Marketing
† ROUSSEEUW and LEROY · Robust Regression and Outlier Detection
* RUBIN · Multiple Imputation for Nonresponse in Surveys
RUBINSTEIN · Simulation and the Monte Carlo Method
RUBINSTEIN and MELAMED · Modern Simulation and Modeling
RYAN · Modern Regression Methods
RYAN · Statistical Methods for Quality Improvement, *Second Edition*
SALEH · Theory of Preliminary Test and Stein-Type Estimation with Applications
* SCHEFFE · The Analysis of Variance
SCHIMEK · Smoothing and Regression: Approaches, Computation, and Application
SCHOTT · Matrix Analysis for Statistics, *Second Edition*
SCHOUTENS · Levy Processes in Finance: Pricing Financial Derivatives
SCHUSS · Theory and Applications of Stochastic Differential Equations

*Now available in a lower priced paperback edition in the Wiley Classics Library.
†Now available in a lower priced paperback edition in the Wiley–Interscience Paperback Series.

SCOTT · Multivariate Density Estimation: Theory, Practice, and Visualization
† SEARLE · Linear Models for Unbalanced Data
† SEARLE · Matrix Algebra Useful for Statistics
† SEARLE, CASELLA, and McCULLOCH · Variance Components
SEARLE and WILLETT · Matrix Algebra for Applied Economics
SEBER and LEE · Linear Regression Analysis, *Second Edition*
† SEBER · Multivariate Observations
† SEBER and WILD · Nonlinear Regression
SENNOTT · Stochastic Dynamic Programming and the Control of Queueing Systems
* SERFLING · Approximation Theorems of Mathematical Statistics
SHAFER and VOVK · Probability and Finance: It's Only a Game!
SILVAPULLE and SEN · Constrained Statistical Inference: Inequality, Order, and Shape
    Restrictions
SMALL and McLEISH · Hilbert Space Methods in Probability and Statistical Inference
SRIVASTAVA · Methods of Multivariate Statistics
STAPLETON · Linear Statistical Models
STAUDTE and SHEATHER · Robust Estimation and Testing
STOYAN, KENDALL, and MECKE · Stochastic Geometry and Its Applications, *Second
    Edition*
STOYAN and STOYAN · Fractals, Random Shapes and Point Fields: Methods of
    Geometrical Statistics
STYAN · The Collected Papers of T. W. Anderson: 1943–1985
SUTTON, ABRAMS, JONES, SHELDON, and SONG · Methods for Meta-Analysis in
    Medical Research
TAKEZAWA · Introduction to Nonparametric Regression
TANAKA · Time Series Analysis: Nonstationary and Noninvertible Distribution Theory
THOMPSON · Empirical Model Building
THOMPSON · Sampling, *Second Edition*
THOMPSON · Simulation: A Modeler's Approach
THOMPSON and SEBER · Adaptive Sampling
THOMPSON, WILLIAMS, and FINDLAY · Models for Investors in Real World Markets
TIAO, BISGAARD, HILL, PEÑA, and STIGLER (editors) · Box on Quality and
    Discovery: with Design, Control, and Robustness
TIERNEY · LISP-STAT: An Object-Oriented Environment for Statistical Computing
    and Dynamic Graphics
TSAY · Analysis of Financial Time Series, *Second Edition*
UPTON and FINGLETON · Spatial Data Analysis by Example, Volume II:
    Categorical and Directional Data
VAN BELLE · Statistical Rules of Thumb
VAN BELLE, FISHER, HEAGERTY, and LUMLEY · Biostatistics: A Methodology for
    the Health Sciences, *Second Edition*
VESTRUP · The Theory of Measures and Integration
VIDAKOVIC · Statistical Modeling by Wavelets
VINOD and REAGLE · Preparing for the Worst: Incorporating Downside Risk in Stock
    Market Investments
WALLER and GOTWAY · Applied Spatial Statistics for Public Health Data
WEERAHANDI · Generalized Inference in Repeated Measures: Exact Methods in
    MANOVA and Mixed Models
WEISBERG · Applied Linear Regression, *Third Edition*
WELSH · Aspects of Statistical Inference
WESTFALL and YOUNG · Resampling-Based Multiple Testing: Examples and
    Methods for *p*-Value Adjustment
WHITTAKER · Graphical Models in Applied Multivariate Statistics

WINKER · Optimization Heuristics in Economics: Applications of Threshold Accepting

WONNACOTT and WONNACOTT · Econometrics, *Second Edition*

WOODING · Planning Pharmaceutical Clinical Trials: Basic Statistical Principles

WOODWORTH · Biostatistics: A Bayesian Introduction

WOOLSON and CLARKE · Statistical Methods for the Analysis of Biomedical Data, *Second Edition*

WU and HAMADA · Experiments: Planning, Analysis, and Parameter Design Optimization

WU and ZHANG · Nonparametric Regression Methods for Longitudinal Data Analysis

YANG · The Construction Theory of Denumerable Markov Processes

YOUNG, VALERO-MORA, and FRIENDLY · Visual Statistics: Seeing Data with Dynamic Interactive Graphics

ZELTERMAN · Discrete Distributions—Applications in the Health Sciences

* ZELLNER · An Introduction to Bayesian Inference in Econometrics

ZHOU, OBUCHOWSKI, and McCLISH · Statistical Methods in Diagnostic Medicine